Biobazaar

Biobazaar

The Open Source Revolution and Biotechnology

JANET HOPE

Harvard University Press
Cambridge, Massachusetts
London, England
2008

Library of Congress Cataloging-in-Publication Data

Hope, Janet, 1972–
Biobazaar : the open source revolution and
biotechnology / Janet Hope.
p. cm.
Includes bibliographical references and index.
ISBN-13: 978-0-674-02635-3
ISBN-10: 0-674-02635-7
1. Biotechnology—Patents. 2. Technological
innovations—Patents. 3. Patent licenses.
4. Biotechnology—Economic aspects. 5. Technological
innovations—Economic aspects. I. Title.

K1519.B54H67 2007
346.04′86—dc22 2007028416

Contents

Abbreviations *vii*

1 An Irresistible Analogy *1*

2 The Trouble with Intellectual Property in Biotechnology *28*

3 Intellectual Property and Innovation *68*

4 Welcome to the Bazaar *106*

5 Open Source Licensing for Biotechnology *142*

6 Foundations of the Biobazaar *188*

7 Financing Open Source Biotechnology *237*

8 Biotechnology's Open Source Revolution *292*

Notes *335*

References *367*

Acknowledgments *391*

Index *399*

Abbreviations

BSD	Berkeley Software Distribution
CAMBIA	Center for Application of Molecular Biology in Agriculture, Canberra, Australia
CGIAR	Consultative Group on International Agricultural Research
EST	expressed sequence tag
FSD	Free Software Definition
GPL	General Public License (formerly GNU Public License)
MTA	material transfer agreement
NIH	National Institutes of Health
OSD	Open Source Definition
OSI	Open Source Initiative
PCR	polymerase chain reaction
SNP	single nucleotide polymorphism
TRIPS	World Trade Organization Agreement on Trade-Related Aspects of Intellectual Property Rights
USPTO	United States Patent and Trademark Office
WTO	World Trade Organization

Biobazaar

1

An Irresistible Analogy

A Different Kind of Scientific Revolution

Early in the new millennium, ten years after completing an undergraduate degree in biochemistry and molecular biology, I returned to the classroom for a refresher course. The 1990s had been a decade of remarkable breakthroughs in the life sciences. First the "worm"—as the diminutive dirt-dwelling nematode *C. elegans* is known to its many enthusiastic devotees—and then the human genomes had been sequenced, ushering in the postgenomic age and triggering a cascade of mysterious new subdisciplines whose names all seemed to end in *-omics*. Having spent these years pursuing a legal career, largely out of touch with scientific developments, I knew I should expect to see big changes as the class I planned to audit moved through its curriculum. But as it turned out, the most striking changes were of a kind I had not anticipated.

Law students are introduced to current legal rules by means of a narrative that traces each line of cases from the earliest decisions through to the present day. This heuristic sends new lawyers a powerful tacit message about the nature of the law—that it is contingent and continuously evolving—and about their own potential role in shaping its future. As trainee biologists at the start of the 1990s, my friends and I had been exposed to a similar style of teaching. Each lecture would bring a new episode in whichever fascinating tale constituted the background to our latest laboratory assignment. The stories usually began with a curious scientist—often an independently wealthy Englishman in periwig and breeches—asking questions about the natural world. In devising and executing some ingenious investigation, this fellow would generate a whole new set of questions, to be picked up by a second protagonist where the first had left off.

As each story progressed, the characters became more diverse and the narrative tension heightened. Restoration courtiers gave way to ambitious female crystallographers and maverick Californian surfer-chemists.[1] We heard of rival theories, personality clashes, and dubious deeds done for the sake of personal prestige at the expense of the greater good. Hoarding data, accepting recognition for work done by junior colleagues, moving in on another group's research after all the hard questions had been answered: all these behaviors were acknowledged. But they were treated as deviations from a general rule of cooperation—enlivened, naturally, by a little fair competition.

Though the stories varied, all were cliff-hangers. That is, they concluded not with answers, but with the latest round of questions. The implication was clear. We novice scientists were being invited to join an epic voyage of discovery, carried on over many generations. Our seniors were handing us the map, showing us the ropes, and imparting to us their code of honor. Our job was to go out and explore new worlds. Whatever we brought back was to be shared with other scientists around the globe, for the good of all humankind. Our reward would be a lifetime of adventure—perhaps even culminating in a walk-on part in the ongoing story of science.

Of course, this pedagogical narrative was largely fictitious. But it was an inspiring fiction, as United States science adviser Vannevar Bush well understood when he gave his famous 1945 funding report the glamorous title *Science: The Endless Frontier.*[2] Moreover, it was backed by many genuinely exciting and creative feats of problem solving.

In fact, so impressive was the science (and so understated the mythology) that my refresher biotechnology course was well under way before I realized that the old familiar story had somehow evaporated. In its place was a rather repetitive refrain that went something like, "Here's a technique. It's owned by Such-and-Such. Here's another technique. It's owned by So-and-So." The professor's oral presentation was accompanied by slick visual aids littered with the names and logos of large corporations: "Expression of proinsulin in *E. coli* (Hoechst and Eli Lilly)"; "Expression of Miniproinsulin in *S. Cerevisiae* (Novo Nordirsk)." Between classes I learned that few of the students aspired to head their own laboratories or conduct independent research. Instead they envisaged careers as technicians in pharmaceutical or biotechnology companies, working to realize someone else's vision. Bright and industrious, they had no trouble decoding the tacit message the life sciences community was now sending its newest recruits. They understood that they didn't need to know where the questions came from. They only needed to know and apply the answers.

Needless to say, these casual observations did not amount to anything like rigorous social science. Even so, they contributed to a hunch that changes in the structure of life sciences research over the past three decades had come to influence even the most basic perceptions scientists hold about their own work, as well as about the nature and purpose of the scientific enterprise overall. In 1962 Thomas Kuhn published a now-famous book called *The Structure of Scientific Revolutions.*[3] In it he described scientific progress as a gradual evolutionary process punctuated by revolutionary "paradigm shifts": profound breakthroughs that require the reconstruction and reevaluation of all that has gone before. The commercialization of life sciences research over the final quarter of the last century can be seen as a different kind of scientific revolution—a paradigm shift in the values underpinning life sciences research.

No less moved than I had been as a teenager by the power and elegance of molecular biotechnology itself, I nevertheless wondered about the consequences of this apparent shift in values. Up on the projector screen, the company logos took on the appearance of "no trespassing" signs along a public right of way. What, I wondered, were the implications of such pervasive rights of private ownership over this remarkable new technology, with all its yet unrealized social and economic potential?

St. Ignucius

Of course, this train of thought was hardly new. From the earliest days of commercial involvement in biotechnology research and development, others had pondered the same question. Specific concerns expressed by scientists and others included the prospect of corporate interests dictating the direction of research, deterioration in the quality of research due to the undermining of traditional peer review mechanisms, exploitation of graduate students and postdoctoral researchers, divided loyalties, financial conflicts of interest, and the danger that academic scientists would lose their credibility as impartial experts on matters of science policy.[4]

Consternation over the effects of privatizing scientific and technological information was not confined to the life sciences. In the late 1970s and early 1980s, the very time when the commercial biotechnology industry was starting to take off, another new industry was emerging out of the academic discipline of computer science. From the point of view of the technical professionals involved, the birth pangs of the biotechnology and information technology industries—together often regarded as defining our current technological era—had much in common.

One who felt those pangs most keenly was Richard Stallman, a member of the Massachusetts Institute of Technology's Artificial Intelligence (AI) Laboratory, a focal point for the "hacker" community through the 1960s and 1970s. (In this context, the term *hacker* does not mean someone who cracks a security system, but—in Stallman's words—"someone who loves to program and enjoys being clever about it.")[5]

I first heard of Richard Stallman when a friend played me a sound recording he had stumbled across when surfing from the XEmacs website.[6] XEmacs is one of two closely related Emacs (*e*ditor *mac*ros) text editors that are especially popular with sophisticated software users and developers. The other is GNU Emacs, from which XEmacs is derived via a code "fork."[7] Forking—the creation of a new branch in the evolutionary tree of a software "genus"—occurs when code from an existing software program is used as the starting point for a new program. With the incorporation of new code into one or both programs, the two code bases ultimately become incompatible. As we shall see, the freedom to fork a software development project is regarded as essential in the world of free and open source software. Even so, anyone whose actions create such a fork experiences considerable pressure to justify that decision, both to other participants in the collaborative development effort and to ordinary users who may be adversely affected by resulting incompatibilities.

The Internet provides a natural forum for such justification. In the case of XEmacs, the personal website of the primary developer, Jamie Zawinski, offered detailed explanations of the reasons for the fork from GNU Emacs, a project started and maintained by Stallman (known to fellow programmers by his initials, RMS).[8] The item that caught my friend's attention was a sort of multimedia supplement to the written explanations: a link to an MPEG file, titled simply "Why Collaboration with RMS Is Impossible."[9] Clicking on the link, one heard Stallman's rather tuneless voice singing, without introduction or accompaniment, the following lyrics:[10]

Join us now and share the software;
You'll be free, hackers, you'll be free.

Hoarders may get piles of money,
That is true, hackers, that is true.
But they cannot help their neighbors;
That's not good, hackers, that's not good.

When we have enough free software
At our call, hackers, at our call,
We'll throw out those dirty licenses
Ever more, hackers, ever more.

Join us now and share the software;
You'll be free, hackers, you'll be free.

This little ditty proved surprisingly catchy, as I found to my irritation over the next couple of days! Curiosity piqued, I did a quick Google search for images of Stallman. The top hit showed him in character as "St. IGNUcius of the Church of Emacs": a pale man with long hair and an unkempt beard, dressed in flowing robes, with a large gold computer disk attached to the top of his head.[11] On one forearm he balanced a closed laptop computer; his other hand was raised, palm open and facing forward. Combined with the makeshift halo, it was a pose brilliantly calculated to evoke a traditional piece of religious iconographic art.

It would be fair to say, on the basis of this and other evidence, that Stallman is a somewhat eccentric person. Yet his achievements, both technical and political, are legendary—as evidenced by the telling and retelling of folk histories describing the origins of the free software movement. Interested readers will find full and fascinating accounts in Steven Levy's *Hackers: Heroes of the Computer Revolution* and, more recently, Steven Weber's *The Success of Open Source*.[12] For our purposes, the following brief history will suffice.

Free Software

In the early days of computer programming, proprietary restrictions on access to and use of source code—the form of software code that can be read and understood by human beings—were rare. Most users did their own programming and exchanged source code according to the collaborative etiquette of a community made up of scientists and engineers employed in academic and corporate laboratories. But in the late 1970s and early 1980s, things began to change. Spin-off companies dedicated to producing proprietary software began to appear, triggering a diaspora of the best programmers from university laboratories and other public-sector institutions.

In the labs, the hackers' sharing ethic had been fairly closely aligned with the institutional missions of their employers. But in the pursuit of private profit, the new companies placed restrictions on sharing. Levy describes the impact of these changes on the hacker community:

> Even if people in the companies were speaking to each other, they could not talk about what mattered most—the magic they had discovered and forged inside the computer systems. The magic was now a trade secret, not for examination by competing firms. By working for companies, the members of the purist hacker society had discarded the key element in the Hacker Ethic: the free flow of information.[13]

Irrespective of discipline, many contemporary scientists would empathize with the personal impact of these new constraints on hackers' professional interactions. Secretive behavior of one kind or another is a fact of laboratory life in all areas of research. In molecular biology, as we shall see in subsequent chapters, the link between commercialization and increasingly restrictive access practices is both predictable and well documented. For example, a series of national surveys in the United States conducted between 1986 and 2002 indicates that a substantial proportion of academic geneticists who withheld data or materials requested by a colleague were motivated, at least in part, by commercial considerations.[14]

With the advent of dedicated software companies, all the hackers felt a disturbance in the force. But Stallman was an extreme case. By Levy's account, he went into deep mourning for the destruction of his beloved AI lab as it had once been—even to the point of telling strangers that his wife had died, leaving them to discover for themselves that he was referring not to a woman but to the old lab culture.[15] As Stallman came to terms with his loss, he remained determined not to go along with what he considered unethical software hoarding. Looking for a way to preserve the possibility of helping one's neighbor in the not-so-brave new world of proprietary software, he hit upon the plan of developing a suite of what he termed "free" software.

The word *free* did not refer to price. Instead, Stallman meant that software users should be at *liberty* to run a program for any purpose, to study how it works and adapt it to specific needs and to redistribute copies, as well as being free to improve the program and release those improvements. In other words, the software was to be "'free' as in 'free speech,' not as in 'free beer.'"[16] For Stallman, this plan represented a way to continue working with computers without compromising his own values. But it was also a radical competitive act. If every essential tool in a programmer's toolkit were to be available "free"—both fully technologically transparent and without legal encumbrance—what might eventually happen to the market for proprietary versions of the same tools? And what new markets might be spawned from the variety of products created with those tools?

The competitive impact of technology freedom is a recurring theme throughout this book. However, Stallman himself did not frame the question in market terms. Instead, he asked whether there was a program or programs he could write so as to re-create a *community* of cooperating hackers.[17] The obvious starting point was to develop a free operating system, partly because that was the kind of work Stallman did best, but mostly because an operating system is the core program without which a computer cannot run. A free operating system—including not just a basic kernel responsible for running other programs, but a full set of features such as the Emacs text editor, compilers, debuggers, and so on—would establish a platform on which other free software could be built, the foundation stone of a rebuilt community.

Stallman called the project "GNU," or "*G*nu's *N*ot *U*NIX." The name was an allusion to the popular UNIX operating system, whose many incompatible forks exemplified one of the adverse practical outcomes of proprietary restrictions on sharing. The GNU project was launched in 1984, accompanied by the "GNU Manifesto," a statement of purpose addressed to fellow hackers to request their participation and support.[18] The Free Software Foundation (FSF), principal organizational sponsor of the GNU project, was established the following year to promote the broader development and use of free software.[19]

Copyleft

One challenge to the success of the FSF's mission was the possibility that free software would be incorporated back into proprietary applications. (*Proprietary* is here used to mean nonfree, not noncommercial: Stallman had no objection to the commercial use of free software, provided it remained accessible and legally unencumbered.) To ensure that state-of-the-art technologies based on free software would continue to be available to the community at large, Stallman devised an ingenious twist on the proprietary approach to software licensing.

To appreciate this idea, it is helpful to first consider the proprietary approach. Many key innovations in biotechnology are protected by patents, but software source code was historically not regarded as patentable subject matter, instead being protected under copyright law as an original work of authorship.[20] Owners of copyrighted works are granted certain exclusive rights, including the right to reproduce and distribute the program and to prepare derivative works. Unlike patent protection, copyright protection applies to unpublished as well as published works, so source code can be simultaneously protected by copyright and as a trade secret.

Vendors of proprietary software typically use both types of protection to stop competitors from imitating their products. Making modifications to a computer program, or using parts of the program code in another program, is very difficult unless a programmer has access to the source code. The buyer of a proprietary software program—technically a licensee—generally receives only the binary or machine code version of the program (the version that is "executable" by a computer); the source code is kept secret. As Bill Gates has explained, "a competitor who is free to review Microsoft's source code . . . will see the architecture, data structures, algorithms and other key aspects of the relevant Microsoft product. That will make it much easier to copy Microsoft's innovations."[21] Even if a licensee does gain access to the source code, he or she is legally constrained by the terms of the copyright license agreement. Under a proprietary software copyright license, the licensor retains the exclusive right to redistribute or modify the program and authorizes the making of only a limited number of copies. Most licenses contain explicit restrictions on the number of users, the number of computers on which the program may be run, and the making and simultaneous use of backups.[22] A typical licensee may not rent, lease, lend, or host products, and may not reverse-engineer the licensed product (convert it into source code) except as expressly authorized by applicable law.[23]

Stallman's idea was to create a license that would emphasize the rights of software *users* instead of software *owners*. He called this type of license "copyleft" because it has effects that are the opposite of those of a conventional copyright license. (The copyleft symbol—a mirror image of the familiar circled "c" of copyright—is often seen accompanied by the caption "all rights reversed," a play on the copyright slogan "all rights reserved.") With guidance from Eben Moglen, now a law professor at Columbia University and pro bono general counsel for the FSF, Stallman drafted the archetypal copyleft license—the GPL or "GNU Public License," later renamed the "General Public License."

Under the terms of the GPL, the copyright owner grants the user the right to use the licensed program, to study its source code, to modify it, and to distribute modified or unmodified versions to others, all without having to pay a fee to the owner. The catch is that if the user chooses to distribute any modified versions, he or she must do so under these same terms. It is this final proviso that makes the GPL a copyleft license, giving it its famous—or, depending on your point of view, infamous—"viral" character. The purpose, according to Moglen, was to "create a hook that gives people access to a commons from which they can't withdraw"—with the user benefits of free software as the bait.[24] Put slightly differently, the role of a copyleft license is to create a collection of usable code that will grow over time as users contribute improvements back to the pool.

The Open Source Revolution

As its name suggests, the General Public License is a template license; it can be applied by any programmer to his or her own code. Linus Torvalds did this in 1991 when he released Linux, an operating system kernel built using tools made available by the FSF. At the time, Torvalds was a graduate student at Helsinki University. He wrote Linux because he wanted a UNIX-like operating system that would run on a PC—a need that was not satisfactorily met by any available commercial product.[25] There was no real prospect that the new software would make any money, even had Torvalds nursed such an ambition. In fact, according to Torvalds himself, the first release of Linux was barely usable: "[It's] a program for hackers by a hacker. I've enjoyed doing it, and somebody might enjoy looking at it and even modifying it for their own needs."[26] Though he did not expect much of a response, Torvalds was also seeking feedback and help building a better version of the program.

The rest is geek history. As Steven Weber tells the story, by the end of the year close to one hundred people had joined the newsgroup, many of them active contributors to Linux's further development. By the end of the decade, GNU/Linux (that is, the Linux kernel together with other operating system elements supplied by the GNU project—hereafter called "Linux") was a major technological and market phenomenon, built from the voluntary contributions of thousands of developers around the world.[27] Another half-decade on, and Linux has become the flagship for an entire techno-social revolution.

Though based on "free" software, that revolution is now generally referred to by a different term: *open source.* In pragmatic terms, free software and open source software are essentially the same thing—although this is an unsatisfying observation, since the main point of dispute between proponents of free and open source software is whether pragmatism should prevail in promoting the use of nonproprietary technology. Stallman wanted his fellow hackers to look beyond short-term expediency in their choice of programming tools, to see that the use of proprietary software raised serious ethical issues and to commit to providing and using an ethically acceptable alternative. By the late 1990s, those who coined the term *open source* wanted to see nonproprietary software more widely adopted, including in commercial settings. They considered the language of "software freedom" to be unnecessarily alienating to businesspeople. Throughout this book, I employ the newer, more widely used terminology.

The Linux project itself predates the term *open source,* but it is regarded as a turning point in the history of the open source revolution. The reason is that before that project began, most people involved in software development—including the free software community—believed that any software as complex as an operating system had to be developed in a tightly coordinated way by a relatively small, close-knit group of people.[28] But Linux evolved quite differently. Almost from the start, it was worked on rather casually by huge numbers of volunteers coordinating only through the Internet, which was just starting to take off around the early 1990s. Quality was maintained not by rigid standards or micromanagement, but by the simple strategy of releasing the code every week and getting almost instantaneous feedback from hundreds of users—a sort of rapid Darwinian selection of the mutations introduced by developers.[29]

Although Linux is often seen as the archetypal open source software development project, in fact it is only one of over 150,000 open source projects now under way, involving more than 1.5 million developers.[30] The number of developers in each project ranges from one or a few to many thousands. Similarly, the number of users of each program produced by open source methods ranges from a mere handful to hundreds of millions.[31] Other measurable characteristics, such as project-level governance and the type of application being developed, also vary widely, so that there is really no such thing as a "typical" open source project. Nevertheless, many projects do have features in common beyond the defining characteristic of code distribution under an open source license. For present purposes, then, the following (drawing on descriptions by innovation management scholars Eric von Hippel, Georg von Krogh, Andrea Bonaccorsi, and Cristina Rossi) is a reasonable approximation of open source software development.[32]

Many open source projects are initiated by an individual or a small group of individuals who are prospective users of the finished program. The intended use is often, though not always, connected with the initial developers' professional activities, which may be carried out in either a commercial or a nonprofit setting. This initiating group may develop a rough version of the program, perhaps with only basic functionality—enough to act as a "seed" for further development. This version is then made freely available for Internet download under a specific open source license, most often through a clearinghouse site such as SourceForge.net. Using tools provided by the site, initial developers may also establish discussion and mailing lists and other project infrastructure.

If this basic version of the program succeeds in attracting interest, some users may create new code and may post that code on the project website for others to use and to generate feedback. This second tier of developers may consist of independent programmers or hobbyists, but also often includes employees of firms that support the project for commercial reasons. New code of sufficiently high quality may be added to an authorized or official version of the program on the say-so of the project maintainers. This core group is often—at least at first—a subset of the initial developer group, though an important feature of the open source approach is that a project's leadership may change over time as participants' needs and priorities evolve.[33]

Even from this brief account, it is clear that voluntary participation and voluntary selection of tasks are central to open source development. Anyone can join an open source project and anyone can leave at any time; each person is free to choose his or her own contribution.[34] This freedom is made possible through a combination of liberal intellectual property licensing (discussed in detail in Chapter 5) and the availability of a core code base that anyone can modify for private use. More generally, open source production is characterized by transparency, exploitation of peer review and feedback loops, low cost and ease of engagement, and a mixture of formal and informal governance mechanisms built around a shared set of technical goals.[35] According to the Open Source Initiative (OSI), a nonprofit advocacy organization and license certification body established in 1998, these characteristics give open source technology development a clear advantage over the now-conventional proprietary approach:

> When programmers can read, redistribute, and modify the source code for a piece of software, the software evolves. People improve it, people adapt it, people fix bugs, and this can happen at a speed that, when one is used to the slow pace of conventional software, seems astonishing. The open source community has learned that this rapid evolutionary process produces better software than the traditional closed source model, in which only a few programmers can see the source and where everybody else must blindly use an opaque block of bits.[36]

In line with this pragmatic view of the benefits of software freedom, the OSI has directed considerable effort since its inception toward promoting the business case for open source software to both users and developers.[37] During that time, public and private sectors have embraced the use of open source software in a variety of forms. Open source software has penetrated government at all levels around the world and is used for major enterprise applications by small businesses through to large corporations.[38] Open source development is championed by IBM and Novell, while Microsoft has identified Linux as a serious competitive threat.[39] Open source software development projects support a wide range of commonly used applications such as OpenOffice, Gnome, and KDE; the database system MySQL; the GIMP, a competitor to Adobe Photoshop; popular programming and scripting languages Perl and PHP; and many others.[40] Open source enterprise has also met with astounding success on the stock market. In August 1999, distributor Red

Hat Linux went public with the eighth-largest first-day gain in Wall Street history.[41] In December the same year, VA Linux trumped that achievement with the most successful initial public offering of all time, its shares rising in value by 698 percent in the first day of trading.[42]

Yet despite the hype, it is still easy to underestimate the degree to which free and open source software influences daily life in our increasingly Internet-driven global society. After all, of the people around the world who actually *have* desktops (a minority, globally speaking), most don't use them to run Linux. The only way to grasp the true impact of open source is to understand that these days, desktop machines are just the tip of the computing iceberg. Tim O'Reilly—a businessman who has long supported, and been supported by, open source software development—uses a neat trick to illustrate the point. Speaking to audiences of computer industry professionals, he will ask how many of them use Linux. When only a very small fraction of the audience raises their hands, he asks, "How many of you use Google?"—and every hand goes up. As O'Reilly explains, "Every one of them uses Google's massive complex of 100,000 Linux servers, but they were blinded to the answer by a mindset in which 'the software you use' is defined as the software running on the computer in front of you. . . . But the operating system [is] only a component of a larger system [whose] true platform is the Internet."[43]

And the Internet is built, overwhelmingly, on open source software. Netcraft's monthly Web server survey for May 2007 shows the open source Web server software Apache continuing to dominate at just over 56 percent market share, compared with the next contender, Microsoft, at just over 31.49 percent.[44] Domain Name System (DNS) software is mission-critical for any firm that uses email or the Internet. The market leader, with more than 75 percent market share of DNS server software globally in mid-2006, is an open source program called BIND.[45] From the late 1990s until around 2001, Sendmail, an open source program whose main task is to handle the interchange and queuing of email messages on outbound and intermediate servers, was estimated to carry approximately 80 percent of the world's email traffic. Although this sector has diversified, so that the picture is now more complex, open source mailers continue to dominate. Market research firm IDC has predicted that the Linux operating system will account for 29 percent of units shipped into the worldwide server market by 2008; in the first half of the current decade, the growth in popularity of Linux servers consistently outran the growth of Windows servers.[46] FreeBSD, another open source operating system that is used by Yahoo! to run its directory services, is also one of the Internet's "killer apps."[47]

Open Source Biotechnology

Unsurprisingly, there have been numerous attempts to explain the astonishing success of open source. One of the earliest and best known is an essay by hacker, self-confessed gun-toting libertarian, and amateur anthropologist Eric Raymond, titled *The Cathedral and the Bazaar*.[48] Together with its sequels *Homesteading the Noosphere* and *The Magic Cauldron, The Cathedral and the Bazaar* gives an account of the open source software development process that emphasizes the distinction between centralized, hierarchical development efforts (the "cathedral") and decentralized, quasi-anarchical development of the kind Raymond claims is typified by Linux and many other open source projects (the "bazaar").[49]

Since 1997, when Raymond's essay was first published online, an enormous amount of ink has been spilt (or bandwidth consumed) enumerating the flaws in this metaphor. In the context of the original essay, the reference to bazaar-style development was a fairly straightforward allusion to the spontaneous, market-like ordering of transactions between leaders and contributors in an open source software project. In a market transaction, decision-making autonomy is key. Participation in the exchange is voluntary, and buyer and seller are separate entities who control their own resources and are not constrained to follow others' orders.[50] Though there now exist more sophisticated analyses of how the open source mode of production differs from *both* firm-based and market-based modes, the image of the bazaar has become an abiding symbol of the open source movement. Hot, dusty and cacophonous, redolent of exotic perfumes and vibrant with color, a bazaar is a place where ideas and cultures recombine like strands of DNA, a hotbed of technological and economic innovation—perhaps even the cradle of a new social order.

Consistent with these revolutionary overtones, the success of open source software development poses some serious challenges to conventional thinking. Why would anyone contribute software code to an open source project for free? How are contributions integrated into the program as a whole? Why don't open source development efforts fall apart before they get started? In the past few years these and related questions have seeded whole fields of scholarship in a broad array of disciplines, including economics, sociology, political theory, law, and innovation management.[51] My purpose is not to add to that explanatory literature—even though, given the cumulative nature of all research, this book could not have been written without the freedom to access and use its ideas. Instead, I want to explore whether and how key open source principles might be translated into a new context: that of biotechnology and its close industrial relations, pharmaceuticals and agriculture. Open source biotechnology would be a manifestation of the bazaar in a bioscience setting: hence, a "biobazaar."

The fundamental reason for undertaking this project is the existence of what seems an irresistible analogy between software and molecular biotechnology. Both technologies have enormous potential to help solve some of humanity's most pressing problems and enrich all of our lives. But their potential will not be realized without further innovation along lines that current industry participants may not yet even be able to imagine. Both industries are highly concentrated: the software industry is characterized by a near monopoly, while the pharmaceutical and agricultural industries, currently the main users of biotechnological innovations, are dominated by oligopolies. Disruptive innovation—the kind that leads to new product types, new industries, and substantial gains in social welfare—threatens the market position of these powerful corporations.[52] From the perspective of society as a whole, it is therefore a Bad Idea to let industry leaders gain too much control over the innovative process. Yet in both software and biotechnology over the past decade, more and stronger proprietary rights have contributed to a decrease in real competition, allowing large corporations—the beneficiaries of the status quo—to gain a stranglehold on the pace and direction of technological progress.

What causes this effect? Intellectual property rights are most often thought of as a way to facilitate bargaining and induce investment in the risky process of innovation. Even on this view, as we shall see in the next chapter, there may come a point at which more and stronger intellectual property rights hinder rather than help the innovative process. But intellectual property rights can also be thought of as private regulatory tools that enable their owners to order the market by fixing prices and controlling the availability of protected goods and services. Seen in this light, intellectual property rights may do less to promote innovation than to encourage rent-seeking via the pursuit of unproductive property rights that are used only to bolster private profits. Holders of intellectual property rights may find it is in their best interests to protect and extend those rights instead of devoting resources to research and development—especially research and development that could generate big changes in the technological landscape. If potentially disruptive innovation occurs elsewhere (for example, in smaller firms), oligopolists may seek to either buy it out and make it serve existing corporate strategies or else suppress it through the use of market power.

Of course, intellectual property rights are only part of the story of market power. But their importance is underlined by the lengths to which multinational corporations—led, not coincidentally, by both computing and pharmaceutical giants—will go to secure a strong global intellectual property regime. Recent scholarship documenting the political maneuvering that preceded the Agreement on Trade-Related Aspects of Intellectual Property Rights (TRIPS) enacted at the 1994 Uruguay round of the World Trade Organization's General Agreement on Tariffs and Trade suggests that the diffuse interests of intellectual property *users* cannot readily compete with the concentrated interests of large-scale intellectual property *owners* at the level of international trade negotiations.[53] Increasingly, the outcomes of these negotiations dictate the content of national intellectual property laws. In consequence, the prospects of achieving domestic law reform to ameliorate the worst effects of proliferating intellectual property rights are also bleak.

A key premise of this book is that open source principles of technology development, licensing, and commercial exploitation offer at least a partial solution to the innovation lock-down caused by extensive private control over scientific and technological information within a highly concentrated industry structure. Open source development shows how groups of volunteers can "collaborate on a complex economic project, sustain that collaboration over time, and build something that they give away freely"—technology that can "beat some of the largest and richest business enterprises in the world at their own game."[54] Because open source licensing makes use of existing intellectual property laws, open source strategies need not rely on domestic or international law reform. Open source is also highly resistant to the kinds of countermeasures traditionally adopted by monopolists and oligopolists when technological innovation threatens their market dominance. As Steven Weber points out, open source software is no marginal phenomenon, but a "major part of the mainstream information technology economy" that increasingly dominates those aspects that are becoming the leading edge in both market and technological terms.[55] It seems natural, then, to ask: Could open source do for biotechnology what it is already doing for software?

A Speculative Model of the Biobazaar

The only incontrovertible proof of the feasibility of open source biotechnology would be a demonstration that it already exists. Chapter 8 describes several examples of real-world open source biotechnology initiatives, but the open-source-inspired biobazaar is still at a very early stage of development. None of the current initiatives constitutes a mature working example of an open source biotechnology project.

Short of proof by example, one might wish to refer to a comprehensive empirical study showing which participants in biotechnology and related industries might choose to adopt the kinds of nonproprietary strategies employed by open source software businesses. It would be interesting to know, for example, how sensitive the trade-off between proprietary and nonproprietary strategies is to such variables as the nature of the specific biotechnology, the industry sector in which the innovator operates, sources of funding, and relationships between the innovator and other industry participants.

Unfortunately, no such study exists. This is partly due to the obvious practical difficulty of gathering data across a range of industries and myriad technologies in enough depth to illuminate not only the details of technology development but also the relative merits of competing business strategies. Just one of the challenges such a study would have to overcome is the extreme commercial sensitivity of the information needed to compare proprietary and nonproprietary strategies in any given business context. While public companies are legally required to disclose some of this information, it is often carefully disguised. The size of the challenge becomes immediately obvious when one reflects that accessing the relevant data would involve asking pharmaceutical companies to disclose the details of their research and development expenditures—something they are notoriously reluctant to do.

Thus, the practical challenges associated with a broad-ranging feasibility study would be substantial. But this is not the main reason why it has been impossible, up until now, to determine empirically whether an open source approach would be rational and practicable for some significant proportion of those engaged in biotechnology research and development. The deeper reason is that, by definition, current industry participants have all succeeded to a greater or lesser degree in engaging with the innovation system in its present form. In consequence, any study that relies on information from current industry participants must exclude those who are logically most likely to be open to unconventional, nonproprietary strategies.

To get a realistic idea of whether open source biotechnology could succeed, we need some way of incorporating the views and experiences of potential participants for whom conventional proprietary strategies have proved too costly or otherwise unworkable, as well as those who have not yet committed to any exploitation strategy. This includes potential innovators who lack the means or incentive to innovate under present conditions but might choose to do so given the opportunity to use and contribute to a cheap, accessible, adaptable, evolving, unencumbered—in other words, open source—toolkit.

Naturally, these potential contributors are much harder to reach for the purposes of empirical study than established industry participants, who may be identified through company websites, prospectuses, and the membership lists of industry networks and associations. Many potential contributors are at the periphery of the industry. Importantly, they include researchers in countries that have little or no biotechnology-related industry. Others are outside the industry altogether, engaged in activities and investments other than biotechnology research and development.

How can these voices be brought into a discussion of the feasibility of open source biotechnology? A detailed, realistic model of open source biotechnology can only be developed by tapping into the ideas and experience not only of those who are already engaged in this field of research and development, but also of those who are not. Clarifying the relationship between generic open source principles and the realities of biotechnology research and development will also help debunk a number of common objections to the feasibility of open source biotechnology that are based on either factual misconceptions or faulty logic.

Thus, a seemingly abstract approach to the question of feasibility is actually the most pragmatic. Not coincidentally, this methodology bears a strong resemblance to open source production itself. Information about whether an open source approach is likely to be rational and practicable in any given biotechnology setting is widely distributed, both within and outside the industry, among people whose identities cannot be centrally determined. Without a shared model of open source biotechnology, there is no common infrastructure on which to build a better understanding of the scope for open source strategies. But once such a model is proposed, it permits the holders of specialized information to contribute to an improved version according to their own interests and capacities. At the same time, it promotes the diffusion of nonproprietary thinking so that more innovators are empowered to experiment with an open source approach.

This is not to say that empirical research on this topic is not important and valuable. Framing interviews and detailed case studies could be particularly useful. Starting in late 2002, I carried out extensive fieldwork on the feasibility of open source biotechnology in major biotechnology research and development centers across the United States. This fieldwork included qualitative interviews with senior executives in agricultural and biotechnology firms; bioscience researchers and managers in public and private nonprofit organizations; fund managers in venture capital firms and major philanthropies; experts in intellectual property law and policy in universities, law firms, and international agencies; technology transfer officials; and leaders in the open source software community and the business community. Company executives were not prepared to open their books for inspection. However, they were willing to speak in general terms about strategic issues. Together with more recent interviews and documentary analysis, this research informs the discussion of open source biotechnology throughout this book.

Nevertheless, for the time being at least, the feasibility of open source biotechnology is a matter for informed speculation, not proof. The larger the number of people who are enabled to speculate on the basis of (1) the model presented here and (2) their own knowledge and expertise, the better.

In this introductory chapter I have highlighted the parallels between software and biotechnology and provided some background to the phenomenon of open source in the software context.

Chapter 2 begins with a brief history of biotechnology commercialization, then introduces the concept of a "tragedy of the anticommons"—underuse of a resource due to excessively fragmented property rights. The chapter goes on to give readers who are not already familiar with the commercial exploitation of biotechnology a sense of the industry setting in which it occurs. The final part of the chapter draws on this industry overview, asking whether anticommons tragedy has actually occurred in different sectors.

Chapter 3 examines the theoretical basis of intellectual property rights. The usual justifications take for granted the need for proprietary exclusivity—either as an incentive for self-interested actors to develop, disclose, or commercialize new technologies, or as a means of coordinating contributions to cumulative and cooperative technology development. But closer examination suggests that these justifications are not convincing, at least with respect to research and development in biotechnology. So why has there been a steady strengthening of intellectual property rights in this and other fields over recent decades? The answer that emerges from empirical research on the globalization of intellectual property law and policy is that major knowledge corporations have been engaged in systematic efforts to extend proprietary exclusivity in order to protect themselves from competition—a phenomenon called "the knowledge game." Chapter 3 concludes with a survey of the adverse structural effects of the knowledge game in biotechnology and related industries.

Chapters 4 and 5 introduce open source biotechnology as a possible antidote to these effects. Intellectual property rights are supposed to enable information to be traded in markets. But markets are only one way of coordinating diverse contributions to technology development. Chapter 4 characterizes open source as an instance of "bazaar governance"—a governance structure with incentives and control mechanisms distinct from those of markets, firm hierarchies, and networks—and describes the opportunities it offers self-interested actors to capture a return on private investments in innovation without relying on proprietary exclusivity.

Chapter 5 extends the discussion of open source as a generic model by articulating the underlying logic of open source licensing. My aim in this chapter is not to provide a comprehensive analysis of all the legal issues surrounding open source licensing or its application to biotechnology. Nor do I offer a ready-made suite of model licenses (or even best-practice guidelines, though I argue in Chapter 6 that this should be on the agenda for anyone interested in promoting the adoption of open source biotechnology licenses). Instead, I aim to formulate the basic principles and purposes of open source licensing independently of features specific to the software industry.

Having generalized the concept of open source to enable its application in the biotechnology context, we move on in Chapter 6 to the question of feasibility. Is there anything about biotechnology research and development that would make it impossible to implement an open source approach in that setting? Crucially, a form of bazaar governance already exists in biotechnology in the conduct of publicly funded, not-for-profit research and development. However, an open-source-inspired biobazaar would differ from the traditional biobazaar in several respects. Chapter 6 canvasses the enhanced use of Internet-enabled peer production methods; it also builds on the discussion of open source biotechnology licensing in Chapter 5 by exploring some of the practical problems and solutions that might arising in formulating working licenses.

Chapter 7 addresses a third difference between traditional and open-source-inspired versions of the biobazaar: the need to effectively integrate commercial as well as noncommercial contributions to open source biotechnology research and development. Chapter 7 addresses this issue by presenting the choice between proprietary and nonproprietary means of exploiting innovation as a strategic trade-off. Such a trade-off must take into account the benefits, opportunity costs, and actual costs of implementing an open source strategy. The chapter concludes with some thoughts on how the costs of open source biotechnology might be met through indirect contributions from a wide range of potential beneficiaries.

The final chapter considers the argument that, even if there is no in-principle obstacle to the implementation of open source biotechnology, the prevailing proprietary culture of the industry presents an insurmountable barrier. I argue that a relatively small number of entrepreneurial actors could catalyze a shift to a new equilibrium in which a substantial fraction of industry participants would no longer operate primarily on the basis of proprietary exclusivity. Several existing open source initiatives that are entrepreneurial in this sense are described. Even if these initiatives do not succeed on their own terms, their efforts to model open source in biotechnology lay the foundations for future change. The chapter ends with some suggestions as to how such initiatives might achieve the scale and momentum necessary to effect a genuine open-source-style revolution.

2

The Trouble with Intellectual Property in Biotechnology

We have seen how Richard Stallman and others fought back against intellectual property owners' interference with the free exchange of information among software programmers. For Stallman, the emotional force of the rebellion derived from his conviction that such interference had the capacity to destroy an entire technical community. But even interference that does not pose such an extreme threat can seriously harm innovation. It is likely, for example, that the Internet would not exist as it does today—as a public good, capable of supporting an enormous variety of next-generation applications—were it not for a continuing commitment to open protocols and standards on the part of those responsible for its infrastructure.

A parallel story can be told about proprietary interference with information exchange in biotechnology, but without the happy ending. The flow of information in academic biology before the advent of intellectual property rights was not, of course, entirely frictionless. But intellectual property has erected new barriers to the access and use of biotechnological information—barriers that add to, instead of replacing, those that were already there.

The impact of burgeoning intellectual property rights in biotechnology is difficult to measure. However, both theory and empirical evidence lead us to expect a range of adverse consequences. From the point of view of current participants in biotechnology research and development, these include frustration, delayed research outcomes, and wasted resources. But the parable of "how the Web was almost won"[1] teaches us that the greatest costs imposed by intellectual property rights in biotechnology may be opportunity costs borne by society as a whole. Could biotechnology's answer to the Internet be a cure for AIDS or malaria or an end to food insecurity? In the absence of an open source movement (or its equivalent) in biotechnology, what great innovations might the world never see?

Clearly, such questions are relevant not only to the future of biotechnology. Any scientific (for that matter, any human) endeavor that depends on information as a primary input is similarly vulnerable to excessive restrictions on its communication and use. Furthermore, the imposition of such restrictions, while it may be cloaked in the seemingly objective language of the courtroom and legislature, is an inescapably political act, in that it tends to promote certain institutional arrangements and patterns of social interaction over others.

A major theme of this book is that the reverse is also true. That is, removing existing restrictions on the exchange of ideas and information introduces new possibilities with respect to the arrangements by which, for example, we feed ourselves or seek to treat and prevent disease. Although it is impossible to predict the precise nature of new arrangements, there is no doubt that small changes in the way society regulates information flow can have big long-term effects.

To illustrate the point: In Chapter 1, I highlighted the role of the seventeenth-century gentleman scientist in the mythology of modern science. From the time of Charles II, the Royal Society of London for the Improvement of Natural Knowledge worked to establish the principle that scientific credibility arises from public scrutiny of experimental methods and results. Its members sought to facilitate such scrutiny through practices such as holding meetings to witness experiments, keeping detailed experimental records, and making them available in scholarly publications like the Society's own *Philosophical Transactions* (now the world's longest-running scientific journal). While early luminaries such as Robert Boyle and Isaac Newton continued to take a keen interest in occult forms of natural philosophy, the lifetimes of these individuals marked the transition from an age in which experimental science was conducted largely within secret societies to the era we now know as the Enlightenment, in which the free exchange of information was acknowledged as central to the scientific endeavor. This is not to say that overly simplistic sociological accounts of the "norms of science" should be resurrected. The point is merely that Western civilization would have developed quite differently, had Scholasticism and alchemy remained its principal means of learning about the natural world.

The originating members of the Royal Society were amateurs—that is, lovers—of natural philosophy. But they were also active in the political and economic realm, influencing the development of new regimes of governance in which citizens "gained standing to evaluate the performance of those in power."[2] That the Royal Society adopted a motto that cautions against the blind acceptance of any kind of authority, scientific or political, was no accident. The motto is "Nullius in Verba": roughly translated, "Don't take anyone's word for it." Nearly three hundred years later, Jewish-born philosopher Karl Popper, under self-imposed exile in New Zealand following the Nazi annexation of Austria, sought to articulate the connection between his understanding of science as a process of conjecture and refutation and the concept of an "open society."[3] According to Popper, people—whether as scientists or as citizens—are engaged in problem solving that must proceed through trial and error, for the simple reason that humans are always fallible. Politics is a matter of trying out tentative solutions to problems; in an open society, those solutions will be open to public criticism so that society as a whole can learn from its errors.

This freedom that scientists and philosophers have valued so highly—the freedom to check an assertion for oneself—also lies at the heart of open source software development, there expressed as the freedom to use, modify, and distribute source code. In this sense, our friend in the wig from Chapter 1 was not only laying the foundations of science as we know it. He was also starting the first, and by far the most successful, open source initiative.[4] In retrospect, it is clear that to describe these efforts as "revolutionary" is no hyperbole. The same may turn out to be true of today's open source movement in software, biotechnology, and perhaps other fields of knowledge production.

I revisit these issues in subsequent chapters. For now, let us take up the tale of intellectual property rights in biotechnology where Chapter 1 left off. How did all those corporate logos, signifying patent rights over basic laboratory techniques, find their way into my new-millennium study materials?

Intellectual Property Rights in Biotechnology

During the Second World War, scientists on both sides of the conflict chalked up a technically, if not always ethically, impressive record of national service. Partly as a result, academic science in postwar Western democracies enjoyed relative independence from external control. Large grants from national governments were distributed by scientists themselves through funding agencies such as the United States' National Science Foundation and National Institutes of Health (NIH), freeing the scientific community from both direct state influence and heavy reliance on industry support.[5]

This postwar relationship between science and the state has been characterized as a simple social contract. In return for money and autonomy, the scientific community was expected to supply a stream of technically trained personnel and discoveries to enhance the nation's health, wealth, and well-being.[6] But by the 1970s, governments were beginning to question the terms of that contract. Policymakers in the United States argued for a new bargain—one that would ensure that discoveries made in university laboratories would find their way out of the ivory tower and be put to work for the benefit of society as a whole.

In 1973, academic scientists Herbert Boyer and Stanley Cohen created the first genetically engineered organisms. Boyer and venture capitalist Robert Swanson founded a firm, Genentech, to commercialize the new recombinant DNA technology. Only five years after Cohen and Boyer's original experiments, Genentech announced the synthesis of human insulin—a lucrative therapeutic commodity—in bacterial cells.[7] This extraordinary early success captured the imagination of investors, and when the company went public in 1980, its stock underwent a more dramatic escalation of value than ever before seen in Wall Street history. By the end of 1981, more than eighty new biotechnology firms had been established in the United States—the start of a multibillion-dollar global industry.[8]

This fairy-tale version of the Genentech story perfectly illustrates the vision behind the revised social contract that was embodied in the United States' Bayh-Dole Act in 1980.[9] Overturning a longstanding presumption that publicly funded research could not be privately owned or exploited, the Bayh-Dole Act authorized recipients of federal funding to patent their research results and to issue exclusive patent licenses. Although the legislation itself was permissive rather than mandatory, it was widely read as imposing a duty on federally funded researchers to commercialize their discoveries. Universities established technology transfer offices to help scientists seek out and strike deals with commercial partners, and university–industry collaboration skyrocketed.[10]

The application of the Bayh-Dole Act was not restricted to the life sciences, but the effects of the new statute were greatly reinforced in that field by judicial decisions expanding the scope of patentability for biotechnology-related innovations. The rule in patent law is that an *invention* can be patented, provided it meets the criteria laid down in the patent statute, but a mere *discovery* cannot. (As we saw in Chapter 1, software was once regarded as unpatentable; this is because it was seen as falling on the "discovery" side of this critical divide.) Hence, before 1980 the United States Patent and Trademark Office (USPTO) had a policy of refusing applications for patents on living organisms. Processes devised to extract products found in nature might be considered inventions, but not the products themselves.

Accordingly, the USPTO initially refused a 1972 application by one of General Electric's employees, Dr. Ananda Chakrabarty, for a patent on an oil-slick-devouring bacterium he had isolated and modified using traditional (nonrecombinant) methods. Chakrabarty appealed, and in June 1980 the United States Supreme Court ruled in his favor.[11] According to the majority of the Court, patentable subject matter included "anything under the sun that is made by man."[12] The result was that living organisms that had been modified by genetic engineering or other means could now be regarded as inventions for the purposes of patent law.

Through the 1980s, further decisions consolidated this reversal.[13] By 1988 the turnaround was complete—as evidenced by the USPTO's willingness to grant a patent to Harvard University on its famous (or infamous) "oncomouse," an animal genetically engineered to be highly susceptible to cancer.[14] Combined with the Bayh-Dole Act, the effect of these judicial developments was that biotechnology patents became easier to obtain just at the time that many more inventors were being encouraged to seek patent protection.

A change to the U.S. legal system completed the pro-patent trifecta of the early 1980s. In 1982 a new specialist court—the Court of Appeals of the Federal Circuit (CAFC)—was established to deal with the growing complexity of patent law. To the satisfaction of those in the pharmaceutical and other industries who had campaigned for an expert court in the belief that it would advance their own interests, the percentage of district court decisions in favor of patent validity that were upheld on appeal more than doubled over the first five years of the CAFC's operation.[15] Penalties for infringement also became much more severe. The court could order "willful and wanton" infringers to pay treble damages and to cover plaintiffs' legal fees, with interest accruing pending any appeal. Business operations could be suspended until the outcome of an appeal was known, so that a defendant's business might be destroyed through loss of revenue—even if he or she ultimately turned out not to have been guilty of infringing behavior.

The establishment of the CAFC made the threat of a patent infringement suit much more effective than it had once been. Defendants were far more likely to lose—and losing could cost a lot. From a patent owner's perspective, of course, these developments made patents considerably more valuable—and not just as a way to protect returns on investments in innovation. Intellectual property owners' ability to use even an invalid patent to sink competitors points to the existence of a whole range of strategies (enumerated in Chapter 3) that extend the power of a sufficiently wealthy patent owner well beyond the basic right to exclude others from making, using, or selling a particular invention.

When patents are used as aggressive weapons, the ability to countersue may be the most effective deterrent. In that case, patent ownership becomes a necessity even for industry participants who have no interest in patenting apart from self-defense. The result is an escalating intellectual property arms race akin to the nuclear proliferation of the Cold War era. Biologists will recognize the pattern in which patenting activity triggers further patenting activity as a positive feedback loop, a common feature of biological systems.

The characteristic indicator of this type of feedback is an exponential growth curve. Patent statistics for the biotechnology industry through the 1980s and 1990s show just such a curve. In 1978 the USPTO granted fewer than 20 patents in the field of genetic engineering.[16] By 1989 the total number of biotechnology patents being granted each year had risen to 2,160, increasing even further to 7,763 new patents in 2002. Despite a flattening out of the curve since 1998, the average remains at more than 7,000 new patents issued per year in the United States alone.[17] Many of these patented technologies are used exclusively or primarily as research tools—that is, as means to the end of further socially and economically valuable innovation.

This brief history of the incursion of proprietary rights into the basic science of molecular biotechnology explains why patent notices are starting to show up in undergraduate teaching materials. But it says little about the potentially adverse impact of intellectual property rights in the life sciences, a multifaceted issue that is explored in the remainder of this chapter and in the next.

Fragment(ed) Ownership

A cover drawing from an issue of the technical magazine *Chemical and Engineering News* dated 12 October 1981—only a few years before Stallman started his GNU project—illustrates early tensions surrounding the prospect of individual scientists profiting from private ownership of research results. In the center, a scientist in a lab coat clutches a test tube, his arms and legs tangled awkwardly in the strands of a giant double helix. The contents of the tube are starting to spill, and the scientist wears an expression of surprise and alarm: he is the object of a vigorous tug-of-war between two other figures, each of whom has a firm grasp on one of his elbows. The prosperous-looking man on the scientist's right is neatly groomed and wears a suit and tie—the archetypal businessman. On the scientist's left, a bearded, bespectacled figure in tweed represents the academy. The two flanking figures glare at each other across the body of their hapless captive, who looks to the viewer in mute appeal for rescue from this unexpected dilemma.

Contrast this image with another, produced nearly twenty years later for the mainstream news press. Gregory Heisler's photograph of Dr. Craig Venter—a leading player in the worldwide effort to sequence the human genome and *Time* magazine's Scientist of the Year for 2000—portrays a single figure standing at his ease.[18] On his right side, Venter wears a white lab coat, stark against a black background. On his left he wears a dark business suit, silhouetted before a white background. The photograph's arresting monochrome composition draws attention to the subject's face, the only splash of natural color, at the focal point of the image. Venter returns the viewer's gaze with absolute composure, arms comfortably crossed, not quite smiling. There is no hint of self-doubt in his expression.

Viewed side by side, these images create the reassuring impression that despite early concerns about the impact of biotechnology commercialization on the integrity of the research process, scientists have since managed to reconcile any conflict between entrepreneurial and academic values. But this impression is false. An example that brings out many of the recurring themes of conflict in contemporary life sciences research is the project to sequence the human genome. The sequencing effort, which so captured the public imagination on both sides of the Atlantic in the early years of the new millennium, was subject throughout the 1990s to sporadic eruptions of controversy over private ownership of research results. In fact, the first such eruption occurred in 1991 and 1992, even as my first undergraduate cohort sat simultaneously learning about osmosis and using it to absorb the precepts of open, curiosity-driven research.

The figure at the center of the storm was none other than Craig

Venter himself, then an in-house researcher at the National Institutes of Health (NIH). Frustrated by the tedium and expense of systematically sequencing DNA from one end of a strand to the other, Venter adapted and automated a technique that used gene fragments to isolate the protein-coding sequences of genes.[19] These fragments, derived from products present in cells when genes are being expressed to make proteins, are known as expressed sequence tags (ESTs). ESTs can help identify genes, but they provide no functional information unless they can be matched to other genes whose function is already known; for this reason it was generally assumed that they were not useful inventions that could be protected under patent law. It therefore came as a shock when, in 1991, NIH lawyers filed applications for patents on several hundred of Venter's ESTs. The patent applications claimed exclusive ownership of not only the gene fragments but also the whole genes they represented and any proteins expressed by those genes.[20]

The genome research community was horrified. As a young man, Jim Watson had helped crack the double-helical structure of DNA.[21] Now head of genome research at the NIH, Watson declared that the decision to seek patents on ESTs was "sheer lunacy"; far from being inventive, their generation via Venter's automated technique could be achieved "by virtually any monkey."[22] Other leaders of the research community refrained from using such strident language, but on the whole they agreed with Watson. Nevertheless, the next year the NIH amended its patent application to include over two thousand more ESTs.[23]

According to observers, the motivation behind these moves was essentially defensive: the NIH was concerned about a patent "land grab" by Venter himself.[24] The possibility arose as a result of conflict between Venter and his employer over genome project strategy that ultimately led to Venter's leaving the NIH to establish his own private research institute, The Institute for Genomic Research (TIGR). Although TIGR itself was nonprofit, it was part of a broader business plan under which a commercial company owned by TIGR investors would be given up to twelve months' exclusive access to TIGR's ESTs, as well as the right to impose conditions on later uses of the data by academic scientists and others.[25]

Naturally, the prospect of obtaining advance knowledge of potential drug targets was extremely attractive to large pharmaceutical firms, and the TIGR business model proved highly lucrative when one such firm snapped up the EST rights for US$125 million.[26] Other biotechnology firms followed suit, selling access to EST databases and imposing reach-through royalties on any commercial developments that might arise from use of the patented sequences. John Sulston, then director of the United Kingdom's Sanger Centre and one of the leaders of the international sequencing collaboration, describes the resulting atmosphere as one of mutual suspicion: "Those who were working to map particular human genes either expected to secure patents on them, or were terrified that someone else would beat them to it."[27]

By 1994, community disquiet over the patenting of gene fragments had reached such a pitch that Harold Varmus, newly appointed director of the NIH, withdrew all the outstanding applications. But the patent race had a momentum of its own: what Sulston calls a "goldrush mentality" had taken hold.[28] As it turned out, the controversy was temporarily resolved by the intervention of a commercial player. Research scientists weren't the only ones appalled at the new genomics companies' attempts to levy a toll on the use of gene sequences, and in 1994 pharmaceutical giant Merck began funding an enormous effort to generate ESTs and deposit the data in public databases.[29]

Relative peace ensued, but not for long. By 1998 Craig Venter headed a new firm, Celera Genomics, soon to become famous as the private company that came close to beating public researchers in the race to complete the draft human genome sequence. When Venter announced Celera's intention of building a proprietary database of a different kind of sequence information, known as single nucleotide polymorphisms (SNPs), a shiver of déjà vu ran through the community. Once again, it seemed that patents would be sought on gene fragments that were not themselves useful or meaningful, but that might be crucial to future medical and other developments.

The Tragedy of the Anticommons

In a much-cited paper in the journal *Science,* academic lawyers Michael Heller and Rebecca Eisenberg put a name to biologists' fears: the "tragedy of the anticommons."[30] A tragedy of the *commons* occurs when a common resource, such as air or water, is destroyed through overuse because no individual user has sufficient incentive to conserve it. A tragedy of the *anti*commons, as the name suggests, is the opposite type of problem.

Economists have long recognized that where property rights on multiple components of a single technology are held by a number of separate individuals or firms, the development and commercialization of new products requires coordination among many different actors. In a world free of transaction costs—that is, a world in which everyone has perfect knowledge and there are no impediments or costs associated with negotiation—this would pose no problem, because property rights would be transferred through private bargaining to those who value them most. But in reality, transaction costs do exist, and they are higher the greater the number and complexity of negotiations needed to bring disparate rights together.[31]

In their article, Heller and Eisenberg explained that a situation in which multiple owners each have a right to exclude others from a scarce resource is a potential tragedy of the anticommons: if owners are unable to negotiate successfully for the bundling of rights so that someone has an effective privilege of use, the resource may be underused. The upshot is that granting too many patents or other intellectual property rights upstream can stifle socially valuable innovations farther downstream in the course of research and product development.

For individual researchers in the field of human genetics, the possibility of anticommons tragedy was a very personal concern: it represented the threat of many years of hard work and hard-won funding leading ultimately into a blind alley of license negotiations. For funding agencies, it meant the potential failure of long-term investments and an inability to deliver useful products to intended beneficiaries. For the big pharmaceutical firms, patents on SNPs meant the prospect of once again having to pay through the nose for the latest information about potential drug targets. Again they chose to intervene. In a deal initiated by Glaxo Wellcome and brokered by executives from Merck who had been involved in establishing the 1994 EST database, ten large pharmaceutical companies—together with the United Kingdom charity the Wellcome Trust, major sponsor of the Human Genome Project—responded by forming a consortium to fund the creation of a free and publicly accessible SNP database.[32]

The SNP Consortium may have averted a potential tragedy of the anticommons with respect to SNP data, but it was only a local solution to a much more widespread problem. We have already seen that an exponentially escalating number of patents on life sciences research tools over the past three decades has made the patent landscape more complex for both researchers and developers. In medical biotechnology, the commercialization of any given invention involves, on average, more patents and more patent holders than ever before.[33] Similarly, research tools in agricultural biotechnology are subject to numerous overlapping proprietary claims.[34] Depending on the complexity of a product, its development may involve the use of dozens of proprietary research tools; an often-cited example is that of GoldenRice, a genetically engineered rice variety developed using approximately seventy different patented technologies.[35]

Clearly, then, innovation in biotechnology often requires the coordination of property rights on multiple technology components owned by a number of separate entities. But this is not enough by itself to bring about a tragedy of the anticommons. Such a tragedy is not inevitable, even in the face of proliferating intellectual property rights, provided transaction costs can be kept low.

Contracting for Knowledge in Biotechnology

What sorts of transaction costs are associated with the transfer of information among participants in biotechnology research and development? According to economists Graff, Rausser, and Small, the economic literature on contracting for knowledge describes the following types of problems: (1) diffuse entitlement problems compounded by poorly defined boundaries among separately assigned rights; (2) value allocation problems resulting from different valuations of the same information by information providers and recipients; (3) difficulties in writing and enforcing contracts to deal with all the technological and commercial contingencies that are bound to arise in dynamic and uncertain environments; and (4) attempts by either or both parties to the exchange to gain a strategic advantage over competitors.[36]

Listed thus, these problems seem rather abstract. To gain a more concrete grasp of the realities they represent, in the discussion that follows it will be useful for the reader to take a mental step outside the well-equipped, well-funded laboratory of a researcher employed by a large company or top-ranking university in Europe or the United States. Let the reader imagine that he or she is a researcher who wishes to commercialize a new medical or agricultural biotechnology, and who works somewhere with limited resources at his or her disposal—a small independent plant-breeding company, perhaps, or a developing-country research institute.

Part of the reason for introducing this added dimension is to help dispel the common (though often unexpressed) assumption that biotechnology research and development is all about inventing new pharmaceutical drugs. While that is certainly its most lucrative application, it is by no means the only, or necessarily the most socially valuable, use to which such a broad enabling technology may be put.

More importantly, the suggested change of viewpoint draws attention to developing-country researchers as an important, though remote, constituency in the development of wealthier countries' policies on intellectual property. Proponents of molecular biotechnology have made much of its promise for the poorest citizens of developing countries in their struggle against extreme poverty, starvation, and disease—but the realization of this promise depends on the ability of researchers in developing countries to work through and around intellectual property barriers that have been erected, on the whole, for the benefit of people in developed nations.

For example, in a 2003 *Science* magazine article,[37] Gordon Conway and Gary Toennissen of the Rockefeller Foundation argue that farm productivity in the poorest parts of Africa would be improved by the use of genetic and agro-ecological technologies that can enhance yields without requiring substantial additional labor or capital inputs. Telling the story through the experiences of "Mrs. Namurunda," a composite character based on numerous real-life farmers, they show how deploying crop varieties designed using cheap biotechnology tools such as sequence-repeat DNA markers and advanced tissue culture techniques to perform well under low-input conditions can help create the conditions required for food security. Clearly, however, if local nongovernmental organizations and research institutes are unable to access such tools on reasonable terms, then the feel-good fiction of Mrs. Namurunda and her rising farm income is far less likely to become fact—something of which the authors are well aware, as demonstrated by their institution's support of open-source-like initiatives in the biology context (see Chapter 8).

The use of a similar narrative device to illustrate the potential hurdles to commercialization associated with intellectual property in biotechnology carries a risk of misrepresenting facts that are true in themselves by placing them in a false context. Software developers have an expression that is apt here: they talk of a "happy path"—the sequence of events one can expect if all goes to plan. In one sense, the description below presents an "*un*happy path," in that it strings together all of the obstacles related to intellectual property that a researcher might expect to encounter on the road to developing and commercializing an innovation. Of course, not all of these need arise in relation to any given project. On the other hand, this description could also be regarded as a "happy path," in that at every turn we assume that our imaginary researcher is able to overcome all obstacles and continue with his or her project. This, too, may be unrealistic.

To begin: Suppose that in order to develop your innovation, you need to make use of several existing technologies. Given the prevalence of intellectual property rights throughout medical and agricultural biotechnology, some of these technologies are likely to be patented. Even if you work in a university or other publicly funded institution, you cannot assume that a research exemption will protect you from expensive and risky patent infringement litigation.[38] The patent laws in your jurisdiction may not contain such an exemption. Even if they do, its scope may not be easy to determine.[39]

Your first step, then, is to conduct what is known as a freedom to operate ("FTO") analysis—a mapping of all the patents and other intellectual property rights, such as plant breeders' rights, that may touch upon the development of your new technology. The purpose of an FTO analysis is to see what sort of path you can forge through the patent thicket in your field. Some areas of research and development may turn out to be wide open in the countries where you intend to operate. In that case, you can innovate freely—in principle, though not necessarily in practice.[40]

In other places the branches of the thicket may be almost close enough to touch, perhaps leaving just enough room for you to slip carefully through. Where your path is completely blocked, and it is either too expensive or frankly impossible to invent around the blocking patent—for example, because it claims a nonsubstitutable tool such as a genome fragment—you will have to think hard about whether and how you can proceed.

It is at the stage of conducting an FTO analysis that you are likely to encounter the problem of diffuse entitlements and uncertain property boundaries. Where there are multiple patents in a given field, the cost of identifying which ones are relevant to a particular avenue of research may itself be prohibitive. To make a thorough search of the patent literature, you need access to sophisticated tools—most of which cost money—and you need to know how to use them. Once you have identified all relevant patents in all relevant jurisdictions, you must determine the scope of the owners' rights to exclude you from exploiting the patented inventions. This is an exacting process that requires careful interpretation and comparison of the highly technical, sometimes conflicting, and often deliberately obfuscating language used in patent claims. If you can afford it, you will pay a specialized lawyer or other professional to do the job. In that case, you will probably be advised that even the most careful and professional FTO analysis can never be conclusive.

The reason for this irreducible uncertainty is that the patent situation in biotechnology is not only highly complex; it is also dynamic. Thousands of new patents are granted in different jurisdictions every year, in some cases after long delays. In mid-2006 the patent on the ubiquitous polymerase chain reaction (PCR) technique invented by Kary B. Mullis in 1983 had still not issued in Australia, a technologically advanced country with a well-established patent system.[41] Meanwhile, technologies are also coming *off* patent in different jurisdictions at different times, sometimes after the standard twenty-year term, but sometimes because they have been abandoned by their owners through nonpayment of maintenance fees. Further, because the grant of a patent creates a presumption, and not a guarantee, of validity, a patent may be "on the books" while at the same time its validity is subject to challenge in litigation that is generally lengthy, unpredictable, and expensive.

Graff, Rausser, and Small give an example in the field of agricultural biotechnology: two early patents originally assigned to W. R. Grace & Co. would as written have given the company control over all genetically engineered varieties of cotton. These authors note that the scope of these patents was eventually narrowed following appeals and protests; such reversals have been relatively common in agricultural biotechnology, where patent litigation has at times been rampant.[42] Similarly, a 2003 survey of intellectual property rights related to Agrobacterium-mediated transformation (a key enabling technology for plant transformation) concluded that ownership of the most far-reaching patents in this area could not be determined because the broadest patents had yet to issue.[43]

Thus, even supposing that available records are complete and up-to-date—not always a safe assumption, given the sheer volume of business now dealt with by patent offices that are often, especially in the developing world, ridiculously underresourced—an FTO analysis can provide only a still snapshot of the ever-changing patent landscape. So even if you are sufficiently wealthy, well informed, and risk-averse to conduct your analysis *before* you invest resources in developing your new technology—also not a safe assumption—you can expect that the situation will be at least somewhat altered by the time you are ready to enter the commercialization phase. As a result, you may yet find that the commercialization of your new product or process is blocked by someone holding a patent over a technology that you have used or incorporated during research and development.

Suppose now that your FTO analysis has uncovered several blocking patents that you cannot, or do not wish to, invent around. You could choose to ignore them and hope that no one will notice or care; this is a common strategy, especially among public-sector researchers, but it is becoming increasingly risky. Your alternative is to negotiate a series of license contracts—written authorizations from the relevant right holders to use the patented technologies for specified acts, in specified markets, and for a specified period of time. Before you can begin negotiations, however, you need to know who has the right to grant a valid license.

Perhaps unexpectedly, patent records provide no more than a starting point for this search. They show the identity of the original patent owner (or owners), but information about any subsequent assignments is not so readily available. A relevant patent may have changed hands more than once, and the rights associated with it may already have been split up among a number of licensees. If you are unlucky, the patent owner—perhaps through a simple lack of foresight, resources, or experience with drafting patent licenses—may have granted an unnecessarily broad exclusive license to the technology to someone else.[44] If you are even less lucky, your efforts to make contact with a prospective licensor may be ignored or summarily rebuffed: outright refusal to grant patent licenses is not unknown, especially where the person requesting a license has nothing valuable to offer the licensor apart from a license fee. As we shall see, in the biotechnology and related industries, networks are important. It is not always worthwhile for key network actors to deal with outsiders merely for the sake of a one-off fee. Clearly, this dynamic tends not to operate in favor of researchers in developing countries or others at the industrial periphery.

Assume now that your FTO analysis is complete, your search for a prospective licensor has been successful, and both of you are prepared to negotiate terms. It is at this point that you may encounter the second and third types of problems mentioned earlier, relating to value allocation, contract writing, and enforcement.

For anyone who is not a licensing professional (or able to afford professional help), entering into a formal licensing agreement can be a daunting prospect. Except in the simplest cases, biotechnology licensing is a multistage process: once the parties have found each other, there may be several rounds of negotiation between first contact and final signature. The process is generally documented in both formal and informal instruments, including confidentiality or nondisclosure agreements, material transfer agreements, option agreements, term sheets or memoranda of understanding, and, increasingly, agreements to negotiate.[45] Key negotiated terms include provisions dealing with definitions of licensed subject matter, allocation of rights in derivatives of and improvements to the licensed technology, the degree of exclusivity of the license (exclusive, sole, or nonexclusive), field of use and territorial restrictions, sublicensing rights, responsibility for maintenance and enforcement of patents, warranties, indemnities and limitations of liability (especially in relation to infringement of any third-party rights in the technology and in relation to product liability), regulatory approvals, term and termination of the license, and, of course, remuneration. Often the license agreement deals with more than one set of intellectual property rights as well as with related trade secrets. Rights in personal property (such as biological materials) are usually transferred under a separate agreement (a material transfer agreement, or MTA).

As a researcher seeking to develop and commercialize a new technology, your anxiety over the technicalities of biotechnology licensing is likely to be compounded by value allocation problems resulting from different valuations of the same technologies by information providers and recipients. Your perception of the worth of a particular technology may differ from that of a prospective licensor because the technology plays a different role in your respective plans. For example, some of the tools you want to use to develop your new technology may be indispensable, while others may be readily substitutable. A tool that is not unique or is needed only to construct an optional extra feature in your new product is not mission-critical as far as you are concerned, and you are unlikely to value it as highly as a licensor whose whole revenue may be derived from licensing out a single technology.

This type of valuation problem is common in biotechnology because the biotechnology and related industries are made up of many different kinds of institutions, each with its own mission, resources, and constraints. In fact, all institutional types—universities, hospitals, private nonprofit research institutions, government agencies, small biotechnology firms, and major pharmaceutical firms or agribusiness concerns—recognize that such differences might sometimes justify asymmetrical terms of exchange. The problem is, they all think the asymmetry should work in their favor.[46]

This last finding points to the existence of irrational as well as rational biases in value allocation. Irrational biases include overestimating the likelihood that salient but low-probability events will occur—for example, that one's own research tool will turn out to be the lynchpin of the next blockbuster invention.[47] The problem of bias is exacerbated in relation to research tools because their ultimate value depends on the outcome of future research—which, of course, cannot be predicted at the time of the license transaction. In addition to this true or irreducible uncertainty, in most transactions licensor and licensee have asymmetrical access to knowledge about the technology. In such a case, the risk of opportunism on the part of the better-informed party makes it harder to agree on value, with both parties systematically overvaluing their own assets while disparaging the claims of competitors.[48]

Problems of valuation are not restricted to the remuneration terms of a license. They can pervade the whole agreement. For example, the field-of-use clause, which defines the purposes for which the technology may be used by the licensee, is often the most contentious part of a biotechnology licensing agreement and the most difficult to draft.[49] One reason is uncertainty as to whether all valuable applications of the technology have yet come to light. Another is that this is an area where the parties' interests are likely to come into conflict: in general, the licensor will want the field drawn narrowly, while the licensee will want it drawn broadly.

The difficulty of drafting field-of-use provisions illustrates the more general problem—number three in our earlier list—of drafting and enforcing contracts in an uncertain and dynamic technological and commercial environment. Uncertainty forces the parties to incorporate terms covering a wide range of possible contingencies, exacerbating the inherent complexity of most licensing agreements. Recall that, in our imaginary scenario, the prospective licensee must negotiate licenses to multiple research tools. Not only will the overall costs be higher the larger the number of negotiations to be concluded, but the complexity of each negotiation will also be increased because you must avoid committing yourself to terms in one contract that would prevent you from fulfilling the terms of any of the others.[50]

Institutional heterogeneity also causes difficulties in drafting and enforcing contracts.[51] For example, participants in a 1997 survey of medical biotechnology industry participants felt that their counterparts in other sectors did not appreciate the difficulties they faced in complying with particular contract terms.[52] But heterogeneity exists within as well as among institutions. In particular, your interests as a researcher may not coincide with those of the lawyers and businesspeople employed to negotiate contract terms on behalf of your institution. Especially if you work for a university, your employer's assessment of your research productivity may not take into account problems of access and freedom to operate; in any case, your main interest will be in acquiring needed research tools as quickly as possible. By contrast, your colleagues across campus in the technology transfer office must protect your employer from incurring obligations that would limit other funding or licensing opportunities or freedom to conduct future research.[53] You may find that the differences between your own sphere of expertise and professional culture and those of your institution's technology transfer professionals leads to serious communication problems, perhaps even mutual hostility.

With respect to transaction costs associated with enforcement of intellectual property rights, litigation is clearly the worst-case scenario. Patent litigation especially is notoriously expensive and complicated.[54] But there are also substantial indirect costs. Even the perception of potential litigation imposes planning costs, while the process of discovery imposes opportunity costs. News of a patent infringement suit generally causes both the patent holder's and alleged infringer's firms' stock values to drop. Where an agreement for the transfer of proprietary research tools establishes an ongoing collaborative relationship, "enforcement" costs might also be thought of as including the often nontrivial costs of monitoring the other party's activities and adjusting the terms of the agreement to changing circumstances.[55]

We are now reaching the end of our imaginary researcher's journey. But there is one last category of transaction costs in our earlier list: those that relate to strategic maneuvering by either party. Patent strategy is discussed more generally in the next chapter.[56] For now, it is enough to note that uncertainty again plays an important role in strategic bargaining over innovation inputs because it induces parties to push for license terms that limit their exposure to risk. A company that owns a research tool you need for your project might be concerned that granting you a license would lead to increased competition, undermine its patent position, or generate data that would trigger a tightening of regulatory requirements for its products.[57] In response, it may try to impose terms requiring you to assign or license any improvements back to the firm on an exclusive basis, requiring you not to challenge the patent's validity, or restricting the publication of your research results. Other types of restrictive license provisions include price and quantity restrictions, territorial restrictions, restrictions on sublicensing, and leveraging arrangements. (The latter include terms that bundle patented and nonpatented products together, extend the license to territories in which the licensor has no intellectual property rights, or oblige you to pay royalties until the last rights in a composite license expire.) Competition laws are designed to prevent these anticompetitive strategies, but if you are operating in a developing country and licensing technology in from overseas, you may not have their full protection.[58]

Unsurprisingly, while researchers in developing countries are particularly vulnerable to exploitative or overly restrictive license terms, all kinds of power imbalances come into play at every stage of the licensing game, not just at the level of strategy. Professional negotiators employ a range of tactics in an effort to dominate the interplay of demands, offers, concessions, and linkages by means of warnings, threats, bluffs, and displays of good faith. Because not all agreements thus concluded are mutually satisfactory, tactical and strategic power plays tend to exacerbate transaction costs associated with monitoring and enforcement.

Is Anticommons Tragedy a Reality?

Any reader in whom this Kafkaesque litany of potential obstacles has induced a strong desire for a couple of generic painkillers and a good lie-down might spare a thought for those determined scientist-explorers who spend many months—even years—hacking through the patent thicket with an increasingly blunt machete in one hand and a sheaf of contracts, overdue grant reports, and attorneys' bills in the other. (And Richard Stallman thought *he* had problems.) But even though intellectual property rights in biotechnology and related fields are stronger and more numerous than ever before, and even though the transaction costs to which they give rise are widely acknowledged to be substantial, these are only the *preconditions* for anticommons tragedy. Whether or not a tragedy of the anticommons has actually occurred in any given field of biotechnology remains a vexed empirical question.[59]

Part of the reason is that it is inherently difficult to conduct rigorous studies of bargaining breakdown in technology licensing markets. Projects may be abandoned without leaving a trace in institutional records, especially if they never get past the "ideas" stage; the reasons for abandoning any given project may be complex and difficult to ascribe to a particular obstacle or class of obstacles; and any records that do exist are likely to be confidential. But a more fundamental reason is that anticommons theory does not take into account the many strategies that industry participants can adopt to keep the wheels of innovation turning, despite the friction generated by intellectual property rights. These include inventing around blocking patents; going offshore; infringement under an informal, legally unsanctioned "research exemption"; and challenging patent validity through litigation.[60] Still other approaches rely on varying degrees of cooperation among industry participants. They include mutual nonenforcement, cross-licensing agreements, patent pooling, mergers and acquisitions, and intellectual property clearinghouses.[61]

Given the range of possible responses by industry participants to high transaction costs associated with proliferating intellectual property rights, it is not surprising that the net effects vary from one industry sector to another. The remainder of this chapter focuses on the two sectors that have so far seen the most significant private investments in biotechnology research and development.[62] These are health care and agriculture, sometimes known as "red" and "green" biotechnology.

The following preliminary overview of these sectors is intended to help readers who are unfamiliar with the business of biotechnology. Others may prefer to skip directly to later paragraphs describing the effects of high transaction costs in each sector.

It is helpful to preface an introduction to the business of biotechnology with a working definition of biotechnology itself. In its original and broadest sense, biotechnology encompasses any application of living organisms or their component parts to industrial products and processes. By this definition, the business of biotechnology is as old as baking, brewing, and the making of cheese and wine.

A more common contemporary usage, adopted throughout this book, is to restrict the term *biotechnology* to the range of tools and products based on molecular biology. Even in this narrower sense, the term refers to an extremely broad enabling technology that affects productivity in a wide range of industry sectors, including health care (drugs, vaccines, devices, and diagnostics); agricultural biotechnology (genetically modified organisms and food safety); industrial and environmental applications (biofuels and biomaterials); biodefense (vaccines and biosensors); and research tools (DNA fingerprinting, bioinformatics, microarray technology, and nanotechnology).[63]

It is therefore quite misleading to refer to "the biotechnology industry" as if it were a single, clearly defined entity.[64] In reality, biotechnology research and development supports a range of business models and noncommercial exploitation strategies whose distribution along the relevant value chain varies from one sector to another. Such an ecological view of biotechnology business models is important to our later discussion of the feasibility and potential impact of open source biotechnology. By contrast, any industry definition that stipulates a particular (proprietary) business model clearly begs the question whether business models that rely on proprietary exclusivity are the only viable means of commercially exploiting biotechnology innovations. Nevertheless, the biotechnology industry is often effectively defined by reference to the business model characteristic of many small dedicated biotechnology firms, sometimes including by extension the for-profit entities with which these firms interact upstream and downstream in their respective value chains.

In this book, the phrase *biotechnology and related industries* has a broader meaning: it includes all for-profit and nonprofit entities engaged in biotechnology research and development.

"Red" biotechnology—the sector related to health care—is not limited to drug development. It also encompasses devices, diagnostics, and "combinations" of two or more components marketed as a single product—for example, surgical mesh with antibiotic coating. Nevertheless, the big money is unquestionably in drugs.

Conversely, there can be no doubt that the development of a new drug—whether a traditional small-molecule drug, such as an enzyme inhibitor, or a product derived from living sources, known as a "biologic"—is both expensive and time-consuming. The process can take anywhere from seven to fourteen years, not including the decades of basic research—and in some cases many generations' worth of traditional knowledge, such as knowledge of the medicinal properties of plants—on which it relies. Estimates of the total cost range from US$100 million to around US$800 million, depending on how the calculation is made.

Because each new drug has its own unique history, any explanation of the drug development process must be a generalization. The process outlined below is also atypical, in that the majority of new drugs that come to market each year are actually minor variations on existing drugs. For these "me too" drugs, the development process is considerably shorter and simpler than for new molecular entities.

Further, in practice, progress from one stage of drug development to the next is not linear. Some degree of iterative learning is inevitable, and the necessary steps are undertaken in parallel as far as possible, in order to cut down on the time taken to bring the drug to market. Conceptually, however, the process may be broken into five stages.[65]

In the first stage of drug development, researchers seek to understand the molecular basis of a disease so as to identify specific molecules that are critically involved in the disease process and would therefore make good targets for intervention. The culmination of this research, which is nearly always carried out at public expense in universities or government laboratories,[66] is the validation of one or more drug targets.

The second major task in drug development is the identification of molecules that interact with these targets, known as drug candidates. It is at this stage of the process, which has been estimated to take one to two years,[67] that for-profit entities generally become involved for the first time.

As patent lawyer Philip W. Grubb explains,[68] the search for drug candidates may involve classical structure–activity correlations, rational drug design, or newer techniques such as high-throughput screening of libraries obtained by combinatorial chemistry or from natural sources. Besides identifying a promising molecule or family of molecules, researchers at this stage must work out how to prepare compounds or samples in the amounts needed for laboratory work, determine test tube or animal models in which to test molecular activity, and establish screening protocols. Screening involves testing the initial candidates for pharmacological and biochemical activity and then selecting and optimizing the "hits." As the name suggests, the outcome of the screening process is a smaller number of more promising drug candidates.

The third step in the drug development process, which typically lasts two to four years, is known as preclinical testing. Only an estimated one in one thousand drug candidates will survive preclinical trials to become "lead candidates."

Preclinical trials involve testing candidates *in vitro* (test tube experiments—literally, "in glass") and on animals. They are conducted in two stages. Stage I involves testing for stability and acute toxicity, conducting detailed pharmacological studies, and developing analytical methods to test for active substances. In stage II preclinical trials, researchers carry out pharmacokinetic studies (that is, studies of the bodily absorption, distribution, metabolism, and excretion of the drug) in animal models and test for subchronic toxicity, teratogenicity (capacity to harm a fetus), and mutagenicity (capacity to damage genetic material in living cells). At this stage, preparations are also made for the next stage of drug development: researchers investigate ways to scale up drug synthesis, develop an appropriate method of delivering the drug to humans, and produce clinical samples.

The fourth stage of drug development involves testing the drug in human volunteers. Clinical trials typically last four to six years and are the most expensive aspect of the overall development process. An estimated one in five lead candidates survives the clinic to be registered as a new drug.

Clinical trials consist of three distinct phases, all regulated in the United States by the Food and Drug Administration (FDA). Before trials can begin, the manufacturer must file an "investigational new drug application." Patent protection is also normally obtained at this stage because of the difficulty of maintaining secrecy once the drug is distributed to doctors and patients. (Drug manufacturers prefer not to file patent applications earlier than necessary because the time taken up by clinical trials eats into the term of patent protection.)

Phase I clinical trials involve testing small numbers of (usually) healthy volunteers to determine their tolerance for the drug and conducting pharmacokinetic studies in humans. Animal pharmacology studies may continue in parallel with these tests.

Phase II marks the beginning of controlled efficacy trials in a population of up to a few hundred patients with the relevant disease;

both doctors and patients are paid for their participation, the former much more than the latter. Usually, the effects of the drug at various dosages are compared with the effects of a placebo rather than with existing drugs—a feature of the process regarded as a flaw by those who lament the number of expensive "me too" drugs on the market. Together with the results of chronic toxicity and carcinogenicity studies in animals, the results determine whether the lead candidate will proceed to phase III.

Phase III clinical trials consist of multicenter patient trials involving hundreds to tens of thousands of patients. They aim to establish the final therapeutic profile of the new drug: indications, dosages and types of administration, contraindications, and side effects. In addition, phase III trials seek to prove the drug's long-term efficacy and safety, demonstrating its therapeutic advantages and clarifying interactions with other medication.

The final major step in drug development is the registration and launch of a new drug. In the United States, registering a new drug entails having a "new drug application" approved by the FDA. With the help of external expert committees, the FDA reviews the relevant data and may grant approval to promote the drug for specified uses and dosages. Generic drugs—drugs that copy a brand-name drug—also require approval, but do not need to undergo clinical trials, provided the manufacturer can show that they are equivalent to an already approved drug.

Having obtained FDA approval, drug manufacturers prepare for launch by providing information for doctors, wholesalers, and pharmacists; training sales staff; preparing packaging and package inserts; and dispatching samples. Following registration and launch, quality control studies and studies relating to new uses continue. These are sometimes referred to as "phase IV" clinical trials.[69]

What of "green," or agricultural, biotechnology? This industry sector exists within the broader framework of agribusiness, which encompasses all entities involved in the production, transformation, and provision of food, fiber, and chemical and pharmaceutical substrates.[70] Some are engaged in the primary production of agricultural commodities; others transform those commodities into value-added products. Both activities are served by suppliers of inputs such as seed and chemicals. Retailers and wholesalers sell both primary commodities and value-added products to consumers. The agribusiness sector also includes entities that support others all along the value chain through services such as education, banking, finance, investment, and legal and technical advice.[71]

Although green biotechnology encompasses animal and microbial biotechnology as well as plant biotechnology, the primary mechanism for capturing private returns on investment in this sector has been the sale of seed in the form of elite proprietary varieties, often coupled with complementary chemical products. For example, a brand herbicide or insecticide may be partnered with seed genetically engineered to withstand application of that particular chemical, as in Monsanto's Roundup/Roundup-ready product pairings. The process of seed production, marketing, and distribution varies from one type of seed, and one market, to another. However, as with pharmaceutical research and development, it can be generalized to a number of distinct stages.[72]

The first stage, plant breeding, aims to produce seeds that embody unique, marketable traits such as higher yields, disease or pest resistance, or traits specific to regional conditions.[73]

The second stage is seed production. Just as pharmaceutical companies engage contract research organizations to carry out clinical trials, seed companies typically outsource production and multiplication of marketable seeds to farmers, farmers' collectives, and private firms. These contract growers receive parent seed stock produced from the "foundation seed" developed by plant breeders. Contract growers are carefully selected, and their activities are closely managed and monitored by seed companies.[74]

The third stage is the conditioning of certified seed for sale to farmers. At this stage seeds are dried, cleaned, sorted, treated with insecticides and fungicides appropriate to the needs of the relevant market, and packaged for distribution and sale. This is also the stage at which seed is inspected for certification in accordance with quality standards, which in the United States are set and enforced by various state agencies.[75]

The final stage of the process is marketing and distribution. Distribution channels vary from one market to another. Large seed firms play a direct role in marketing and distribution in regional, national, and international markets. They also frequently outsource these tasks to wholesalers and retailers, farmers' collectives, and individual farmer-dealers.[76] Agricultural extension services perform an equivalent role with respect to niche crops and crops for which there is no market because of farmers' and consumers' limited capacity to pay. As public-sector extension services come under increasing financial pressure in both developed and developing countries, distribution of these crops has diversified to include for-profit firms, nongovernmental organizations, quasi-commercial government bodies, and various types of partnerships and cost-sharing arrangements. At an international level, International Agricultural Research Centers, members of the Consultative Group on International Agricultural Research (CGIAR), work to harness the potential benefits of biotechnology for the poor in developing countries and to preserve germplasm diversity.

In practice, of course, there is no clear distinction between red and green biotechnology. Indeed, the potential breadth of application of many biotechnologies, extending across a number of sectors, helps explain why intellectual property rights that may be reasonably well tailored to encouraging innovation in one sector can wreak havoc elsewhere. Nevertheless, this somewhat artificial distinction is helpful as a means of grasping the complex effects of increasing transaction costs.

What are some of those effects? In the context of red biotechnology, a 1997 study by Rebecca Eisenberg, commissioned by the NIH in response to the EST patenting controversy discussed earlier, suggested that in many areas of medical biotechnology, bargaining failure had become a reality. For scientists, bargaining breakdown was evidenced by significant delays attending the outcome of negotiations over material transfer agreements (MTAs), patent license agreements, and database access agreements. For university technology transfer officials, it was evidenced by resource problems arising from the need to renegotiate previously routine agreements and the need to resist attempts by outside parties to impose increasingly onerous terms. For private firms, the clearest sign of bargaining failure was the growing administrative burden of conducting negotiations and increased delays in research.[77]

More recent studies in the United States, Europe, and elsewhere suggest that an outright tragedy of the anticommons has not, in fact, occurred in most areas of medical biotechnology, largely because the value of many individual transactions is perceived to be high enough to outweigh the costs.[78] Yet the overall progress of research and development depends heavily on low-value exchanges of methods, materials, and data that allow incremental innovation. Even if the value of each individual exchange forgone due to bargaining failure is low, the aggregate social value of these exchanges may be considerable.[79]

One area of medical biotechnology where many agree that anticommons tragedy may be imminent is the field of molecular diagnostics. The human genome is continually evolving through mutations introduced by replication errors. These mutations contribute to human diversity, but some are also linked to disease. Genetic testing using basic molecular biology techniques can therefore be used to help diagnose a large number of pathologies, including infectious diseases such as HIV and hepatitis, genetic and neoplastic diseases such as cystic fibrosis and some cancers, and to help predict an individual patient's likely response to particular drugs—an emerging science known as "pharmacogenetics" or (more or less interchangeably) "pharmacogenomics." Molecular diagnostic techniques are also used outside a clinical setting in forensic identity testing (DNA fingerprinting), in detection of bio-threats, and in a range of agricultural and livestock applications.

Genetic testing patents exist in relation to a number of diseases, ranging from the relatively common (including breast cancer) to the very rare. Some patents cover techniques, while others cover the genes themselves. Gene patents can be extremely broad. For example, a claim that covers the observation for diagnostic purposes of an individual's genetic makeup at a locus associated with a particular disease effectively bars others from looking at that locus, irrespective of the method used. In the absence of any generally applicable defense to patent infringement, a particular concern to many practitioners in this field is the possibility of encountering a "submarine patent"—that is, a patent that is not issued or enforced until after the relevant test has been widely adopted, leaving users in a weak bargaining position with respect to the patent owner.[80]

Patent owners who are effectively free to dictate the terms of use of an established test may choose to license it broadly, asking only a reasonable royalty from all laboratories that offer the test to patients. On the other hand, they may choose to restrict the license to selected laboratories or even to a single test provider. In the latter case, the consequences can be severe. Monopoly control over a particular test tends to limit accessibility and increase costs, with obvious negative implications for the equitable provision of cheap health services. Quality assurance is also compromised: with only a handful of test providers, regulators may find it is not cost-effective to develop adequate proficiency testing. A lack of comparative data may allow systematic errors to go undetected—no small concern, given the seriousness of many of the relevant diseases and the expense and trauma associated with their treatment. Restricting the number of licensed providers also means fewer trained practitioners and reduced opportunities and incentives to develop improvements and advance the field generally.[81]

Given the close interaction between research and clinical practice in molecular diagnostics, clinical restrictions can have substantial indirect effects on research. If the rights to perform a particular test are held by a particular company and that company is one step removed from actually dealing with patients, the cross-fertilization that is crucial to the further development of this field may be lost.[82] The research thereby undermined ranges from the basic to the applied: for example, from the study of genotype–phenotype correlations to the development of drugs or gene therapy based on the gene sequence. Thus, ironically, private ownership of genetic tests may allow patients to find out that they have a disease but at the same time block the possibility of any treatment.

These problems relate to the exclusive or narrow licensing of genetic testing patents—the "single supplier," or "lock-in," problem that is so familiar in the software world and underpins the business case for using open source software instead. But even if all gene patents were broadly licensed for a reasonable fee, along the lines adopted by Stanford University in relation to the Cohen-Boyer recombinant DNA patent, the potential for anticommons tragedy would still exist. This is because the twenty-year term of a patent may be many times longer than the cycle of innovation in immature fields such as molecular diagnostics and many other areas of biotechnology research and development that do not share the long product cycles of new drug development. New technologies are constantly emerging that suggest new uses for existing patented products and processes, and fees that are perfectly reasonable for existing applications may be high enough to render new ones totally unfeasible. For example, it is already technically possible to screen many thousands of different genes or gene products on a single chip, but royalties set with single-gene applications in mind may make such high-throughput screening prohibitively expensive.

What about the effects of high transaction costs in green biotechnology? As I explain in the next chapter, agricultural biotechnology differs from medical biotechnology in that, until recently, almost all agricultural research was conducted in the public sector—and in developing countries, still is.[83] Nevertheless, ownership of much of the intellectual property generated by this research has over the years been transferred to the private sector. Although the remaining public-sector intellectual property portfolio in agricultural biotechnology is still strong when taken as a whole, its ownership is highly fragmented among different institutions. These now seem to show the classic symptoms of an anticommons.

As in the case of red biotechnology or biomedical research, the problem affects different types of institutions differently. With respect to the licensing out of technologies in exchange for revenue, some institutions own very little intellectual property, while others may own substantial portfolios but face difficulties in relation to effective management and marketing. With respect to licensing in (that is, gaining access to research tools owned by others) the clearest distinction is between research institutions in developed and developing countries.

Researchers in developed countries are frequently under the misapprehension that they do not need to obtain permission to use other people's technology, on the basis that they and their institutions are protected from any infringement action by a research exemption. In fact, patent laws do not always contain any substantial research exemption, and although the actual risk of being sued is in many cases still low, it is likely to increase as public and nonprofit institutions form closer relationships with industry. In the United States, for instance, legal precedent "does not immunize any conduct that is in keeping with the alleged infringer's legitimate business, regardless of commercial implications. . . . The profit or non-profit status of the user is not determinative."[84]

By contrast, researchers in less developed countries are inclined to overestimate the risks associated with using other people's technology, which are often not patented in the researcher's own jurisdiction. The perception that a particular technology is owned by someone else who would object to its use can be as effective in constraining researchers' conduct as the legal reality—although in any case, perceptions and reality are likely to converge as developing countries implement their obligations under international trade agreements to protect intellectual property rights, discussed in Chapter 3.[85]

Meanwhile, the private-sector agricultural inputs industry has undergone a startlingly rapid and comprehensive restructuring over the past two decades. Chemical giants like Dow and DuPont have moved aggressively into plant biotechnology, buying up all the larger national seed firms in North America and acquiring most surviving start-ups in the research-intensive agricultural biotechnology sector by the end of the 1990s.[86] The industry structure that has emerged is characterized by a "small number of tightly woven alliances, each organized around a major life sciences firm and vertically integrated from basic research and development through to marketing."[87] In this environment, new agricultural biotechnology start-ups are quickly integrated into the worldwide oligopoly once the promise of their technical innovations has been demonstrated.[88] The pattern is strongly reminiscent of the proprietary computer software industry.

There is evidence that this consolidation has been driven primarily by the need to avoid high transaction costs associated with multiple intellectual property rights.[89] But whatever the causal explanation for the dramatic series of mergers and acquisitions that took place in agricultural life sciences during the 1990s, the outcome is that most key enabling technologies are now owned by one or another of a small handful of firms. Agrobacterium-mediated transformation—a widely used method of integrating foreign genes that code for desired traits into a plant genome, allowing the regeneration of whole genetically engineered plants from the transformed tissue—is a case in point. In 2003 the Center for Application of Molecular Biology in Agriculture (CAMBIA), a nonprofit institute based in Canberra, Australia, conducted a survey of intellectual property relating to this important tool. Although most of the research that led to its development took place in public-sector institutions, the survey showed that of twenty-seven key patents in the crucial "vector" category, twenty-six were owned by a mere three private institutions. Further, all of the patents on binary vectors (which largely supersede earlier vector technologies) were held by a single multinational company. The same company held a dominant position in the category of dicotyledonous plants, which covers most commercially important crop plants apart from cereals.[90]

More broadly, by 1999 the top seven firms in the agricultural industry in terms of intellectual asset holdings controlled three-quarters of patents on transformation technologies and genetic materials, together with nearly all germplasm patents.[91] Moreover, although consolidation of intellectual property ownership appears to have reached its limit in relation to current technologies, it is likely that consolidation will increase rather than decrease with the emergence of new technologies.

Thus, in contrast to the public sector, no tragedy of the anticommons can be said to have occurred in private-sector agricultural biotechnology. Instead, private industry has side-stepped this outcome by forming institutions that lower transaction costs. Self-described anticommons optimist Robert Merges argues that in some contexts where there are multiple owners and transaction costs are high, an anticommons tragedy may be avoided if communities of intellectual property owners develop collective institutions to lower the transaction costs of bundling multiple licenses. Such institutions include copyright collectives in the music industry and patent pools in the automobile, aircraft manufacturing, and synthetic rubber industries and, more recently, the consumer electronics industry.[92] Merges sees these institutions as beneficial in their own right, observing that they provide a framework for standardizing techniques and for institutionalizing the exchange of unpatented technical information—advantages that might not be realized in the absence of strong property rights.[93] As we shall see in the next chapter, others are less sanguine. One plant biotechnology executive describes the outcome of a decade of mergers and acquisitions as "complete and total constipation."[94] Certainly, longer than expected development times for biotechnology-related products have been cited as one reason why, by the beginning of the current decade, oligopolists were starting to scale down their long-term investments in biotechnology-based crop improvement.[95]

This chapter began by describing the dramatic privatization of biotechnology research and development that has come about since the early 1980s through the legal mechanism of intellectual property. Intellectual property rights, especially patents, have been crucial in the development of today's biotechnology industry. But this development comes at a price, including the danger that industry participants may encounter so many legal obstacles that promising avenues of research and development remain unexplored—a "tragedy of the anticommons."

In this chapter, we have seen that the two essential preconditions of anticommons tragedy are (1) proliferating intellectual property rights, leading to fragmented ownership of complementary technological assets, and (2) high transaction costs associated with the transfer of those rights via licensing or other exchange mechanisms. These conditions are generally acknowledged to be widespread throughout biotechnology and related fields. Debate continues as to whether a tragedy of the anticommons has actually occurred in biotechnology. But such debate presupposes that in the absence of anticommons tragedy, intellectual property rights are good for innovation. Otherwise, why be concerned about anticommons tragedy *per se*? Unless there are convincing arguments in *favor* of intellectual property rights, evidence of increased transaction costs should be sufficient on its own to indicate a problem.

Earlier in this chapter I touched briefly on the conventional rationale for extending intellectual property protection to biotechnological innovations. The next chapter describes that rationale in more detail. It then offers an alternative analysis from which, in Chapter 4, I will build the case for open-source-style intellectual property management in biotechnology. In the meantime, the history of proprietary interference with scientific and technical exchange in both software and biotechnology is placed in context as part of a global agenda that has been dubbed "the knowledge game."

3

Intellectual Property and Innovation

The potentially adverse effects of intellectual property rights in biotechnology research and development can also be viewed with a wider lens—one that takes in not only the plight of current industry participants caught up in a potential tragedy of the anticommons, but also the broader structural effects of an overwhelmingly proprietary approach to innovation.

This chapter begins by outlining some established justifications for extending intellectual property protection to biotechnological innovations. Comparing these justifications with what is known about the nature of biotechnology research and development explains why intellectual property rights are often a hindrance rather than a help to the innovative process. But if theoretical arguments in favor of protecting intellectual property are incompatible with the realities of innovation, why are intellectual property rights so important to the biotechnology industry? We shall see that at least in some cases, intellectual property may have little to do with innovation and much to do with the ordering of markets to maximize private benefits. This insight reveals the breadth of problems that could be associated with excessive intellectual property protection; at the same time, it begins to suggest the shape of a possible solution.

My purpose at this stage is to give the reader an overall sense of the risks to innovation posed by excessive intellectual property protection. Importantly, I am not claiming a precise fit between the problems described here and the promise of open source biotechnology: the nature of the relationship between problems of anticommons tragedy and intellectual property–related barriers to competition, on the one hand, and open source as a possible solution, on the other, is explored in Chapter 4.

Neither is the picture presented here of the "knowledge game" intended as a complete and balanced portrait of biotechnology industry dynamics. The various mainstream uses of intellectual property rights in commercializing biotechnology are discussed in Chapter 7. My present goal is to articulate the thinking that increasingly leads scholars, activists, and entrepreneurs to resist (or try to circumvent) the current global trend toward stronger and stronger intellectual property protection. That thinking is sometimes radical. Much of it is unlikely to resonate with industry incumbents or the professionals who serve their intellectual property needs. Nevertheless, we shall see in later chapters (especially Chapter 8) that even if these dissident voices are irrelevant to the *present* structure of the biotechnology industry, it would be unwise to dismiss them as irrelevant to its future.

Instrumental Justifications for Intellectual Property Rights

Property rights are complex social technologies.[1] Understanding whether and how they should be applied to the results of scientific research—itself a complex social technology—is far from straightforward. But the fundamental tension is easily articulated. Intellectual property rights, including patents and copyright, are supposed to encourage innovation by various means that all involve allowing owners to restrict the flow of innovation-related information. Yet some degree of freedom of information exchange is necessary to the conduct of scientific research and technology development as social activities. Hence, property rights that are intended (in the words of the United States Constitution) to "promote the progress of science and the useful arts" can sometimes block such progress.[2] In the discussion that follows, my aim is to unpack the theories that underpin this basic tension.

All property laws have a distributive impact that tends to favor some groups at others' expense. Consequently, it is no surprise that the extension of legal protection to new forms of intellectual property has always been controversial. Patents—the dominant form of intellectual property in biotechnology—originated in Britain several hundred years ago, ostensibly as economic instruments to encourage the introduction of new technologies from other countries. In the sixteenth and seventeenth centuries, controversy surrounded the Crown's abuse of patents as a source of patronage and revenue.[3] The anti-patent movement reemerged during the British industrial revolution of the late eighteenth and early nineteenth centuries, and again in the 1860s, when the main focus was on the restraining effects of patents on industry and free enterprise.[4] Patent-based cartels were the targets of early twentieth-century "trust-busters" in the United States. At the end of the twentieth century, the shift toward an information-based global economy again sparked fierce debate over the nexus between trade and intellectual property policy.[5]

This ongoing debate about the legitimacy and scope of intellectual property rights has provided both sides with plenty of opportunity to hone their arguments. Proponents of patent rights have elaborated a range of justifications with both moral and economic bases. Moral arguments—including the Lockean notion that mixing one's labor with an idea should generate some right of ownership—carry some intuitive weight, and have helped shape the law in some areas.[6] Nevertheless, economic arguments have always been more influential. Certainly, as we saw above, the United States Constitution incorporates an exclusively instrumental justification for intellectual property legislation.

Economists Richard Nelson and Roberto Mazzoleni have identified four distinct rationales for granting patent protection to inventors.[7] First is the "invention inducement" theory: the anticipation of receiving a patent motivates useful invention. Second is the "disclosure" theory: patents facilitate wide knowledge about and use of inventions by inducing inventors to disclose their inventions when otherwise they would rely on secrecy. The third rationale, which featured prominently in discussions leading to the passage of the Bayh-Dole Act, is the "development and commercialization" theory: patents induce the investment needed to develop and commercialize inventions. Finally, the "prospect development" theory is that patents facilitate the orderly exploration of broad prospects for derivative inventions.

The first three of these arguments—invention-inducement, disclosure, and development and commercialization—all treat patent protection as a way of preempting market failure resulting from the "free rider" problem.[8]

A free rider is someone who imitates an invention and thereby gets the full benefit without having made any significant investment of time, effort, skill, or money. Assuming that inventions are easier to copy than to make, the theory is that a rational actor would not choose to invest the resources necessary to make a new invention, or to disclose or develop an existing invention, without some means of protecting that investment. Consequently, by conferring on the patentee or his or her assignees the exclusive right to commercially exploit an invention for a limited time, patent rights create a needed economic incentive to engage in the relevant phase of the innovation process.

The fourth theory—prospect development—has a slightly different emphasis. For a long time, economic discussion of patent rights centered on their role in facilitating product markets by allowing owners of goods to sell the goods separately from the associated intellectual property. But in 1962 Kenneth Arrow observed that patents and other intellectual property rights can be seen as a way to facilitate markets in information.[9] The integration of valuable information from a range of sources requires industry participants to bargain for the transfer of that information. But in the absence of patents, such bargaining runs into difficulties. If the owner of information discloses it to a prospective buyer, the buyer has obtained the information for free. On the other hand, if the owner does not disclose the information, the buyer will be unable to judge its value and will therefore be unwilling to pay the asking price. A patent allows the owner of the information to disclose it to prospective buyers without losing its value.

The prospect development theory builds on this view of patent rights as a way to facilitate markets in information (as, in fact, does the invention-inducement theory). Its first proponent, Edmund Kitch, was concerned with the inefficiency of permitting patent holders to capture larger returns on their investment in innovation than is actually necessary in order to induce the investment (a gap known to economists as "rent"). He postulated that granting broad patents on early-stage inventions allows patent holders to coordinate subsequent research and development within the limits of the patent claim (the "prospect").[10] If the patent holder has an exclusive right to exploit the new technological prospect, later arrivals will be unable to derive economic benefit from developing the prospect unless they negotiate directly with the patent holder to obtain licenses to the underlying technology. Thus, the patent holder becomes a link among all those working to develop the prospect, preventing wasteful duplication of effort and facilitating the transfer of information.

All four conventional economic theories of the patent system are routinely used, usually in combination, to justify intellectual property rights in biotechnology. For example, Sheila Jasanoff contends that patents played a foundational role in the development of the biotechnology industry in the following ways. First, the extension of patentability to inventions in the life sciences created "property rights in things that were previously outside the realm of what could be owned," so that "these objects became commodities that could have value, be exchanged, circulate in markets, and foster productivity." Second, in the early stages of technology development, biotechnology companies had no marketable products, so patents served to convince venture capitalists that there "would be something of future value to justify present investment." Third, patents provided some assurance to investors that they would not be swamped by legal disputes over rights to any future products. Finally, patents offered a way to make sense of the "competing claims of participants in an increasingly complex web of invention."[11]

Of the four theories listed earlier, the only one not incorporated into this fairly typical analysis is disclosure theory. However, disclosure theory makes frequent appearances in discourse justifying intellectual property rights in biotechnology inventions. For example, *amicus curiae* ("friend of the court") briefs submitted by Genentech and the Pharmaceutical Manufacturers Association in the *Chakrabarty* case argued that allowing patents on living organisms would keep genetic engineering research "out in the open" because patents compelled publication of the means and methods that led to a patentable product.[12]

The four conventional theories of intellectual property described in this section embody three basic assumptions.[13] The first is that technological information is easy to copy, resulting in a lack of incentive to invest in innovation. The second is that the market is—or should be—the primary mechanism for the exchange of technological information among industry participants. These first two assumptions together give rise to the perception that patents are necessary in order to overcome market failure due to the inappropriability of investment in technological innovation. The third assumption, which underpins the prospect development theory, is that central coordination of research and development activity by a patent holder is more efficient than decentralized innovation carried on in the absence of patent rights. As we will see, all of these assumptions are open to question in the context of biotechnology research and development.

Intellectual Property and the "Story of Science"

Ever since the days of Plato and Aristotle, the pursuit of truth has been regarded as a communal activity. The very concept of "scientific progress," which dates from the sixteenth and seventeenth centuries, has long been linked with (1) an ideal of free and open dissemination of scientific information and (2) the notion that scientific knowledge should be public knowledge.[14]

For example, nineteenth-century political philosopher John Stuart Mill, in laying the foundations of our modern notion of freedom of speech, took it for granted that unfettered communication and disputation were essential to the extension of human knowledge. "The only way in which a human being can make some approach to knowing the whole of a subject," he wrote, "is by hearing what can be said about it by persons of every variety of opinion, and study all modes in which it can be looked at by every character of mind. No wise man ever acquired his wisdom in any mode but this; nor is it in the nature of human intellect to become wise in any other manner."[15] Twentieth-century philosophers held similar views. Karl Popper believed that the growth of scientific knowledge depends on individuals making guesses or conjectures that are then subjected to communal criticism.[16] Michael Polanyi, a physical chemist turned philosopher of science, likened scientists to a group of people working on a jigsaw puzzle, who cooperate by "putting the puzzle together in sight of the others so that every time a piece of it is fitted in by one helper, all the others will immediately watch out for the next step that becomes possible in consequence."[17]

In the 1940s, pioneer sociologist Robert K. Merton theorized that a norm of common ownership of research results—the norm of "communism," or "communalism"—functioned together with other scientific cultural norms to align the interests of individual scientists with the overarching institutional goal of scientific progress.[18] Although Merton's theory is no longer current, having been superseded by constructivist accounts that are both more nuanced and more thoroughly grounded in empirical observation, it deserves attention for two reasons apart from mere historical interest. First, its influence on legal scholarship concerning intellectual property in biotechnology continued well into the 1980s. Second, Merton's account resonates strongly with hacker lore seeking to explain the success of open source (for example, Eric Raymond's essay *Homesteading the Noosphere*).[19]

Consistent with both earlier and later views, Merton's theory reflected an understanding of scientific research as essentially cumulative and cooperative in nature. In order to collaborate and build on each other's work, scientists needed access to a common fund of knowledge. The norm of communism was supposed to encourage scientists to contribute to this common fund by communicating the results of their research to other scientists; it ensured that secrecy was condemned, while timely, open publication was rewarded.[20]

The norm of communism was also supposed to preserve scientific knowledge as public knowledge—that is, knowledge that is able to be freely used and extended in the public interest. Merton wrote: "The substantive findings of science . . . constitute a common heritage in which the equity of the individual producer is severely limited. An eponymous law or theory does not enter into the exclusive possession of the discoverer and his heirs, nor do the mores bestow upon them special rights of use and disposition. Property rights in science are whittled down to a bare minimum by the rationale of the scientific ethic. The scientist's claim to 'his' intellectual 'property' is limited to that of recognition and esteem."[21]

As we saw in Chapter 2, even if this was true under the original social contract between science and the state, negotiated in the United States by Vannevar Bush after the Second World War, it has not been the case in biotechnology since at least the 1980s. Throughout that decade, when the boundaries between commerce and the academy were being progressively eroded, concerns about the impact of proliferating proprietary rights were expressed in terms of Merton's theory of scientific norms.[22] Merton himself had warned that the scientific ethos can be subjected to serious strain when the larger culture opposes a scientific norm. In the case of the norm of communism, he saw such conflict arising out of the incompatibility of the scientific norm with the definition in capitalist economies of technology as private property. The prospect that commercialization of university research might tip individual scientists' balance of incentives away from contributing to a common fund of knowledge and toward restrictive communication practices was exactly the kind of strain Merton had in mind. In fact, he specifically referred to patents, with their exclusive rights of use (and, he remarked, often nonuse) and to the suppression or withholding of knowledge—for example, through trade secrecy—as being opposed to the rationale of scientific production and diffusion.[23]

The norms Merton described were not supposed to be codified or necessarily explicit. Rather, they operated as "prescriptions, proscriptions, preferences and permissions . . . legitimated in terms of institutional values . . . transmitted by precept and example and reinforced by sanctions." Their existence could, it was argued, be inferred from a moral consensus among scientists expressed "in use and wont, in countless writings on the scientific spirit and in moral indignation directed toward contravention of the ethos."[24]

Of course, as later sociologists have pointed out, the fact that scientists (and by the same token, open source software programmers) talk a lot about what constitutes appropriate conduct does not mean there is any such thing as a "norm of science" in an objective sense. Disagreements among scientists cannot really be treated as minor deviations from a consensual norm: as Harry Collins, Trevor Pinch, and others have shown, controversy is ubiquitous in science.[25] Empirical research demonstrates that violations of Merton's norms are frequent, often rewarded, and sometimes even important for scientific progress: for example, Ian Mitroff has presented substantial evidence of successful "counternormal" behavior.[26] More subtly, scientists who are doing their best to follow norms of disinterestedness, objectivity, and rationality may find themselves led to very different conclusions about what constitutes conformity with these norms. Since no rule can specify completely what is to count as following or not following that rule, we cannot assume that any norm can have a single, unambiguous meaning independent of the context in which it is applied.[27] Nevertheless, normative talk does appear to have some persuasive force, especially in forming the sensibilities of newcomers—a hypothesis borne out by my own experiences as a novice scientist, described at the start of Chapter 1.

Be that as it may, by the time Cohen and Boyer made their historic discovery of recombinant DNA techniques in 1973, most sociologists of science had begun to doubt the reality of normatively controlled behavior, preferring instead to treat references to norms in the course of scientific debate as rhetorical tools or rationalizations for self-interested behavior.[28] Influenced by parallel developments in the history and philosophy of science, including the work of Thomas Kuhn,[29] sociologists questioned the existence of any distinctive scientific ethos, at the same time becoming more skeptical of absolutist claims about scientific progress. According to contemporary sociologists of science, the process of conjecture and refutation does not necessarily bring scientists closer to objective truth—or even, for that matter, to consensus. Instead, science and technology are seen as historically situated, culturally contingent, dynamic constructs generated within social parameters that reflect multiple identities and interests and are characterized by a high degree of conflict and competition.[30]

It does not, of course, automatically follow that scientists construct their account of the natural world in a manner indistinguishable from any other form of storytelling. Most scientists would agree that Mertonian idealism no longer explains how science gets done in practice—if it ever did. Despite the caricature that unfortunately infects some sociological discourse, most natural scientists would also agree that science and technology are social creations and that the pronouncements of scientists are far from infallible. However, many would contend that science differs from other ways of knowing, especially in that it incorporates mechanisms for continuously improving its "stories" as reliable explanations of natural phenomena.[31] While there may be some genuine divergence between this point of view and the views of sociologists practicing in the sociology of scientific knowledge (SSK) tradition that began in the 1970s, the differences have been vastly exaggerated.[32] Like their colleagues in the natural sciences, most SSK practitioners care deeply about science; many describe themselves as employing the scientific method in their own research, and some are highly knowledgeable about the content of scientific theories in their field of empirical study.[33] Their work exposes uncertainties and failures in the practice of science as a social activity, but most thoughtful sociologists of science nevertheless believe that science is the best method we have for finding out about both the natural and the social world.[34]

How does the constructivist view of science differ from the Mertonian view in relation to intellectual property rights? Constructivist theories suggest that patents and other intellectual property rights need not be always and everywhere inimical to scientific production and diffusion. On the other hand, neither is private ownership of scientific data—broadly defined to include findings, results, samples, materials, reagents, laboratory techniques, protocols, know-how, experience, algorithms, software, and instruments—necessarily benign. Nowadays, if a sociologist or philosopher of science were asked whether intellectual property rights obstruct scientific progress, the response would be something along the lines of "It depends." But on what does it depend? Partly, to be sure, on what is to be counted as scientific progress—which in contemporary sociological theory is no longer a given. But for any definition of scientific progress, it also depends on the specific access practices that prevail in the relevant field.

To understand why, consider Stephen Hilgartner and Sherry Brandt-Rauf's "data stream" model of scientific research.[35] In keeping with its constructivist roots, the data stream model subjects the concept of "data" to social analysis, treating data not as well-defined, stable entities—the end products of research—but as elements of an evolving data stream composed of heterogeneous networks of information and resources. Hilgartner and Brandt-Rauf describe these elements as ranging from mundane components of the ordinary social infrastructure (such as water, electricity, or computers), through elements that are widely available, though they may be specific to a particular research area (such as journal articles or assay kits), to specialized elements that are not available through public channels but may be disseminated through personal contacts, and finally to novel or scarce elements available only via one-off arrangements. Clearly, data access problems are more likely to arise in relation to the last two categories of data than the first two.[36]

Different elements in a data stream have different information status. According to Hilgartner and Brandt-Rauf, at one extreme the elements of a data stream may be generally accepted as reliable and valuable, while at the other they may be so uncertain that even the scientists who produce them doubt their credibility or usefulness. Data are constantly interpreted and reinterpreted through the research process, so that scientists' perceptions of the reliability and value of particular parts of the data stream vary over time. This can be important in decisions about access, as scientists ask themselves whether data are "ready" for dissemination, or how much data are "worth."[37]

Finally, data streams are composed of chains of products. Scientists record data using primary inscription devices, such as X-ray film or electrophoresis gel, then convert the data into second-, third-, or fourth-order inscriptions. Materials may be processed and purified; electronic information may be subjected to a series of manipulations; and so on. These translations and conversions affect access practices because, besides altering the information content and material form of the data, they change the purposes for which data can be used.[38]

Not only are data streams themselves heterogeneous, but transactions involving data are negotiated within complex research networks. The transactions themselves help to construct and maintain such networks, in which each actor is linked with many other people and organizations. Hilgartner and Brandt-Rauf point out that a decision to grant access to data may involve many parties, including scientists, sponsors, and university or corporate bureaucrats, all with different goals and differing claims to portions of the data stream, who may disagree about the best way and time to disseminate the data. Similarly, they note that audiences or markets for data are heterogeneous, perhaps including competing research groups, potential collaborators, authors of studies with conflicting results, gatekeepers who control key resources, potential markets for research-based products, or venture capitalists.[39]

While traditional models of scientific exchange emphasize peer recognition as a scientist's primary reward for discovery, with publication as the primary legitimate means of achieving recognition, the data stream model acknowledges that open publication is only one of many possible mechanisms for disseminating portions of a data stream. Data may also be bartered in negotiations with prospective collaborators or sponsors, distributed to selected colleagues, patented, transferred by visitors being trained in new techniques, provided to a limited group on a confidential basis, bought and sold, prereleased to existing sponsors, or kept in the lab pending future decisions about disposition. Each such mechanism is associated with particular incentives and strategic considerations, which are specific to particular areas of research. For example, scientists can choose to exploit the competitive edge conferred by possession of unique data by restricting access and using it to produce more data or by providing carefully targeted access. Alternatively, they may choose to provide widespread access in order to enhance their scientific reputations. Other relevant factors include timing, the portion of the data stream to be made available, and the costs and logistics associated with different modes of access.[40]

This brief introduction to a constructivist account of scientific exchange helps explain why extending the scope of intellectual property rights to biotechnological inventions has caused transaction costs to skyrocket. In the absence of detailed legal regulation, scientific exchange is governed in ways that reflect, because they arise out of, the continuity and variety of data streams and research networks. (This is not, of course, to say that they are functional in the sense of promoting any overarching goal.) In contrast, the legal approach to data ownership involves "plucking items from the data stream and placing them into discrete categories" so as to "designate an end product" that qualifies for a much more rigid and general type of protection—patent, copyright, trade secrets, misappropriation, contract, or conversion.[41] As noted in Chapter 2, the application of intellectual property laws to subject matter previously regarded as falling outside the limits of legal protection complicates transactions by introducing an additional layer of regulation. But the disparity between reductionist legal rules and the complex, messy reality of scientific research implies that the effect on transaction costs is not merely additive. Applying legal rules also creates a selective pressure that favors some kinds of transaction over others and can lead over time to the kind of paradigm shift described in Chapter 1.

Intellectual Property and Information Markets

To understand the nature of this selective pressure, it is helpful to revisit the idea of intellectual property rights as facilitating a market in information. This time, however, we take a more nuanced view of the nature of information itself, in line with both constructivist theories of scientific "data" and insights from transaction cost economics.

Information differs from material goods with respect to both production and dissemination. Information is not consumed by use, as in the case of goods that economists designate "rival." Rather, it grows with use, and its social value is enhanced through dissemination. The cost of *producing* information is independent of the scale on which it is used, and much of the cost of *transferring* information is incurred by the recipient in absorbing the information and allocating scarce resources to its use. All these factors affect the efficiency of market mechanisms for transferring information, but their impact depends on the type of information being transferred.

Technological information exists on a continuum of what economists call "codification." For example, economist Thomas Mandeville notes that although some technological information is codified into machines, blueprints, technical and trade journal articles, patent specifications, and the like, much exists in less codified form. Codification is formalized learning: it represents knowledge "organized into a pattern" and, ultimately, embodied in a tangible object. A technique is not codified unless it consistently yields the same output—in other words, unless it is reliable. For this reason, highly codified or tangible information appears only after substantial learning has already taken place. At the other extreme, uncodified information consists of undeveloped ideas and unarticulated know-how—it is "pure," intangible information.[42]

Although most real-world information lies somewhere in between these two extremes, argues Mandeville, the bulk of economic phenomena associated with innovation occurs toward the uncodified end of the continuum. The proportion of technology that remains uncodified in any given field at a particular time is determined by both the technology's newness and its inherent complexity. Generally, the older or more mature the technology, the more it has been codified. A new industry based on a new technology—such as biotechnology—is in a fluid situation where most relevant technological information has yet to be codified.[43]

The ease, speed, and mode of diffusion, transfer, or imitation of technological information all depend on its degree of codification. Highly codified information can be communicated without the need for personal interaction. But uncodified information is best communicated in person, through practice and "learning by doing." Because transfer costs are higher the less codified the information, the conventional assumption about ease of copying—the assumption that underpins the free-rider argument—holds only for the highly codified end of the information spectrum. For uncodified technology, information and user costs inhibit imitation even in the absence of patent protection.

As an example, when I worked in a molecular biology laboratory in the early 1990s, there were a number of important techniques that could be reliably performed only by people who possessed a certain specialized knack; these people were said to have "magic hands." The only way to acquire the magic seemed to be through a process akin to apprenticeship. But personal instruction combined with hard work were not always enough to reproduce the necessary skill: some people just seemed to have a talent for certain jobs that others did not. In other words, their competence was highly uncodified. Interestingly, the same tasks that required magic hands a decade or so ago are now routinely and reliably performed in undergraduate laboratory classes. Gradually—in some cases, even imperceptibly—these processes have become more codified over time.

The fact that uncodified information is more costly to transfer than codified information suggests that markets may not be the most efficient means of coordinating the production and distribution of uncodified information. Nonmarket mechanisms—transfer via hierarchies within firms, personal communication networks and personal mobility, open publication, collaboration among technology suppliers and between users and suppliers—are often more efficient. (We will return to this point in the next chapter.)

The conventional economic view of innovation emphasizes the role of the individual innovative firm. But an information perspective on innovation—like both the traditional story of science and more recent constructivist accounts—highlights the interactive nature of the process. Innovation is cumulative, in that the existing stock of technology is a crucial input in the production of new technology. It is also collective, in that it relies on dealings among many participants. Cumulative innovation depends on information flow between present and future innovators; collective innovation depends on information flow among current participants, including—perhaps especially—among competitors.

On this view, even unauthorized copying among competing firms can be beneficial to overall technological innovation (and hence to society) because it is part of a process of transfer and learning. But patents block other firms from freely adopting, imitating, or improving on patented ideas without the consent of the patent holder. As we saw earlier, the prospect development theory assumes that patents encourage the diffusion of ideas by giving patent holders an incentive to sell the patented product or license the patented technology. However, this diffusion occurs via market mechanisms. Patents may aid the market exchange of highly codified technology, but they also discourage the flow of associated uncodified information via nonmarket mechanisms.[44] Absent the patentee's consent, that is exactly what they are designed to do.

One way to conceptualize the superposition of proprietary access restrictions on existing barriers—relating, for example, to competition among academic scientists for recognition, funding, and other traditional rewards, or among commercial rivals struggling to retain a competitive edge—is to draw a (limited) analogy with real property law. As sociologist Stephen Hilgartner has observed, the politics of intellectual property often involve conflicts between the new claims of intellectual property holders and the established practices of other parties. In keeping with the notion of an endless scientific and technological frontier, intellectual property rights are often treated as if they were novel holdings staked out at the leading edge of knowledge production. It is assumed that because intellectual property emerges only at the frontier (represented in patent law by the novelty requirement), conveying rights to previously unexplored territory, it cannot impinge on earlier rights. But, says Hilgartner, the frontier metaphor should give us pause; for the history of colonialism shows that the land that distant powers perceive as uninhabited is sometimes already occupied. He argues that in fact, intellectual property does not simply grant rights over virgin territory, but also curtails existing rights and transforms social practices.[45] Thus, we saw in the previous chapter that slick contemporary images of the successful scientist-entrepreneur conceal a persistent tension between traditional and proprietary practices surrounding the exchange of innovation-related information in biotechnology.

The conventional counterargument, that the blocking effects of patents can be overcome through licensing and other contractual arrangements, is also less convincing with respect to uncodified technology—that is, most technology. Even if a patent holder were willing to license the technology to all comers, we have seen that license agreements between arm's-length agents in the marketplace can be a much slower and more costly form of information transfer than nonmarket mechanisms. The more uncodified the technology, the higher the transaction costs associated with such arrangements. While conventional theory supposes that the restrictive effects of patents can be justified if they ultimately encourage the production of new information, such a trade-off makes no sense in the realm of uncodified information because there is no clear distinction between production and use. As Mandeville points out, stifling the flow of information automatically stifles its production.[46] Patents may not do much harm in a mature industry where much of the relevant technology has already been codified. But in new, highly innovative industries like biotechnology, where a greater proportion of technology remains uncodified, patent rights are likely to have a significant negative impact on innovation.[47]

Intellectual Property and Innovation

As we saw earlier in this chapter, conventional economic theories of intellectual property rights embody three basic assumptions. The first is that technological information is easy to copy, resulting in a lack of incentive to invest in innovation. We have now seen that this is not necessarily true of any information, and certainly not of uncodified information, which makes up a larger fraction of all technological information—especially in immature fields such as biotechnology—and is more important to innovation.

The second assumption is that the market is, or should be, the primary mechanism for the exchange of technological information among industry participants. Again, it turns out that market mechanisms are a more costly way to transfer information than many existing nonmarket mechanisms; this is especially true of uncodified information.

The third assumption is that central coordination of research and development activity by a patent holder is more efficient than decentralized innovation carried on in the absence of patent rights. We have seen that early- to mid-twentieth-century philosophers and sociologists of science believed that the most efficient possible organization of scientific research involves independent initiatives by competing scientists working with awareness of each other's achievements.[48] These scholars argued that even where imperfect knowledge of competing initiatives leads to duplication of effort, such duplication may be valuable: for example, multiple overlapping research efforts may improve the impact and accessibility of new research claims or help establish their validity, while different researchers may make different mistakes, interpret results differently, or perceive different implications of the same results, thereby achieving greater overall understanding.[49]

More recent work by legal scholar Yochai Benkler affirms that under appropriate conditions, "commons-based peer production"—the mechanism by which productive activity is coordinated in traditional academic science—has systematic advantages over methods that rely on management directions within firms or price signals within markets.[50] Peer production involves self-selection of individuals for particular tasks in generating, checking, and integrating contributions to an overall project. Self-selection is better than firm- and market-based methods at identifying and assigning human capital to information and cultural production processes because it loses less information than either management directions or price signals about who the best person for a given job might be. Further, peer production allows large clusters of potential contributors to interact with large clusters of information resources in search of new projects and opportunities for collaboration, a process that is inherently more efficient than employing property and contract as organizing principles of collaboration.[51]

This analysis might appear to contradict the premise of Kitch's prospect development theory of intellectual property rights. In fact, there is no contradiction. Kitch himself explicitly stated that the prospect function performed by the patent system in relation to applied research could not apply to basic scientific research. He believed that in that context, coordination depended instead on peer review and other mechanisms of the kind described by Benkler and by earlier scholars—precisely because it was not possible to fashion a meaningful property right around a mere discovery or explanation of scientific phenomena.[52] Of course, in formal terms this distinction is built into patent law itself in the form of the discovery–invention divide mentioned in Chapter 2. But as we have seen, the substance of this distinction has been subject over the past several decades to a full-scale assault—and not just in the field of biotechnology. Software is now regarded as patentable, as are many business methods.

More broadly, we are living in an age in which new copying technologies and new means of communicating ideas and knowledge are consistently accompanied by renewed efforts to subject information to private control—efforts that, across the board, have largely been successful. Legal scholar Jamie Boyle describes the effects in terms of a "second enclosure movement," a sequel to the profitable but socially disruptive enclosure of common lands in eighteenth-century Britain.[53] We have seen that none of the instrumental arguments for intellectual property outlined earlier in this chapter is particularly convincing with respect to innovation in biotechnology. The same is true in many other fields where intellectual property rights and other forms of proprietary information lockdown have been steadily strengthened in recent years. What, then, is the rationale behind this trend?

The Knowledge Game

In their book *Information Feudalism,* Peter Drahos and John Braithwaite document a phenomenon they call the "knowledge game." According to their analysis, the knowledge game predates the biotechnology industry by many decades, helping to shape it from the start.[54]

Drahos and Braithwaite suggest that the first institutions capable of playing the game were major corporations in a range of industries in the United States and Europe at the dawn of the twentieth century. Leaders of these corporations understood the importance of managing knowledge, and invested heavily in research and development. Seeking to ensure that knowledge generated in large-scale industrial laboratories stayed under private control as long as possible, they realized that knowledge about patents, trademarks, copyright, and trade secrets was just as important as technological knowledge. Accordingly, they established in-house legal departments staffed with patent attorneys to build patent portfolios and gather intelligence about the strengths and weaknesses of competitors' patent holdings. In the hands of these professionals, the knowledge game developed both competitive and cooperative dimensions—and patents acquired a strategic value that was increasingly independent of their supposed role in encouraging innovation.

In this account, the competitive aspect of the knowledge game entails acquiring large numbers of patents while employing elaborate strategies to neutralize the value of competitors' intellectual property. The acquisition of large numbers of patents is now a way of life for many companies, with management setting higher and higher patent quotas from year to year. Such "ramping up" is not usually accompanied by a corresponding increase in the company's research and development budget—rather, employees are expected to "harvest" a greater proportion of existing inventions for patent purposes.[55] This is no surprise, given that the point of having a large patent portfolio is not to facilitate one's own innovation so much as to (1) generate licensing revenue and/or (2) discourage innovation by others through various "blocking" strategies that involve seeing the direction a competitor is headed and deliberately creating patentable roadblocks along that path.

Tricks for increasing licensing revenues include patent "stacking"—taking out many patents on different aspects of a single innovation—and the use of "submarine" or "stealth" patents that are not deployed until large numbers of users have become dependent on the patented technology. Specific blocking techniques include "clustering" (patenting around one's own core patents), "bracketing" (patenting around a competitor's core patents), "blitzkrieg" (patenting a large number of similar or related devices, such as molecules in chemical or pharmaceutical settings), "blanketing" (mining every step in a manufacturing process with patents claiming minor modifications), "flooding" (acquiring many patents on minor or incremental variations on technology developed by another company), "fencing" (blocking certain lines or directions of research and development using a series of patents), and "surrounding" (enclosing a key patent with minor patents that collectively block its effective commercial use).[56] None of these strategies requires the patent owner to actually use or permit others to use the patented technology. Indeed, the aim in some cases may simply be to ensure the quiet death of a potentially disruptive innovation. In some fields, including software, it has been argued that the majority of patented technologies are never used because people who actually make things can't afford the risk.[57]

The strategic use of patents in the competitive knowledge game suggests, as we saw in Chapter 2, that at least some of the recent exponential increase in patenting can be explained by the increase itself: the more patents others are pursuing, the greater the perceived need for defensive patenting. However, the disconnect between intellectual property rights and innovation is even more starkly illustrated by the emergence of companies that defend themselves from infringement suits not by threatening to countersue but by avoiding productive activity altogether. These dedicated "intellectual property portfolio holders"—known pejoratively as "patent trolls"—thrive by suing or threatening to sue other companies. Some are the last remaining shell of formerly innovative companies; others acquire patents at rock-bottom prices when other companies go under. Either way, patent trolls are effectively immune to reprisals because they have no other business—no products, or even necessarily any employees. These companies are loudly condemned by other players of the competitive knowledge game, but they are only taking the game to its logical extreme—openly demonstrating, to the discomfort of other players, how little it need have to do with innovation.

Meanwhile, the cooperative dimension of the knowledge game described by Drahos and Braithwaite relates to the use of patent rights to dominate markets, in particular by establishing cartels.[58] A cartel is an arrangement among industry competitors to fix the price or limit the production of a commodity. Of course, such arrangements are highly profitable, but historically they have been vulnerable to both defection by individual members and legal challenge by competition authorities. According to Drahos and Braithwaite, players in the knowledge game as it developed between the two world wars recognized that intellectual property rights offered a solution to both problems. Complex patent-licensing agreements were much harder for competition lawyers to attack than straightforward articles of association, partly because such an attack would be seen as an interference in the use of private property. At the same time, patent licenses could be used to impose legally enforceable restrictions on price and production and to divide territories among cartel members.[59]

For a time, networked intellectual property portfolios became the outstanding characteristic of global knowledge cartels in a range of industries, especially the chemical and pharmaceutical industries. But after the Second World War, players of the cooperative knowledge game found themselves on the defensive.[60] Everywhere, the costs of research and development were rising. In the United States, cartels faced aggressive enforcement of antitrust laws that also made it risky to acquire smaller innovative companies. In the pharmaceutical industry, these problems were compounded by competition from generic manufacturers and by a decline in the rate of drug discovery by synthetic chemical means. It was time for a change of strategy. Drahos and Braithwaite describe how both chemical and pharmaceutical industry giants turned to the life sciences, including the newly emerging science of biotechnology, in the search for continued oligopolistic profits, and to universities and smaller companies to bear the risks of expensive research and development (universities having the added attraction of providing indirect access to public funds).[61] The results, they argue, are those we saw in Chapter 2—namely, the strengthening of intellectual property rights in biotechnology through the 1980s and beyond, and the incorporation of universities and other public-sector institutions into the knowledge game.

Winners and Losers

Obtaining a single patent on a single invention is not cheap: an international filing costs on the order of ten thousand dollars, including government and private fees. But it is nothing compared to the knowledge game buy-in. Only the largest players can afford to own and manage thousands of patents across the globe and to wager on the outcome of litigation worth hundreds of millions of dollars. In the intellectual property skeptics' view, other industry participants may be likened to bystanders at a corrupt poker game who are forced to join in for survival's sake, with no real chance of winning.

Like a poker game that takes over a whole town, the knowledge game arguably diverts resources from other more constructive pursuits—including innovation. We have seen how the proliferation of patents and other intellectual property rights can lead to high transaction costs in biotechnology research and development. Drahos and Braithwaite's account suggests that this may be not a mere side effect, but rather a deliberate strategy. The argument would be that the knowledge game is profitable to industry giants precisely because it is so expensive. High costs keep potential competitors out of the market and help insure against disruptive innovation by making smaller firms and public-sector institutions dependent on alliances with larger companies. For the bigger players, the cost of participation is a small price to pay for the rewards of oligopoly.

Chemical and pharmaceutical industry lobbyists claim that strong intellectual property protection is essential to bring more and better biotechnology products to market. But, as noted in Chapter 1, the kind of innovation that goes hand in hand with oligopoly control is often not the kind that serves as a basis for new technologies, products, services, and industries. Whoever is winning the knowledge game at any given time will be inherently averse to innovation that causes old inventories, ideas, technologies, skills, and equipment to become obsolete—what early twentieth-century economist Joseph Schumpeter called "creative destruction."[62] Instead they will tend to favor incremental or evolutionary innovations, preferring to extend and combine already successful products and brands rather than invest development and marketing dollars in the risky business of creating entirely new offerings.[63]

As for innovation that takes place elsewhere, in smaller firms and public-sector institutions, oligopolists may typically respond to potentially disruptive innovation in one of three ways: copy it, buy it, or crush it.[64] Arguably, a company with a team of expert lawyers and a huge intellectual property portfolio need not be unduly concerned about the legality of copying technology from smaller players. Any objections are likely to be met with, "Oh, did we infringe your patent? Well then, let's just back up the truck and see what *you* might be infringing." Acquiring innovations, with a view to either exploiting them in the competitive knowledge game or stifling them for the sake of cooperative market domination, is also straightforward. Neither small firms nor universities are generally capable of wearing development, regulatory, distribution, and marketing costs on their own. For many small innovative firms in the biotechnology industry, being bought out by a multinational is a highly desirable exit strategy—and the sooner the exit, the higher the return. Similarly, the market for most university biotechnology patents is made up of large companies with interests in chemicals, pharmaceuticals, and agriculture.[65]

In any case, there is not a great deal that innovators who have a problem with oligopoly control can do to fight back. The winners of the knowledge game hold the high ground in relation to intellectual property. They have enormous market power that allows them to do all the bundling tricks, threaten to withdraw products, distribute efficiently into all the nooks and crannies, buy advertising at competitive rates, and gain leverage with downstream distributors.[66] Further, major knowledge companies are widely acknowledged to be accomplished political operators, capable of rigging the rules to favor their own interests.[67]

The foregoing characterization of twentieth-century developments in global intellectual property law and policy as the product of a deliberate strategy on the part of a handful of multinational companies strikes a jarring note, especially when heard in conjunction with the much more familiar arguments put forward by the companies themselves. (These are along the lines of the mainstream economic justifications for intellectual property rights described earlier in this chapter, together with a healthy dose of moral justification in the case of copyright-protected information goods like software, movies, and music.) Some readers may therefore be inclined to question the credibility of the knowledge game account.

In fact, this account is supported by an unprecedented breadth and depth of qualitative research in the field of global business regulation, being part of a study by Braithwaite and Drahos based on over five hundred interviews with international business and government leaders.[68] Some of these interviewees were remarkably candid in their responses. In many cases, the explanation is straightforward: their part in establishing and then winning the knowledge game has been a source of personal pride, not shame. One key player *not* interviewed in the study was Edmund T. Pratt Jr. (now deceased), chairman and CEO of Pfizer in the crucial years leading up to the negotiation of the World Trade Organization Agreement on Trade-Related Aspects of Intellectual Property Rights (TRIPS). Pratt's term as CEO ended before the agreement was concluded. Nevertheless, he described the fight for global intellectual property protection as one of the highlights of his career.[69]

Further, Drahos and Braithwaite are by no means the only scholars who have reached these conclusions. In his 1998 book *Knowledge Diplomacy: Global Competition and the Politics of Intellectual Property*,[70] Michael Ryan describes the policy advocacy, tactics, and strategies adopted by major corporate players throughout negotiations to forge international trade and intellectual property policy in the 1980s and 1990s as a new form of "knowledge diplomacy" that, he argues, sets the rules governing the exploitation of innovation and expression in a digital, twenty-first-century global economy. Duncan Matthews's 2002 study of the TRIPS agreement confirms that it was largely the result of an initiative by a handful of multinational companies based in both the United States and Europe who sought to protect their own intellectual property (and hence their markets) through international law.[71] Similarly, Susan Sell's account of the origins and subsequent evolution of TRIPS strongly supports the thesis that international intellectual property law and policy are clear examples of governance by nonstate actors on a global scale.[72]

This is not to deny that the knowledge game account of contemporary intellectual property politics is strong stuff. Despite the weight of evidence supporting this account, some readers may find it *too* strong to swallow as an accurate historical explanation for the present structure of the pharmaceutical, chemical, and biotechnology industries. It is therefore important to note that the appeal of the open source approach described in subsequent chapters does not depend on the reader accepting the above version of events as objectively true. Even if everything that has been written about the knowledge game were to be dismissed as nonsense, the fact remains that there are many who agree with that analysis and are looking for ways to turn the tables on the perceived winners. The lesson of the broader open source movement is that, despite long odds, they may ultimately succeed.

Leaving aside the question of credibility, the foregoing account may be difficult to grasp in that it is rather abstract. A more concrete appreciation of the structural impact of intellectual property rights can be gained by returning to look at the situation in red (medical) biotechnology and green (agricultural and environmental) biotechnology. Our earlier discussion was confined to the question whether anticommons tragedy had occurred in either context. We now look more broadly at the global social and environmental implications of strengthening intellectual property rights in biotechnology.[73]

Red Biotechnology

For most people in the world today, health and life expectancy are affected by a range of complex issues having to do with poverty, food insecurity, and limited access to medical treatment. The topic of food insecurity is subsumed in the next section, and most poverty-related health issues lie beyond the scope of this book. Hence, the focus here is on access to medical treatment; but note that although access to newly developed medicines is important for health outcomes, it is often less important than other factors. A well-fed person with access to clean water and living conditions and to information about how diseases are spread is less vulnerable to most diseases even if there is no actual treatment in existence. He or she is also, of course, less likely to die of simple starvation or exhaustion. To illustrate the point, the rise and fall in infant mortality in Brazil over the past century has been correlated with fluctuations in real wages rather than with the introduction of new medical treatments.[74]

Disease is an enormous and growing problem in today's world. Together, tuberculosis, malaria, and HIV/AIDS claimed nearly 6 million lives in 2002 and led to debilitating illness for millions more.[75] Even from a purely medical perspective, the problem is multidimensional: for example, at least one-third of the 39.5 million people living with HIV around the world are likely to be coinfected with tuberculosis.[76] Africa remains the global epicenter of the AIDS pandemic, according to a 2006 report by the Joint United Nations Programme on HIV/AIDS. By the end of 2005, South Africa's epidemic (one of the worst in the world) showed no evidence of a decline: an estimated 18.8 percent of South Africans fifteen to forty-nine years old were living with HIV, almost one in three pregnant women attending public antenatal clinics were infected in 2004, and trends over time show a gradual increase in HIV prevalence.[77] While HIV/AIDS is the biggest single cause of mortality in developing countries, a number of less common diseases such as measles, sleeping sickness, leishmaniasis, and Chagas disease also collectively affect large numbers of people, with the disease burden falling disproportionately on developing countries.[78]

For diseases in developing countries that are also prevalent in developed countries, such as HIV/AIDS and diabetes, research directed at markets in developed countries may produce appropriate treatments. In such cases, the challenge is to ensure widespread access to those treatments outside the rich world. For diseases in developing countries that either are not prevalent in developed countries or commonly take a different form there, so that treatments designed for patients in developed countries would be ineffective in developing countries, there is a need not only to provide access to existing treatments, but also to mobilize the resources required to develop new treatments.[79]

Empirical research has demonstrated that intellectual property rights have little positive effect with respect to mobilizing resources for this kind of research and development: less than 5 percent of worldwide expenditure for pharmaceutical research and development goes to finding treatments for diseases in developing countries. In the private sector, this is because research and development activity is driven by the need to make a profit. From this perspective, the high risk of failure at each stage of the drug development process, from identification of molecular targets to clinical trials, means that the market for any new drug must be significant for it to be worth the investment. In the standard "blockbuster" business model, a few enormously profitable drugs effectively subsidize all the others.[80] Because the product life of each blockbuster can be extended only so far beyond the period of patent protection, and the outcome with respect to each new candidate is unpredictable, firms are unwilling to take on a project that does not offer at least the possibility of huge commercial success. But the market for medicines for diseases that are specific to or concentrated in developing countries is small, because although there are many sufferers, they have little capacity to pay.

As for the public sector, public-sector institutions in the developing world have little capacity to conduct pharmaceutical research and development. In richer countries the priorities of public-sector institutions are determined principally by domestic considerations.[81] In any case, as we saw in the previous chapter, any publicly funded research on diseases in developing countries that may be undertaken in developed countries may be vulnerable to anticommons effects resulting from restrictions on access to proprietary research tools. Thus, even if there is no patent on a particular research tool in the relevant developing country, intellectual property rights in developed countries may constrain research and development on diseases in developing countries.

Hence, empirically, intellectual property rights appear to have no positive effect on innovation relating to the most urgent health needs of the majority of the world's citizens. There is considerable evidence, though, that they do have a negative effect on access to treatments once developed.[82]

Access to the final products of biomedical research and development depends on two factors: (1) affordability and (2) the existence of a health service infrastructure that can support delivery. The second factor has little to do with intellectual property rights, at least in the short to medium term.[83] Not so the first.

In developed countries, generic competition causes drug prices to fall sharply, especially if the market is large enough to support a number of generic competitors, indicating that patents keep the prices of drugs higher than they would otherwise be—unsurprisingly, since that is their function according to the conventional theories described earlier.[84] Prices matter even more in the developing world because most poor people pay for their own drugs, instead of having the costs of treatment met by the state or through insurance.[85] Yet prices for identical drugs are often much higher in the developing world than in the developed world because of the practice of transfer pricing, in which intellectual property is sold into a tax haven at a low price and out again at a high price, shifting taxable profits to the haven. Most low-income developing countries rely on imports for their supplies, so the existence of patents in potential supplier countries allows patent holders to control price and availability even in countries where there are no patents. Since 2005, transitional provisions of TRIPS relating to the patenting of pharmaceutical products are no longer available, cutting off the supply of cheap generic drugs from countries such as India.[86] Importing countries are unlikely to be able to grow their own generic industries because their small markets and lack of indigenous technological, productive, and regulatory capacity make it difficult to create a competitive environment for patented and generic products.[87]

Green Biotechnology

What about agricultural biotechnology? As in biomedicine, the social welfare stakes are high. About 800 million people in the world today are suffering from malnutrition, many of them farmers who cannot grow or sell enough food to make ends meet.[88] Of course, agricultural innovation on its own cannot solve this problem. But it could help, both by generating gains in production and productivity and by stimulating broader economic growth that can help break the cycle of poverty and food insecurity.[89] Innovations in biotechnology may be uniquely well suited to address problems of soil management, crop production, and environmental sustainability in poor rural areas, because they can be embodied in seeds, which are familiar to users and can be distributed cheaply via established channels.[90]

The kinds of agricultural biotechnology research and development that are most likely to benefit poor people in developing countries involve traits and crops that are useful to small subsistence farmers. Relevant traits include those that increase yield potential, increase the stability of yields through resistance to biotic and abiotic stress, or enhance farmers' ability to grow subsistence crops in difficult conditions (such as drought and salinity). Relevant crops are the basic staples of the poor: rice, wheat, white maize, cassava, and millet.[91] As in biomedicine, however, private research and development priorities tend to reflect the needs of large commercial operations targeting big markets. The private sector has little interest in developing crops for which there is no substantial market, whether they be minor specialty crops in the developed world or major staple crops in the developing world.

Until recently, the private sector mattered less to agricultural research and development than did the public sector. But public-sector funding for agricultural research has been stagnant or declining since the 1960s, a trend that affects research directed at both developed and developing countries. For example, funding for the Consultative Group on International Agricultural Research (CGIAR), derived from predominantly first world donors, has fallen in real terms since 1990 to the extent that both its research efforts and its ability to maintain valuable gene banks are now under threat.[92] In developed countries, especially in Europe and North America, the decline in public funding for agricultural research has been accompanied by a rapid growth in private investment. Private-sector research, supported by intellectual property protection and sustained by demand from farmers in developed countries and the commercial sectors of a few of the richer developing countries, is now the dynamic element in agricultural research and development.[93]

The impact of this shift is exacerbated by the effects of the knowledge game. Few developing countries possess any significant capacity for biotechnology research and development, but those that do are hampered by problems of access to proprietary research tools. This is not so much because of patents in their own countries as because of a tendency to overestimate the risks of infringement action, combined with a desire to export to developed countries where the relevant technologies *are* patented. Even in the absence of a patent covering the relevant technology in any jurisdiction, one effect of high overall barriers to entry is to restrict the avenues available to researchers seeking to obtain crucial know-how with respect to the technology. The smaller the number of people working in any given technology area, the harder it is for new users to acquire the skills, materials, and other uncodified information necessary to adopt even those techniques that are not protected by intellectual property rights.

As we saw in the previous chapter, empirical evidence suggests that public-sector researchers in developed countries are affected by a tragedy of the anticommons, exacerbating the neglect of agricultural research and development conducted for the benefit of the world's poor. Meanwhile, the private sector has avoided a tragedy of the commons only at the cost of a radical restructuring that has dramatically reduced competition within the industry. Reduced competition means higher barriers to entry for players who might be interested in serving markets that are too small to be attractive to agronomic systems giants like Monsanto. It also means higher prices due to customer lock-in—that is, dependence on particular suppliers or distributors—for both research tools and agricultural inputs. Effective biosafety regulation is undermined by concentration of the capacity to conduct biotechnology research in the hands of those who stand to gain from minimizing regulatory interference. Finally, the legacy of the Green Revolution includes widespread adoption of monoculture-based farming practices that threaten the diversity of genetic resources.[94] This problem is exacerbated by the development of intellectual property fences and technological locks that are necessary in order to permit intellectual property owners to sell essentially the same technology year after year, but make traditional agricultural practices such as the saving of seeds difficult or impossible.[95]

In Chapters 2 and 3 I have argued that, while intellectual property rights have been crucial to the development of an independent biotechnology industry based on proprietary business models, they tend to backfire as a regulatory strategy for encouraging innovation in biotechnology. Part of the reason is that intellectual property law fails to take into account the reality of scientific research and development as a cumulative, cooperative social activity fueled by the exchange of largely uncodified information. Another part of the reason is that the politics of intellectual property are dominated by actors for whom innovation—other than in the narrowest sense of generating incrementally improved products for established markets—is not, in fact, the name of the game.

All of the problems catalogued in these two chapters can be summarized as follows. In both medical and agricultural biotechnology, the ability to conduct socially valuable research and development depends on access to a full set of enabling technologies, analogous to the basic toolkits needed for cooking, gardening, sewing, or any of a thousand other familiar productive activities. In biotechnology, key elements of these toolkits are protected by intellectual property rights. Not only does this tend to increase the costs of assembling a complete toolkit for any given task; it also tends to concentrate the *capacity* to innovate in the hands of actors that have little *incentive* to innovate, except in ways that promise to generate substantial profits while sustaining the structural status quo.

For reasons that are explained in Chapter 7, public perceptions are extremely important to the biotechnology industry.[96] In consequence, every new development, positive or negative, is subject to a considerable degree of spin. Biotechnology companies themselves are not the only ones responsible for all the hype surrounding this field of research and development. Investors, professional advisers, governments—indeed, any group that supports the industry in its current form—all have an interest in projecting an image of the biotechnology industry as vibrant, innovative, and successful. Yet among the more thoughtful and objective industry observers there is now a strong sense that biotechnology has failed to live up to its potential.[97]

This is not to say there have been no successes. There have, some spectacular. The point is not that biotechnology has failed to deliver any social benefit—which is untrue—but that it should have delivered more. It is not the absolute scale of achievement that is in question, but the gap between promise and reality. Of course, this gap may appear larger than it really is. After all, the purely technical promise of biotechnology as a broad enabling technology has itself been the subject of hype—not only by industry and its supporters, but also, and for longer, by scientists with their own set of incentives to downplay failure and exaggerate success. Nevertheless, it is surely reasonable to ask whether current institutional arrangements are fully optimized to deliver the maximum amount of socially valuable biotechnological innovation. It appears from our discussion so far that the answer is no.

If that is correct, then the current state of affairs is not merely inefficient. To the extent that would-be innovators are excluded from the opportunity to contribute to research and development and to influence the trajectory of a powerful new technology, it also has broader negative implications for autonomy, justice, and development—a point not lost on those responsible for carrying forward the foundational science of biotechnology. In 2003, Nobel laureates and other scientists attending the Nineteenth International Congress of Genetics declared:

> Fifty years since the double helix structure of DNA opened our eyes to new means of using genetics to contribute to human wellbeing, we are increasingly faced with the challenge of ensuring that the next fifty years delivers these benefits to all people. . . . We see democratization of innovation, including genetic modification, to be essential. . . . It is essential to empower innovators everywhere, small and large, in public and in private sectors, by ensuring their access to enabling technologies. . . . The answers are . . . in encouraging local capacity to innovate and respecting local choice of technologies. The freedom to innovate must not be hindered by barriers imposed by any interest group. To do so would be disrespectful of the legitimate drive of all people to solve their own problems. These barriers include . . . restrictive ownership of enabling technologies. . . . The right to innovate must not be the sole province of the highly capitalized, nor of a few owners of key intellectual property, who could thereby control, direct or limit innovation globally. . . . The tools of innovation must not be withheld.[98]

Earlier in this chapter, I noted that both economic and sociological theories of innovation emphasize the value of nonmarket exchange mechanisms in the production of information goods. However, as we have seen, intellectual property rights create a selective pressure in favor of market mechanisms at the expense of nonmarket information exchange. Before the encroachment of intellectual property rights, the mechanisms governing knowledge production in biotechnology were essentially those of traditional academic research. It might be supposed that the best way to undo the adverse effects of excessive intellectual property protection would be to revert to this system by somehow dismantling the property rights that encroach upon it. But this logic is flawed. Accounts of scientific norms and the nature of the scientific enterprise offered by Merton, Polanyi, and other early- to mid-twentieth-century observers took for granted an entire political and economic framework that acted as a life-support system for "public science." For better or worse, that framework no longer exists: the knowledge game is now entrenched. If Drahos, Braithwaite, and others are correct in their assessment of the game as a way for big business to discipline states as well as markets—through a combination of threats and regulatory capture at both domestic and international levels—then we should not expect the sheriff along to break it up anytime soon.

The alternative is to look for other ways to "democratize innovation" in biotechnology research and development. If building local capacity to innovate is important, the ideal solution would be to provide an alternative to the use of proprietary tools—a toolkit for biotechnology innovation that is affordable, accessible, and unencumbered. The rest of this book is about one possible means of generating such an alternative: the "biobazaar."

4

Welcome to the Bazaar

Chapters 2 and 3 showed how the radical propertization of biotechnology research and development that has occurred over the past three decades introduces inefficiencies that limit its social and economic potential. Nevertheless, as explained in Chapter 3, it is generally assumed that strong intellectual property protection in biotechnology is essential in order to (1) secure investment in innovative activity and (2) permit coordination of contributions to cumulative and/or cooperative technology development. Part of the underlying logic of this assumption is that intellectual property rights can facilitate arm's-length exchanges of information by conferring a value on information that outlasts its disclosure. By enabling technology and knowledge to be treated as tradable commodities, property rights facilitate the development of markets for intermediate technological inputs.

The reason such a mechanism is considered indispensable to biotechnology innovation in particular is that the knowledge base of biotechnology research and development is so diverse that not even the largest players can build a sufficiently strong research base to cover all technology areas. Nor is it generally possible to assemble under one roof the full range of skills required to get new products to market. To compensate for this lack of internal capability, firms that might otherwise have integrated research and development with activities such as production, marketing, and distribution seek to "outsource" innovation through relational contracting with universities, technology-based start-ups, and other established firms. Such contracting takes a variety of forms; sociologist Woody Powell lists joint ventures, research partnerships, strategic alliances, minority equity investments, and licensing arrangements.[1]

Thus, in biotechnology and related industries, both technical and commercial considerations make it necessary to achieve some division of labor among participants in the innovative process. Converting intermediate technologies into salable commodities by way of intellectual property rights permits a division of labor based on market transactions. But the market is not the only possible means of coordinating innovative activity that crosses boundaries between firms.

This chapter introduces the concept of open source as an instance of a nonmarket form of governance, here termed "bazaar governance." In the course of this discussion, we shall see that the success of open source challenges not only the assumption underpinning the prospect development theory (that proprietary exclusivity is necessary to permit coordination of cumulative and cooperative innovation), but also the assumption behind the theories of invention-inducement, disclosure, and development and commercialization—namely, that proprietary exclusivity is needed to induce investment in innovation.

Market, Hierarchy, Network, Bazaar

Discussions of technological innovation frequently draw on the branch of economic theory known as transaction cost economics. Transaction cost theorists generally recognize two main types of governance structure that operate to reduce the uncertainty—and therefore the costs—associated with any economic transaction. These are the firm and the market.[2]

In this context, a firm is any hierarchical structure—not necessarily a for-profit company—in which economic activity is centrally coordinated through the authoritative directions of managers or leaders. As we have seen, this mode of production is referred to in the software context as "cathedral building." According to Raymond, proprietary and free software cathedrals are characterized by the dominance of a single architectural vision.[3] The word *architect,* literally meaning "chief builder," itself implies a hierarchy. In a market, by contrast, economic activity is coordinated in a decentralized manner, through price signals.

Scholars in other disciplines have suggested that markets and firms or hierarchies are not the only possible means of coordinating economic activity. For example, empirical research has shown that the biotechnology and pharmaceutical industries are characterized by intensive relational contracting among firms (including public and private nonprofit organizations) that leads to the formation of observable networks.[4] Sociologist Woody Powell argues that the network is a form of governance distinct from both firms and markets. In a network, economic activity transcends organizational boundaries but is not coordinated solely by firms' or individuals' perception of price advantages within individual transactions. Instead, it is influenced by their consciousness of the value of long-term relationships. Members of a network forgo the single-minded pursuit of their own interests at others' expense for the sake of indebtedness and reliance over the long haul.[5] As a network evolves, reputation, interdependence, trust, and even altruism may become integral parts of its constituent relationships.[6] Network relationships are still characterized by direct reciprocity or conditional action, but the assessment of reciprocal fairness extends over a longer time frame than that of a single transaction. In Powell's words, "the books are kept open" in the interests of building on satisfactory results.[7]

Both traditional scientific research and open source software development are examples of yet another mode of production, variously termed "horizontally networked user innovation,"[8] "commons-based peer production,"[9] and "collective" or "bazaar" production.[10] Management scholars Benoît Demil and Xavier Lecocq argue that bazaar production is based on neither firm, nor market, nor network governance. They describe the points of distinction as follows.

First, no formal fiat can enforce decisions within the bazaar.[11] No one is obliged to perform any particular task, and work cannot be imposed or mandated by a leader. In this respect, the bazaar differs from a firm; like the market, it is an example of spontaneous, decentralized ordering of transactions. Unlike a market, however, the bazaar coordinates economic activity, not by means of extremely reduced abstract information as provided by price signals, but via concrete information about the production system's subject matter.[12]

To grasp this point more clearly, recall philosopher Michael Polanyi's conception of scientists as being like a group of people working on a jigsaw puzzle.[13] If you were to observe such a group, you might notice that no one appears to be in charge. The participants may not even look at each other; instead, they are focused on their own pieces and on the pieces of the puzzle that are laid out for everyone to see. Similarly, in a bazaar setting, participants' primary relationship is not with each other but with the project. The properties of the project itself coordinate their contributions by dictating the possibilities for action within a framework of generally accepted rules, just as jigsaw players accept that pieces must fit together in two dimensions instead of three. In the software context, shared notions of technical validity are established through fundamental programming conventions. Indeed, Andrea Bonaccorsi and Cristina Rossi argue that "software itself is a convention or a common language, in which errors are identified and corrected through the mechanism of compilation."[14]

Not only is communication within the bazaar not mediated by price signals, but pecuniary self-interest is not the only motivation of agents operating within this governance structure. Later in this chapter I draw on insights from a field of management studies known as user innovation theory to show that extrinsic rewards coexist with intrinsic or process-oriented rewards as drivers of investment in innovative activity. The bazaar is further distinguished from both firm- and market-based governance in that it does not formally differentiate agents' roles as users and producers: any given agent may be both producer and user of the same technology (hence the concept of users as innovators).

Bazaar governance is also distinct from network governance. Network-based production depends on relational ties between specific participants; bazaar production does not. The norm of direct reciprocity, whether assessed in the short term (as in the market) or the long term (as in networks), does not prevail in the bazaar. Instead, reciprocity is diffuse or generalized.[15] Rishab Ghosh has explained this idea using a "cooking pot" analogy: you put in ingredients and you take out stew, but no one minds whether the stew you eat is made from the same ingredients you contributed.[16] Similarly, while networks call for long-term engagement in order to minimize opportunism, a high level of commitment is not a prerequisite to participation in bazaar-style production. In a bazaar setting, as in social life, free riding is tolerated within the limits of collective resources: provided there is enough stew for everybody, the guest who forgets to bring a contribution need not be excluded. Free riding with respect to a particular type of investment may even be encouraged for the sake of other benefits. The forgetful guest may be a great storyteller, enhancing everyone's experience of the shared meal. By the same token, the value of a knowledge product may be enhanced by positive network effects (as in the case of a fax machine that is more valuable to each user the larger the number of others who use the same technology) or through quality checking by users who are not themselves developers.

Thus, bazaar governance can be distinguished from firm, market, and network governance in a variety of ways. There are two key points of distinction, though, that separate the bazaar from all three previously recognized governance structures.

The first relates to the radical use of property in bazaar settings to promote distribution and sharing of the object of production. In firm, market, and network structures, property rights are used to control assets, either individually or jointly. The purpose of intellectual property and quasi-intellectual property rights in these contexts is that of exclusion, whether the protected innovation is exploited within a firm through trade secrecy, exchanged in a market, or used as a bargaining chip to extract concessions from other members of a network. By contrast, the bazaar is predicated upon a formal or informal contractual framework—in the software case, open source licensing—that seeks to prevent, not to facilitate, appropriation.[17] We return to this point in Chapter 5.

The second point of distinction between the bazaar and previously recognized governance structures is that membership in the bazaar is open to anyone who wishes to participate: there is no formal barrier to entry. Firms select their members. In a market, contracting parties select each other. The direct links between agents within a network ensure that its membership is limited, though a network's boundaries may sometimes be "fuzzy." In the bazaar, members select themselves for productive tasks and decide for themselves the nature and extent of their contributions. No one can prohibit access—for example, in order to prevent free riding.[18]

How does the bazaar stack up against other governance structures as a means of reducing the uncertainty associated with transactions? Transaction cost theorists identify two means by which governance structures "infuse order" into transactions that might otherwise degenerate into conflict, undoing the opportunity for mutual gains.[19] These are (1) control and (2) incentives. "Control" refers to the capacity of a governance structure to contain opportunism and align the behaviors of parties to a transaction. "Incentives" regulate parties' motivation to be effective in their production functions.[20]

Initial theoretical characterizations of firm and market governance structures conceived of these two mechanisms as being in tension with one another. Firm-based modes of production entail a high degree of control and relatively low incentives. Conversely, market-based modes entail a low degree of control, but high incentives. Network governance appears to display an intermediate intensity of both control and incentives. According to Demil and Lecocq, bazaar governance scores low on both dimensions. In the bazaar, neither formal/legal nor informal/social controls exert a strong influence on parties' conduct. At the same time, while the incentives associated with the bazaar—that is, with "free revealing" or nonproprietary exploitation strategies, discussed below—may be compelling, they generally do not apply to more than a small proportion of members. As a result, free riding is prevalent.[21]

If both control and incentive mechanisms for ordering transactions operate only weakly within the bazaar, we might expect bazaar governance to be relatively unsuccessful at reducing transactional uncertainty compared with firm, market, and network governance. Yet the bazaar's weaknesses are also its strengths.

According to transaction cost theorists, the best governance structure for a particular transaction or set of transactions is the one that minimizes the sum of production and transaction costs.[22] To see how bazaar governance can do both, consider first its weak control mechanisms. This weakness is attributable primarily to two factors: (1) self-selection of contributors and (2) the unconventional use of property rights as means of preventing appropriation (that is, private control) of the object of production, which gives rise to a "norm of non-excludability."[23] How do these factors contribute to the strength of bazaar governance?

We saw in Chapter 3 that Benkler claims that the bazaar or "commons-based peer production" has advantages over other governance structures when the object of production is information or culture and the physical capital necessary for production is widely distributed.[24] He gives two reasons. First, self-selection is a better way to identify and assign human capital to specific tasks in information and cultural production processes than either selection by superiors in a firm hierarchy or selection through price competition in a market. This is because potential contributors themselves have the best information about their own capacities; self-selection loses less information than other methods about who the best person for a given job might be. The second reason is that removing exclusive proprietary rights as the organizing principle of collaboration substantially reduces transaction costs of the kind described in Chapter 2. As Benkler points out, there are substantial increasing returns in terms of allocation efficiency to "allowing larger clusters of potential contributors to interact with large clusters of information resources in search of new projects and opportunities for collaboration."[25]

Brian Behlendorf, cofounder of the highly successful Apache open source software development project, describes the advantages of the bazaar approach in terms of a "spectrum of involvement."[26] In the world of closed source or proprietary software, this spectrum consists of a smaller number of discrete levels of engagement with the development process. For example, a help desk staffer may act as the only intermediary between dedicated programmers and end users who are obliged to treat the software as a "black box." Open source development methods, by contrast, allow people to find intermediate levels of engagement that precisely match their own interests and capacity. Potential contributors differ from one another with respect to the resources they are willing and able to devote to looking inside the black box. Further, the situation of any particular user-developer is likely to change over time. An unskilled user may acquire programming skills; conversely, a onetime project leader may want to scale down his or her commitment to make room for other priorities. Given this variety and variability, it makes sense to allow people to settle at their own point along the spectrum and to facilitate the flow of information all along the spectrum, from dedicated designer to black box user and back again.

The same logic underpins the characterization of open source development as a "horizontal user innovation network" in the literature on user innovation,[27] which has its roots in the fields of innovation management and industrial organization.[28] Like Benkler's "commons-based peer production," the term *horizontal user innovation network* is broadly synonymous with the bazaar; but the emphasis on innovation by users highlights the fact that bazaar governance tends to blur the distinction between those who develop or produce new technologies and those who use them, thereby breaking down the barriers that prevent potential contributors from engaging with the object of production in the most efficient manner. Of course, not every user innovates. For example, it is possible to treat an open source operating system like GNU/Linux as if it were Windows—to install the software and take no further interest in it. Many users do just that, and a large number of those who consider themselves developers do not much more. The essence of user innovation is that *some* users do choose to innovate—and that all are both technically and legally empowered to do so.

To further understand why the bazaar's weaknesses are also its strengths, consider the question of incentives. Incentives are weak in the bazaar, in the narrow sense that the private rewards associated with contributing to the production process tend to influence only a small proportion of potential contributors. For example, a survey of open source projects conducted in 2000 found that 10 percent of the developers were credited with producing more than 70 percent of the code.[29] Compared with the universal appeal of monetary rewards, which generate such strong incentives in the market, the value of nonmonetary rewards depends on agents' peculiar characteristics or circumstances.

What kinds of incentives are associated with bazaar governance? To answer this question, it is helpful to consider some insights from user innovation theory. The user innovation model of open source analyzes innovators' motives in terms of rational self-interest, but avoids falling into the trap of equating either self-interest with pecuniary interest or profitability with exclusivity. This approach goes some way to counter a common objection to the feasibility of open source biotechnology, which is that nonproprietary exploitation strategies—called "free revealing" in the user innovation model—are the exclusive province of hobbyists and ideologues.

Bazaar Incentives and Business Models

According to user innovation theory, rational self-interested actors often innovate in response to an expectation that they will derive some benefit from the existence of a new technology, but not always. At least some innovative activity is driven primarily by benefits the innovator expects to derive from the *process* of innovating. In the open source software context, empirical research indicates that these include fun, learning (such as acquiring new programming skills), and the sense of belonging to a community. Another possibility is that the innovative activity may be incidental to the pursuit of some other goal. In that case, innovation is serendipitous—a side effect rather than a return on any conscious investment. An example is Alexander Fleming's discovery in the late 1920s of penicillin mold growing as a contaminant on deliberately cultured bacterial colonies.

Thus, not all innovators innovate with the intention of exploiting the innovation. But even those who do may expect to benefit from its existence in any of a range of different ways.[30] In the user innovation literature, a person or a firm that benefits primarily from using an innovation is called a *user;* someone who expects to benefit primarily by selling (or licensing) the innovation to users is a *manu-*

facturer. Suppliers of goods or services needed to produce or use an innovation and wholesale or retail *distributors* benefit indirectly, through increased demand resulting from the adoption of the innovation by users. Others—for example, insurers or providers of professional services such as lawyers, accountants, or fund managers—benefit from increased activity in the sector as a whole. In the biotechnology context, we might add a category of *consumers*. Consumers are those who benefit from the end products of biotechnology research and development—drugs, diagnostic services, agricultural products, environmental management tools—irrespective of the specific technologies used to generate those products. A premise of user innovation theory is that any of these "functional classes" may be sources of innovation under appropriate conditions.

The reason the above categories are so designated is that they describe the functional relationships through which firms and individuals derive benefit from a given product, process, or service innovation.[31] It follows that they are not fixed, but depend on the particular innovation being examined. For example, Boeing is a manufacturer of aircraft, but a user of machine tools.[32] Similarly, a large pharmaceutical firm is a manufacturer of drugs, but a user of research tools; a dedicated biotechnology firm might be a manufacturer of one research tool and the user of another. Thus, while the concept of a functional class may seem to suggest a structural relationship between innovator and innovation, in fact the relationship is a matter of strategy: with respect to any given innovation, a firm may adopt a manufacturing strategy, a user strategy, and so on. In addition to combining different strategies with respect to different innovations, as in the examples just given, a single business model may combine different strategies with respect to the same innovation. For example, a firm may derive some of its revenue from using a particular technology in-house and some from providing services in relation to that technology.

So, logically, there are many different reasons why a rational, self-interested actor might invest in innovative activity. According to the "invention-inducement," "disclosure" and "development and commercialization" theories encountered in Chapter 3, proprietary exclusivity is essential. But which, if any, of the above motivations actually depends on the innovator's gaining exclusive proprietary control over the innovation? Investment that is motivated primarily by an expected process benefit does not rely on proprietary exclusivity. In that case, the act of innovating is its own reward. Similarly, in the case of innovation that is incidental to the pursuit of some other goal, the incentive relates to the innovator's primary purpose; it has nothing to do with any proprietary rights he or she might obtain over the innovation itself. What of private investment in innovative activity that is motivated by an intention to exploit the innovation? In that case, the salience of proprietary exclusivity depends on the innovator's chosen exploitation strategy.

Proprietary exploitation strategies include both in-house use and licensing of innovations protected by intellectual property rights, trade secrecy, or personal property rights. The essence of any proprietary strategy is exclusion. So, for example, a pharmaceutical firm typically uses patents on drugs to exclude competitors from the market, thereby keeping drug prices high enough to recoup the costs of research and development. Similarly, a dedicated biotechnology firm might use a research tool patent to protect its licensing income by excluding nonlicensees who would otherwise erode potential licensees' incentive to enter a paying license agreement.

In the user innovation lexicon, both of these situations would be described as "manufacturer innovation." Clearly, a manufacturing strategy lends itself to proprietary exclusivity because the ability to exclude competitors increases the price that a manufacturer-innovator can charge for the innovation in the marketplace. Firms that are able to team a manufacturing strategy with strong intellectual property rights may find it highly lucrative to "invent one and sell many" copy-protected items: indeed, this is the archetypal knowledge game strategy.

Nevertheless, it is perfectly possible to adopt a manufacturing strategy that is nonproprietary, and in fact that is the norm in most industries. Even in technology-intensive fields, while firms may acquire intellectual property rights for various purposes, empirical studies since the early 1980s have consistently shown that firm managers regard the proprietary exclusivity conferred by patents as being less important in generating returns on R&D investment than other expected benefits such as reduced cost or improved quality of manufacture, being first to market, and developing strong marketing relationships.[33]

Perhaps because proprietary manufacturing strategies have been so spectacularly profitable—if only for some companies in some industries—they tend to eclipse other potential strategies and, hence, other incentives to innovate apart from those that depend on proprietary exclusivity. Nevertheless, the benefits that firm managers in the above-mentioned studies perceived arising from strong research and development point to the real-world value of other strategies for generating private rewards from investment in innovation (user, supplier, distributor, and others).

For our purposes, what matters most about these other strategies is that they are likely to benefit from a nonproprietary or "free revealing" approach, even where strong proprietary protection is available. The reason is that, while proprietary exclusivity raises the expected returns from information production, thereby helping (in theory, at least) to induce investment in the current round of innovation, it also decreases the availability of the information as an input to further rounds of innovation. Whereas a manufacturing strategy depends on the "sale value" of the innovation in question (its value as a product, final good, or output), other strategies rely to a greater extent on maximizing the innovation's "use value" (its value as a tool, intermediate good, or input). In the latter case, benefits to the innovator may actually be enhanced if he or she follows a nonproprietary strategy—that is, if the innovation is freely revealed.

"Free revealing" is defined in the user innovation literature as voluntarily giving up all intellectual property rights to an innovation so as to allow free and equal access to all comers.[34] In fact, as we shall see in Chapter 5, contributors to open source software projects do retain ownership of their intellectual property, but because they use it to facilitate free distribution and improvement, open source licensing is considered a form of free revealing.

Free revealing does not mean that recipients necessarily acquire and use the revealed information at no cost to themselves. For example, they may have to pay for an Internet connection or a field trip to acquire the information being freely revealed; alternatively, they may need to obtain complementary information or other assets in order to fully understand the information or put it to use. However, if the information provider does not directly profit from any such expenditures made by information recipients, the information itself is still freely revealed.[35]

Free revealing may, of course, be the default if an innovator does not take any steps necessary to keep an innovation secret or to obtain intellectual property protection. But empirical research indicates—as does the phenomenon of open source software itself—that some innovators adopt free revealing as a deliberate strategy, spending significant resources to ensure that innovation-related information is effectively and widely disseminated. In the software context, writers of computer code may work hard not only to write the code but also to eliminate bugs and to document it in a way that is easy for potential adopters to understand. Active efforts to diffuse proprietary information without any expectation of direct compensation have also been observed in a wide range of other contexts—including among profit-seeking firms for whom innovation is far more capital-intensive than it is in software development.

Thus, Eric von Hippel notes that routine intentional free revealing has been observed in relation to the development of methods for processing iron ore and pumping water out of mines in nineteenth-century Britain; the design of semiconductors, automated clinical chemistry analyzers, and lithographic equipment; and the design of sporting goods, including mountain bikes and kite surfers.[36]

Why would innovators intentionally give away for free what they have invested private resources to develop? The main premise of a free-revealing or nonproprietary exploitation strategy is that the use value—simply, the *usefulness*—of a given technology will be enhanced if it is made available in a form that is easy to understand and modify, on terms that allow users to make changes and to use or distribute the resulting modified versions as they see fit. Having brought about such enhancement through free revealing, innovators then seek to convert users' gain to private economic benefit.

At least three aspects of a technology's use value may be enhanced by free revealing. First, the value of a tool to its user is often higher if the user is able to understand fully how the tool works. Technical transparency and the legal freedom to use the tool in any way the user desires both facilitate such understanding.

Second, the usefulness of a technology depends on its quality, which in turn depends on accuracy, reliability, versatility (or, conversely, specificity), interoperability with other tools, and robustness to changes in the use environment. The open source approach of permitting anyone who is so motivated to become involved in technology development contributes to quality improvements in two ways: it expands the number of people who can help eliminate design flaws and introduce enhancements ("given enough eyes, all bugs are shallow");[37] and overall development efforts are more likely to be directed toward improvements that really matter to users.

Third, the usefulness of a technology depends on its accessibility. Accessibility is a function of (1) availability and (2) affordability.

The terms on which open source technologies are distributed are calculated to improve both availability and affordability by making it possible for anyone at all to become a distributor and by ensuring that anyone who happens to gain access to the technology is legally free to use it. In the software context, for example, developers know that open source code will always be available to them; they need not be concerned about becoming dependent on suppliers who might go bankrupt, disappear, or change the terms of access.

In some cases, the mere fact that a technology has been freely revealed and is therefore accessible to a larger number of users is enough to enhance its overall use value even in the absence of tangible improvements. This is true for any technology whose use is subject to positive network externalities, or "network effects" (the example given earlier was that of a fax machine). In the software context, network effects are largely driven by the need for interoperability. Free revealing via open source licensing promotes interoperability because open source code effectively establishes an open technical standard. But technical interoperability is not the only potential driver of positive network externalities. "Certification signal" or "peer review" effects that enable users to treat a particular technology as reliable on the basis that it had been tested or checked by other users also come under this rubric.[38]

In addition to the enhancement of use value associated with widespread use of a technology, free revealing may trigger a series of cumulative transactions that improve the technology itself. In the user innovation literature, this is known as "collective invention," and open source software development is a prime example. In the open source context, the release of a technology under an open source license implies universal permission to use, redistribute, and modify the technology. Some licensees may take up the opportunity to replicate and use the technology; others may go a step further and improve upon it or apply it in an entirely new setting. If a sufficient number of follow-on innovators also adopt a free-revealing strategy, a cycle may emerge in which subsequent incremental improvements to a technology are also freely revealed, triggering new rounds of innovation.[39]

Just as open-source-style innovation does not imply that every software user becomes a developer, open-source-style collective invention does not depend on the free revealing of every innovation relevant to a particular field of activity, either in the initial round or in subsequent rounds. Like river water that can be diverted to irrigate surrounding farmland provided diversions remain below the threshold needed to sustain the health of the river system as a whole, a collectively produced data stream can sustain both proprietary and nonproprietary uses, provided proprietary diversions do not dry up the supply for downstream users. We shall see in Chapter 5 that open source licensing schemes help insure against this outcome.

Assuming that free revealing on the part of an innovator can help generate a more readily available, cheaper, better, and/or more transparent technology, how can the innovator—or, for that matter, anyone else—turn this enhanced use value into private rewards? In what follows I answer this question first in general terms, in order to lay the foundation for Chapter 7's exploration of nonproprietary strategies in biotechnology, and then by reference to actual software business models.

For users of a freely revealed technology, capturing its enhanced use value is straightforward. Obviously, individuals, companies, or other institutions (including universities and public or private nonprofit research institutions) that use a particular technology in a research program or as a component of a production process benefit directly through cost savings or efficiency gains.

For innovators who are not themselves users of a freely revealed technology, one way to capture improvements in use value is to provide services. In this model the technology is distributed on an open source basis in order to grow the market for the technology itself and associated offerings. Revenue is generated by selling the technology in a form that is easier or more convenient to use than the freely available version and by providing services such as training, consulting, custom development, and after-sales support or accessories such as user manuals.

A second possibility for nonusers is to employ the freely revealed technology as a market positioner. For example, an open source product that itself generates little or no revenue could help build the firm's overall brand and reputation, add value to conventional products, and increase the number of technology developers and users who are familiar with and loyal to the product line as a whole. Brand licensing and franchising are business strategies that depend on good market positioning: both involve charging a fee for the right to use brand names and trademarks associated with technology that is itself nonproprietary.

A third possibility for innovators who cannot benefit directly from the improved use value of a freely revealed technology is to leverage that value to enhance the appeal of a complementary product. So, for example, a company primarily in business to sell hardware might distribute enabling software such as driver and interface code at no charge along with the hardware; the hardware is more valuable the better the data and the better the tools for manipulating the data. More generally, complementary products include enabling or "platform" technologies and the niche or application technologies that are designed to be used with a particular version of the enabling technology. In the short term, proprietary licensing of platform technologies can be very lucrative. However, the broader the platform, the greater the incentive for others to invent around it—which makes the proprietary strategy vulnerable to ongoing innovation in the field. In the longer term, therefore, a better strategy may be to freely reveal the platform or enabling technology in order to encourage its adoption as an industry standard and concentrate on building proprietary products that sit on top of the platform. A bazaar-style or open source approach to technology development promotes the adoption of useful standards because the transparency of open source tools means it is obvious which technology is the best for any given platform function.

Finally, free revealing may be used as a way to restructure the competitive landscape in an industry—one's own or someone else's. We will return to this point in Chapters 7 and 8.

As a foil to the somewhat abstract nature of the discussion so far, let us now consider how these various free revealing or nonproprietary strategies have been incorporated into real-life business models—and with what success. While the empirical research cited earlier shows that nonproprietary strategies are not limited to the software context, software provides a convenient focus for this brief overview.

An informal review of open source software business models conducted in 2005 by Bruce Perens, author of the official Open Source Definition, enumerates a number of different classes of contributor to open source software development.[40] They are (1) volunteers (whom I will designate "hobbyists," since the term *volunteer* could be interpreted much more broadly to include anyone who contributes to bazaar production); (2) academics; (3) distributors; (4) dual licensors; (5) vendors of proprietary add-ons; (6) service providers; (7) hardware manufacturers; and (8) end users.

Because the first two classes of contributor are not operating in a commercial setting, they cannot be said to be following a business model. However, they may still expect some private return on their investment. For hobbyists, contributing to technology development generates process benefits: it is a way of learning new skills or having fun and connecting with other lovers of the art of software programming. Another motivation for hobbyists is that of the prospective user: for example, by participating in open source development, one may ultimately obtain a specialized tool for private use that would be beyond one's own resources to create.

"Academic" contributors are paid by government or philanthropic grants; Perens also includes in this category programmers who are employed directly by government. What members of this class have in common is that the resources they devote to open source software development are public, not private, and therefore do not need to be justified by reference to self-interest. (We return to this important point in Chapter 6.) In fact, of course, the distinction between public and private resources is often blurred in an academic context. Either way, probably the most common "strategy" (and hence incentive to innovate) relevant to this class of contributor is that of a user.

The next class of contributors comprises businesses engaged primarily in distribution of open source technologies; Red Hat Linux is perhaps the best-known example. Distributors charge customers for the convenience of receiving a packaged version of software that, given time and expertise, they could put together for themselves by downloading the relevant bits and pieces of open source code from the Internet. Depending on the technical proficiency of the customer base and the degree of user-friendliness of generic, unpackaged versions of the technology, such companies may face an uphill battle in differentiating their offerings from generic versions. To achieve this goal they use various techniques, including branding, certification by other vendors, and what Perens (somewhat provocatively) terms "sequestration" of parts of the program, so that users are charged a fee to access features that are not available in the generic version. As with proprietary software, most of the money to support this business model comes ultimately from sales of packaged software to IT cost centers.

The next class of contributors to open source development includes companies whose product is a single open source program; Perens points to MySQL and Trolltech as examples. These companies make their technology available for free under copyleft-style licenses, thereby taking advantage of other users' contributions to improve their basic product. At the same time, they offer their customers identical technology under a proprietary license that charges a fee but does not require the licensee to copyleft any derivatives he or she may develop. In other words, these companies survive through "dual licensing." According to Perens, these contributors' primary customers are companies that want to be able to sell downstream technologies to their own customers, such as enterprise users, embedded device developers, and software application developers. Their ultimate source of income is sales to IT cost centers based on the number of users covered by the proprietary license—again, the same sales approach as that employed by mainstream software vendors. In the case of dual licensors, sales income is often supplemented by additional revenue from training or development services.

The next contributors in Perens's list are essentially proprietary software vendors who contribute to open source development in order to expand the market for proprietary add-ons to the open source platform. Perens notes that this has been quite a successful approach for companies such as Sendmail.

Service providers are the next class of contributors. According to Perens, service providers create solutions for their customers by integrating multiple open source programs with customized "glue" software. They may also provide support and maintenance for a collection of open source programs. According to Perens, an older version of this approach was to specialize in servicing a single open source program, but this has turned out to be less successful than the multiple-program model, presumably because keeping up with a single program is a sufficiently manageable task that potential customers do not need specialist help, preferring to do the job either in-house or through an existing service provider.

Perens's next-to-last classification is hardware vendors. In the software context these include businesses like IBM and Hewlett Packard, whose main profits are in the sale of hardware or other nonsoftware products. These companies are among the largest investors in open source software development because releasing software products under an open source license helps promote the sale of complementary hardware without substantially threatening the company's ability to differentiate its hardware products. At the same time, packaging hardware products with nonproprietary software saves hardware manufacturers the cost of producing proprietary software from the ground up in-house.

To appreciate the success of open source in this particular commercial niche, consider the case of IBM. As Benkler points out in *The Wealth of Networks,* this technology giant—according to Drahos and Braithwaite, one of the original architects of the knowledge game—has been the most patent-productive firm in the United States, amassing in total more than 29,000 patents between 1993 and 2004.[41] Yet during the same period it has been aggressively engaged in adapting its business model to the emergence of free software. In the year 2000, according to Benkler, activities described by the firm as "Linux-related services" accounted for practically no revenues. By 2003 they provided an astonishing *double the revenues* derived from all patent-related sources.[42]

Perens's final class of contributors, end users, corresponds roughly to the functional class of "consumers" suggested earlier as a potential source of investment in biotechnology innovation (I expand on this suggestion in Chapter 7). Besides private individuals, this category includes companies that make use of open source technologies in their own operations; the most prominent examples involve the use of open source Web applications by online traders such as eBay and Amazon. According to Perens, these companies—like hardware manufacturers—generate a very significant proportion of contributions to open source projects. They also contribute indirectly, as ultimate customers for many of the other business models just described.

Thus far, our discussion of the incentives associated with bazaar-style production has shown that self-interested actors may engage in innovative activity for a range of reasons. Of these, some relate to anticipated "process" benefits, as distinct from benefits relating to the existence of a new technology or other innovation; others are unrelated to either process or outcome, as in the case of a serendipitous scientific discovery. As we have seen, in neither of these cases is proprietary exclusivity needed to induce investment in innovation. Even when motivated by the expectation of deriving benefit from the innovation itself, an innovator need not adopt a proprietary strategy in order to capture the rewards of private investment. It is true that an innovator who plans to capture the value of an innovation by selling or licensing it in the marketplace is likely (though not bound) to pursue a proprietary strategy, because, other considerations aside, this is likely to be the most lucrative approach. However, an innovator whose strategy depends on maximizing the use value of the innovation may prefer a nonproprietary or "free revealing" approach. Free revealing promotes widespread use of an innovation and increases the likelihood that it will be improved upon in subsequent rounds of innovation. That free revealing is a viable business strategy in practice as well as in theory has been empirically demonstrated in a range of industry contexts, including but not limited to software and other information goods. Finally, a business model—or, in the nonprofit context, a strategic plan—may be composed of a variety of strategies pursued (ideally) in a coordinated fashion; the foregoing survey of open source business models confirms this, at the same time demonstrating that nonproprietary strategies can survive even in an intensely proprietary industry environment. (Note that, to this point, I have not attempted to argue that nonproprietary strategies can or will succeed in a biotechnology setting. That question will be addressed in Chapter 6.)

One task remains before we return to our more general discussion of bazaar governance as a generic model of open source technology development. This is to clarify how the range of incentives to innovate described above relates to the bazaar as one of several possible governance structures.

Of the several different categories of private rewards associated with free revealing, both process benefits and direct user benefits are available in an ideal or "pure" bazaar mode of production. However, the incentives associated with free revealing from the point of view of (nonproprietary) manufacturers, suppliers, distributors, professional services providers, end consumers, and others who are not themselves users of the relevant technology all depend on the coexistence of the bazaar with other governance structures, especially the market. This point requires some explanation.

In their introduction to bazaar governance, Demil and Lecocq point out that the conceptual delimitation of different governance structures is heuristic; in reality, agents are free to choose different governance structures for different transactions. Taking examples from open source development, they note that there may be strong network ties among the founders of an open source project; some members of open source communities are employees of firms; and firms often use open source software to develop products or services that they then sell on the market.[43]

Thus, the bazaar coexists with other governance structures; and to the extent that bazaar production can be parlayed into private rewards within firms, markets, or networks, the coupling of bazaar transactions with transactions governed by other structures may boost incentives to contribute to bazaar production. The classic example is that of a programmer who parlays a reputation gained through participation in open source software development into enhanced employment prospects or better pay—rewards normally associated with participation in a firm hierarchy. Other examples involve using bazaar-style technology production to grow the market for complementary goods and services or, conversely, to damage the market for competing proprietary versions of the relevant technology. In this way, manufacturers of related innovations as well as suppliers, distributors, service providers, and consumers can all profit from bazaar-style development through the operation of markets alongside the bazaar.

Bazaar Production and the Open Source "Community"

User innovation theorists Georg von Krogh and Eric von Hippel characterize open source software development as a network of innovative users supported by a community.[44] Others have also emphasized the role of community in open source development,[45] and the term is used liberally by open source developers themselves.

Nevertheless, it has been suggested that there is no such thing as a community in open source software development. Some supporters of open source development claim that talk of a "community" creates the misconception that a core group of ideologues is responsible for the continued success of open source software, thereby leading potential licensors and adopters of open source technology to be too wary of falling foul of particular individuals.[46]

If there is a community in the open source software context, it need not be an essential element of the open source approach. For example, in an analysis based on case studies of distributed computing and carpooling, Benkler demonstrates the existence of sharing practices that "are not limited to tightly knit communities of repeat players who know each other well and interact across many contexts," but are either "utterly impersonal" or take place among "loosely affiliated individuals."[47]

These apparent contradictions imply a need to address the role of community in open source software development and, more broadly, in the bazaar.

Like other governance structures—the firm, the market, and the network—the bazaar is an ideal type. Real production systems contain a mixture of mechanisms for coordinating transactions. Thus, as we saw in the previous section, most open source development projects do not display pure bazaar characteristics throughout their life cycle or at every level of organization. Many open source projects begin as primarily network structures, scaling to bazaar governance as they grow larger or reach higher levels of technical maturity. Large-scale open source development projects often evolve hierarchical administration mechanisms.[48] Contributors to open source development include firms as well as individuals, and the incentives of potential contributors to participate in bazaar-style technology development are heavily influenced by the dynamics of markets in related goods and services.

The relative proportions of different governance structures in real-life open source development projects vary from one project to another—and also, presumably, from one technology context to another. This is because the relative effectiveness of each form of governance as a means of reducing transactional uncertainty depends on the nature of the transactions that are necessary for production to take place in any given setting. For example, it is generally argued that both networks and bazaars, with their open-endedness and relative absence of explicit quid pro quo behavior, enhance the ability to transmit and learn new knowledge and skills compared with markets and firms.[49] Which of these two modes of production predominates may depend on the type of information being exchanged. Powell suggests that the exchange of uncodified information—know-how, technological capability, a particular approach or style of production, a spirit of innovation or experimentation, or a philosophy of zero defects—is more likely to occur in networks.[50] Benkler's analysis implies that bazaar governance may work best where innovation-related information is highly codified;[51] hence, perhaps, the shift from network to bazaar governance as open source software development projects mature. Given the uncodified nature of much innovation-related information in biotechnology, we might speculate that real-life open source biotechnology development projects would be characterized mainly by bazaar governance, but with a bigger dash of network governance than is usual in the software context.

Thus, the existence of network governance even within production systems that are dominated by the bazaar helps account for the importance of relational ties in open source software development, and hence partially explains how community support may be seen as a key element of open source production without necessarily being an essential feature of bazaar governance. So far, so good; but there is another piece to this puzzle. Up to this point we have considered governance structures that operate in different ways to maximize *private* rewards for investment in technology development. What happens when private rewards provide insufficient incentive to innovate? For example, what if a particular open source development project involves tasks for which all private incentives—not just those of the bazaar, but also firm, market, and network incentives to the extent that they are also present—prove inadequate?

Of course, one solution to this problem is for the government to levy taxation in order to supply monetary or other subsidies to those who contribute to the production of public goods. According to Adam Smith, one "duty of the sovereign or commonwealth is that of erecting and maintaining those public institutions and these public works, which, though they may be in the highest degree advantageous to a great society, are, however, of such a nature, that the profit could never repay the expence to any individual or small number of individuals, and which it therefore cannot be expected that any individual or small number of individuals should erect or maintain."[52] Today, however—especially in United States—the rise of a conservatism that finds Adam Smith too much a liberal makes it increasingly difficult to advocate a leading role for government in the production of public goods, and this is clearly a factor in the relative decline of public funding for biotechnology research and development.

In the absence of monetary or other subsidies, a collective action model of technological innovation relies on the characteristics of *community* to supply the necessary incentive to participate.[53] Production systems that are governed predominantly by the rules of the market, firm, network, or bazaar therefore often also display some of the attributes of collective action communities—including active leadership, recruitment, and retention of participants[54]—and rely to some extent on collective-action–style incentives, such as enhanced reputation. This effect is particularly pronounced in bazaar-style production systems because, as we have seen, the bazaar presupposes private incentives that are independent of proprietary exclusivity and therefore tend to resemble collective action incentives more closely than the private incentives associated with firms, markets, or networks.

Thus, Steven Weber has observed that coordination of contributions to real-life open source projects depends on *both* the code architecture—the technical design of the object of production, critical to bazaar-style coordination—and on aspects of community (collective action) governance such as public shaming mechanisms (flaming and shunning), de facto constitutional documents (open source licenses), exhortation by project leaders to join or stay involved in particular projects or to undertake particular tasks, and a range of formal and informal institutions designed both to facilitate collective decision making and to promote the code transparency needed for individual decision making according to the bazaar model.[55]

We have seen that although direct ties between participants are not essential to ideal bazaar governance, they may enhance productivity in real-life open source development projects both as aspects of network governance and as a way to reinforce private incentives in the face of residual collective action problems. But there is yet another sense in which community matters to open source technology development. According to the user innovation literature, the "user community" that supports open source production is not just a network of relational ties or a set of collective-action–style incentives; it is a complete community in that it provides sociability, support, information, a sense of belonging, and a social identity.[56]

The existence of such a community does support knowledge production, but it also extends beyond user-innovators to include other users who are neither actively engaged in knowledge production nor necessarily adopters of the technology but are loyal to open source as a brand or social movement. An example is the SourceForge website, which supports development by allowing open source developers to see what projects exist in different technical areas and also provides a platform and tools for collaborative code development. Not everyone who visits SourceForge regularly is a registered user; it is also a kind of "virtual hangout" for people who want to see what is happening in the community.[57] The readership of Slashdot.org ("News for nerds, stuff that matters") is broader still.

Such "brand communities," which often (though not always) coexist with innovative user communities, are well known to marketers because they carry out the important functions of sharing brand information, perpetuating the history and culture of a brand, providing assistance to other users, and exerting pressure on members to remain loyal.[58] Thus, although this aspect of open source community is not essential to knowledge production within the framework of any particular open source project, it is still a real phenomenon with significant consequences for the brand as a whole—a point to which I return in Chapter 8.

Advantages of Bazaar Production

Earlier I noted that bazaar incentives are "weak" in the sense that not every agent who has the capacity to contribute to bazaar production will be in a position to take advantage of process and user benefits. At the same time, many who do not have the capacity or inclination to contribute may be in a position to free ride with respect to all but process benefits. Nevertheless, when bazaar incentives do happen to "click" with the needs or priorities of a particular agent, they can be as powerful as, or more powerful than, those associated with other governance structures; and the weak control mechanisms ensure that agents for whom these incentives *are* compelling are not prevented from acting on them.

How do weak bazaar-style incentives strengthen the bazaar as a governance structure? Primarily by lowering production costs. In the case of contributions that are motivated purely by process benefits, the contributor anticipates (by definition) that the net cost of generating a contribution using his or her private resources will be negative. In the case of contributions motivated by the prospect of obtaining user benefits by improving the usefulness of a technology, productivity is its own reward. In neither case is it necessary to attach a sale value to each contribution and rely solely on price competition to keep costs low. Instead, bazaar production is driven by use value, which is maximized in the bazaar because each contributor is able to reap both positive network externalities associated with widespread adoption of a technology and the benefit of others' contributions, whether in the form of new features or quality control. Where the contractual framework underpinning bazaar production is not just nonproprietary but also incorporates a copyleft-style commitment to make downstream technologies freely available, this already-expanded use value expands further to include the use value of any future technologies that incorporate either the agent's own contribution or any other contribution made under the same licensing regime. The bazaar is particularly effective at lowering production costs where the object of production is knowledge or knowledge goods, because *it allows contributors to be paid in kind out of a resource that is both nonrivalrous and inexhaustible.*

Thus, when like transactions are compared, bazaar production is potentially both cheaper and more efficient than other modes of production. How does this conclusion relate to our earlier discussion of intellectual property rights in biotechnology?

We saw in Chapter 2 that one consequence of proliferating property rights in biotechnology is an increase in the transaction costs associated with assembling proprietary inputs to research and development. By definition—all else being equal—a freely revealed or nonproprietary technology offers greater freedom of access and freedom to operate for prospective users than a proprietary technology. It is true that in practice, open source only approximates the "free revealing" ideal of perfectly unrestricted access and use. This is because the open source version of free revealing relies on intellectual property protection as a prerequisite to open source licensing. Nevertheless, open source licenses are more likely to promote widespread access and use of the licensed technology than other uses of intellectual property rights, because, unlike those other uses, that is what they are designed to do.

Of course, the disparity (in terms of "openness") between a given open source license and a given proprietary strategy depends on the details of each. A nonexclusive proprietary license that makes the licensed technology available on the same terms to all comers at minimal cost and does not include any reach-through or grant-back provisions comes close to offering the same degree of user freedom as an open source license (and supports a similar range of business strategies). Nevertheless, such an approach differs in principle from open source licensing, in that the latter is designed to maximize use value even at the cost of eliminating sale value.

The rationale for prioritizing use value is that the greatest economic impact of any technology lies in the secondary markets and nonmarket activities that it enables, not the income it may generate for its owner in markets for the technology itself.[59] In the biotechnology context, as we have seen, attempts to balance these competing values have arguably led to underexploitation and underdevelopment of a technology whose great promise, like that of software, is strongly linked to the breadth of its potential applications—that is, to its power as an enabler. The open source software experience suggests that in the right circumstances, wholehearted commitment to fostering use value may produce more satisfying results.

Begging the question, for the time being, whether open source is actually feasible in biotechnology, an open source or bazaar-style mode of production as described in this chapter would go a long way toward solving many of the problems highlighted in Chapters 2 and 3. When anyone who wants to contribute to technology development is given both the legal freedom to do so and the ability to exercise that freedom—in the software context, access to source code—the technology can be adapted to users' needs, irrespective of the size of the market those needs represent. It also tends to evolve more quickly, becoming progressively more robust and reliable. Further, because distribution is unrestricted and subject to free competition, it is generally both more affordable and more readily available.

One question that has not yet been explicitly addressed is the extent to which an open source approach to biotechnology research and development could relieve or reverse an incipient tragedy of the anticommons. We saw in Chapter 2 that anticommons tragedy occurs when the transaction costs associated with assembling a "toolkit" composed of multiple proprietary elements escalate to the point that the effort is abandoned (or never begun). If the owner of one of these proprietary elements were to adopt an open source strategy, then the overall transaction costs would, in general, be reduced because an open source technology is available for use, modification, and distribution by "anyone, anywhere, for any purpose," without payment of royalties.[60]

Note that an open source approach does not eliminate the costs to innovators of obtaining and maintaining intellectual property rights. Nor does open source eliminate *all* of the costs to prospective users of obtaining permission to use the licensed technology, though open source licensors strive to keep these costs to a minimum. Further, copyleft-style open source licenses place conditions on the distribution of downstream technologies that, despite their nonproprietary purpose, do limit freedom to operate with respect to those technologies. The rationale for adopting an open source license, as distinct from simply making one's technology available with no strings attached, relates to the overwhelmingly proprietary setting in which open source licenses are designed to operate. This issue is discussed in some detail in the next chapter; but it is true that in itself, straightforward free revealing is more effective in reducing transaction costs than open source licensing.

Whether open source licensing of one element of an otherwise proprietary toolkit would reduce overall transaction costs sufficiently to reverse an imminent tragedy of the anticommons depends on the circumstances. Free revealing of a single toolkit element does not, of course, automatically dissolve other proprietary barriers: assuming independent ownership, the owner of one technology has no direct control over the terms on which other technologies are made available. Open source has no magic power to change this; on the other hand, a switch in strategies on the part of one owner does alter the environment in which others must make their own decisions. For example, a researcher considering whether it is worth investing resources to "invent around" a proprietary technology will take into account the availability of other toolkit elements needed for the relevant research. If access to all of them is blocked, he or she is less likely to consider the investment worthwhile than if the proprietary tool in question is the last holdout in a toolkit that is otherwise fully free or open source. In fact, this was the dynamic at work in the development of the GNU/Linux operating system. Although Richard Stallman set out to create an entire free operating system, it was not necessary for him actually to succeed at this task. He only needed to be somewhat successful, because the more he achieved, the more clearly others saw that it would be worthwhile to devote their own resources to knock over the remaining obstacles (specifically, the missing kernel).

The point is that incremental reductions in transaction costs related to intellectual property can have a disproportionate effect on anticommons conditions; there is a kind of tipping point beyond which the payoff for developing an open source substitute for any given toolkit element is high enough for the task to attract focused effort on the part of all who would like to see a fully open collection of tools. Nevertheless, while open source licensing may have some effect on existing anticommons conditions, it is not primarily a means of dealing with existing anticommons tragedies. Rather, as we shall see in the next chapter, it is a way to preempt such tragedies by establishing a robust commons with respect to basic or fundamental technologies whose value is likely to be enhanced by cumulative innovation.

Another of the problems specifically related to intellectual property canvassed in earlier chapters is, in a sense, the opposite to that of the anticommons. What might open source biotechnology do to address the issue of overly broad patents on basic (platform or infrastructure) technologies? The answer depends on whether the platform is entirely nonsubstitutable, as in the case of the human genome sequence. If it is, then open source can do little once a proprietary choke hold has already been established. On the other hand, free revealing is an excellent way to *prevent* this from happening, as public-sector human genome researchers realized: we return to this example in Chapter 8.

If the platform or infrastructure technology in question is *not* impossible to invent around, then open source has something to offer even after the platform has fallen under proprietary control. Although it is possible to build nonproprietary alternatives by means other than bazaar-style production, the bazaar has the advantages of distributing the associated cost and risk as broadly as possible and harnessing the creativity of a wider range of potential contributors than is feasible under other governance structures. We shall see in the next chapter that an open source version of bazaar production uses reciprocal or "self-enforcing" licenses to establish and maintain cooperation among diverse players for whom problems of trust and confidence might otherwise be insurmountable obstacles to collective action. At the same time, bazaar-style production offers a range of opportunities for innovators to capture returns on private investment in innovation through the kinds of business models we see in the world of open source software. The provision of nonproprietary alternatives to proprietary platform technologies therefore need not—in principle, at least—require huge injections of public funding.

This last point warrants further comment. Like traditional academic research, open source is an example of bazaar production. However, we have seen that the legal and economic life-support system that sustained traditional academic research as "public science" has been gradually dismantled over the past several decades. Through a clever adaptation of conventional intellectual property licensing strategies and the development of new business models, open source production has managed to integrate itself into contemporary mainstream legal and economic structures in a way that preserves nonproprietary, peer-based knowledge production alongside proprietary research and development. Contributors to open source software development may be private individuals or corporations, for-profit or nonprofit, driven by commercial or noncommercial considerations or by a mixture of both. In the biotechnology context, the involvement of corporations in the bazaar is important because the minimum capital costs of research and development are generally (though not always) beyond the reach of private individuals. But more than this, the interdependence of for-profit and nonprofit innovators has become integral to modern biotechnology. Clearly, the biotechnology industry could not have come into being without access to foundational research conducted in a bazaar setting; less widely understood is that the private sector is still profoundly dependent on the public as a source of knowledge assets. Conversely, basic scientific research has come to depend heavily upon industrial innovation. Most bioscience research would today be inconceivable without centrifuges, PCR machines, microscopes, commercial antibodies, digital computers, and other tools that initially arose as user innovations in academic laboratories. In commercializing these instruments and selling them back to the academy, manufacturers relieve scientists of the burden of having to build their own equipment from scratch. Once upon a time, university science departments had their own workshops, employing specialist tradespeople for this type of work. But the sophistication of modern instruments and the economies of scale associated with their production are such that this approach is often no longer viable.

All of this implies that the challenge is not merely to sustain a scientific commons, but to do so in a way that preserves the capacity for commercialization on which the commons has come to depend.[61] Far from being an argument against open source biotechnology, this is a point in its favor. Open source business models show how commercialization of biotechnology research and development could continue without stifling the flow of innovation-related information. At the same time, the weak control mechanisms associated with bazaar governance help facilitate productive collaborations across the private-public divide. This is because where contributions are coordinated via the object of production, as in the bazaar, rather than through relationships between collaborators, as in network governance, many sources of potential conflict are removed.

In the next chapter, we shall see how these strengths of bazaar governance are reinforced by the characteristics of open source licenses.

5

Open Source Licensing for Biotechnology

The last chapter characterized open source software development as a special case of a broader phenomenon, bazaar production, that has existed in a range of commercial and noncommercial settings since well before the advent of the Internet. A key element in this mode of production is the practice of free revealing, in which innovators voluntarily give up all intellectual property rights to their innovations and allow everybody equal access to innovation-related information. One reason open source is a special case is that authors of open source software do not, in fact, give up their intellectual property rights—they just exploit them in a way that supports free revealing instead of proprietary exclusivity. The lynchpin of this maneuver is the open source license.[1]

So far in this book we have encountered several layers of meaning to the phrase *open source*. It simultaneously denotes a set of licensing criteria (introduced briefly in Chapter 1), a development methodology, and an open-ended yet characteristic approach to the commercial exploitation of technological innovations (both discussed in Chapter 4). Of these, licensing is fundamental. Open source exploitation strategies rely on open source collaborative technology development, which in turn relies on open source licenses, both as legal instruments and as embodiments of a specific social covenant among contributors. In consequence, the success of modeling open source in biotechnology depends to a large extent on the "devil in the details" of incorporating open source principles into biotechnology licensing.

For many important biotechnology tools that also happen to be software programs, open source licensing poses no special challenge. Some of the most powerful and commonly used bioinformatics tools—that is, tools for handling the data sets produced by molecular biologists—are already open source in the strictest sense.[2] Given the importance of bioinformatics in the contemporary life sciences, this is no trivial case of open source biotechnology, except in the mathematical sense of needing no further proof. However, the focus of this chapter is on extending the principles of open source licensing to *non*software biotechnologies.

The starting point for this discussion is a brief exploration of the various sources available to guide the translation of open source licensing from software to other fields, including biotechnology. These include individual open source software licenses, licensing standards, and statements of principle that are intended to capture the spirit of this innovative and unconventional approach to intellectual property management.

The next step is to articulate the underlying logic of open source in terms that make sense outside the software context. The chapter concludes with some basic principles of institutional design that may help biotechnology licensors implement this logic.

Drafting an Open Source Biotechnology License: Where to Begin?

Both software programs and biotechnology tools are generally made up of a number of different components. In software, these components are relatively homogeneous, in that they all consist of highly codified information in the form of software code or written documentation. By contrast, any given biotechnological innovation may be made up of a combination of data, software, hardware, "wet-ware" (including living biological materials), documentation, and specific know-how—all in varying degrees of codification.

This technological heterogeneity gives rise to heterogeneous patterns of ownership. Thus, although patents are the most important type of legal protection for biotechnology innovations, other forms of ownership, including copyright, trademarks, and data protection, are also relevant. Despite increasing use of patentable biotechnologies in plant breeding, classical methods are still in common use; many plant biotechnology licenses must therefore deal with plant variety rights. Biotechnology licenses also commonly transfer personal property rights in biological materials (usually by means of a separate material transfer agreement) and make provision for the protection of trade secrets and the transfer of nonproprietary information that would be expensive or difficult for the licensee to obtain via other channels. Each technology is thus covered—often incompletely—by a patchwork of different protections.

These technical and legal characteristics of biotechnology innovations make biotechnology licensing—whether proprietary or open source—inherently more complex than open source software licensing. Open source software licenses are primarily copyright licenses, though some also contain a limited patent grant.[3] By contrast, the precise mix of proprietary and quasi-proprietary rights transferred under license in the field of biotechnology varies from one technology to another, even within the same data stream. In fact, defining the technical and legal subject matter that is owned and intended to be transferred by the licensor is often the most difficult aspect of drafting a biotechnology license agreement. This is especially true for biological materials because of their inherent ability to replicate without human intervention, to exist in different forms, to be modified and used to generate completely different substances, and to be transferred from one organism to another.[4]

The challenge, then, of modeling open source licensing in biotechnology is to create new licenses that can accommodate the complexity and variety of biotechnology transfer agreements, yet remain faithful to the underlying logic of open source. It might be supposed that the most efficient way to tackle this brief would be to choose a representative open source software license and make the minimal modifications necessary for it to function in a specific biotechnology setting. But the existence of dozens of other open source licenses renders the choice problematic.

The Open Source Initiative (OSI) website lists close to sixty approved open source licenses.[5] Of these, only a subset are in general use, and some are more popular than others. Nevertheless, there are enough distinct open source licenses in circulation to give rise to what some in the open source community call "the combinatorial problem"—the confusing and potentially conflicting interaction of multiple different licenses when several open source modules are incorporated into a larger program.[6] Given the proliferation of open source licenses, what does it mean for a particular license to be "representative"? Should we choose the most commonly used license? The most professionally drafted? The most widely applicable? Should we choose the longest established, or the most recently revised?

Of course, prospective open source software licensors are confronted with exactly the same problem; the OSI's approach to the issue of license proliferation is discussed further in Chapter 6. But even supposing the choice of a starting text were straightforward, any prospective licensor who set out to tweak a particular open source *software* license for use in *biotechnology* would soon come up against another obstacle.

Anyone who has ever adapted a published experimental protocol for use in a new setting, or ported a software program from one operating system to another—anyone, in fact, who has ever undertaken any work of translation or adaptation—will be familiar with the problem of finding substitutes for original elements that have no direct counterpart in a new setting. What is the nearest equivalent—and given real-world constraints, how close is close enough? In the case of open source licensing, it is impossible to identify a precise biotechnology equivalent for every concept, term of art, or drafting device that is used in a software copyright license. Some creativity is required—and where there is room for creativity, there is room for distortion.

Faced with a choice among several imperfect translations from one language to another, an interpreter minimizes distortion by looking for clues as to the author's overall style and intent. An engineer in a similar predicament might refer to some kind of standard specification—essentially, a written description of what the technology is supposed to *do*. For a would-be open source biotechnology licensor, one obvious place to seek guidance in drafting a biotechnology license that conforms as closely as possible to open source principles is the text of the Open Source Definition (OSD), the latest version of which can be viewed on the OSI website, www.opensource.org.

The OSD lists the criteria for approval of a license by the OSI board. Software distributed in source form under an approved license is "OSI-Certified open source software." Such certification carries considerable weight within the open source community: licenses that fail to meet the requirements set out in the OSD are not accepted by key organizations such as the project clearinghouse SourceForge and tend not to be widely adopted.[7] Thus, the OSD is a voluntary nontechnical standard: software licensors who wish to take advantage of open source certification can choose to comply, just as a white-goods manufacturer can choose to comply with energy efficiency standards in order to sell more washing machines or a fruit grower can choose to comply with organic farming standards in order to access a more lucrative market.

The OSD originated as a set of guidelines developed in consultation with project contributors by Bruce Perens, leader of the Debian

Linux project, in mid-1997.[8] At that time, a number of what are now approved open source licenses were already in use, along with several others that purported to be "free" but were felt by the Debian community to be too restrictive to merit that label. The distinction mattered because Debian contributors wanted to build their own distribution of Linux entirely out of free software. The Debian Free Software Guidelines were developed as a yardstick against which to measure existing licenses; when the OSI was established, these guidelines were adopted as the OSD with only minimal changes.[9]

Despite its success with both licensors and licensees, the OSD was not written as a formal standard, and its language is not always easy to interpret. In fact, it reads like it is: a freeze-clamp specimen of ongoing debate and discussion among hackers about what it means for software to be "free." Unsurprisingly, the OSD is less popular with lawyers than with programmers. Littered with the artifacts of ancient controversies, the text is often vague and sometimes downright ambiguous. Worse, applying the ordinary rules of interpretation that guide lawyers in constructing precise meanings from imprecise language gives results that do not gel with interpretations forged over time among open source developers.

This is no criticism of the author, whose decision to favor textual stability over flawless drafting was deliberate and has no doubt contributed to the very high level of acceptance the OSD enjoys within the programming community.[10] Nobody could have foreseen the demands that are now routinely placed upon the current text of the OSD, which has been only slightly modified from its original version.[11] Arguably, its ambiguities have had a positive effect on community and consensus building in the software context because they force prospective licensors into a dialogue with the OSI board—and thereby indirectly with members of the community who subscribe to the OSI's license approval discussion list.[12] Nevertheless, the OSD on its own does not provide a clear set of specifications that can be conveniently generalized and applied to the drafting of open source licenses in the biotechnology context.

Where else might we look for clues about what it means for a license to be "open source"? Although the OSI is the only body that administers a formal certification system for open source licenses, the Free Software Foundation (FSF), established by Richard Stallman in the early 1980s, maintains its own list of "free" licenses, distinguishing between those that are compatible with the GPL and those that are not.[13] The FSF decides which licenses should go on its list by reference to the Free Software Definition (FSD), published on its website at http://www.fsf.org. This definition is authoritative partly because of the FSF's continuing influence in the community and partly because the GPL (Version 2) is the single most commonly used free and open source software license: even programmers who do not share the FSF's commitment to avoiding proprietary software need to know which other licenses are GPL-compatible.

Leaders of individual open source development projects likewise exert substantial influence by setting one-off standards to which licensors must conform if they want their contributions incorporated into the project's official code base. For example, the Debian project continues to use the Debian Free Software Guidelines to determine the status of contributions covered by different licenses.[14]

In addition to these de facto standard-setting activities, there exist numerous written opinions on the subject of open source licensing. Like the OSI, Debian maintains an email discussion list specifically to facilitate community consultation on licensing issues;[15] these lists tend to cross-fertilize through overlapping membership. The public archives of these and other free and open source discussion lists constitute a detailed record of the development of open source licensing principles over time. Other documentary sources seek to distill concise principles from evolving practice. For example, Lawrence Rosen, founding general counsel of the OSI, has proposed a set of five principles intended to encapsulate the "key things to look for in open source licenses, and the key things missing in nonopen source licenses."[16] Thoughtful commentators from outside the community, notably Steven Weber, have made similar attempts to capture the essential features of open source licensing.[17]

These various statements of principle, policy, and opinion—discussed further below—reveal a high degree of consensus about the characteristics a license must possess in order to be considered open source. Agreement is not absolute, however. Like any group that values active participation and deliberation, the open source software community subjects its own core values to constant debate and reinterpretation; reducing this multifaceted, dynamic discourse to clear, stable standards for the purpose of modeling the open source licensing approach outside software is not easy. But in any case, even the clearest such standards relate to the licensing of software, with all of its technical and legal peculiarities. To formulate a model of open source licensing for biotechnology, it is therefore necessary to approach the problem at a yet higher level of abstraction.

Software Freedom

What is open source licensing all about? Contrary to one common misconception, open source licenses are not inherently anti-intellectual-property. It is true that for some open source programmers, the assertion of proprietary rights in support of a nonproprietary strategy is a deliberate attempt to subvert a legal regime they consider to be harmful or even immoral. Just as insurgent forces use captured weapons to arm themselves, these programmers employ intellectual property rights to undermine the use of intellectual property as a means of achieving exclusive proprietary control over technical information. But there are also many open source licensors—including that patent colossus and co-architect of the knowledge game,

IBM—who emphatically have nothing against intellectual property rights, regarding open source licensing as merely a useful addition to a legitimate but limited repertoire of proprietary strategies for exploiting intellectual property ownership.

Similarly, although it is true as a general rule (see the next paragraph) that open source licensing does not itself generate significant revenue, it would be a mistake to conclude that open source licenses are inherently anti- or noncommercial. We saw in Chapter 4 that the primary commercial application of open source licensing is to support a free-revealing or nonproprietary business strategy. Such strategies sacrifice proprietary exclusivity for the sake of larger economic gains. In other words, open source licensing is a form of investment, akin to more conventional commercial investments in research and development or marketing. Whether the investor is a for-profit firm, a government funding agency, or a philanthropic foundation, such investments are always made in the expectation that they will ultimately generate more gains than losses.[18]

In any case, the blanket perception that open source licensing does not generate any revenue is not accurate: there is room in the open source model for licensors to recover at least some of the up-front costs of protecting intellectual property and disseminating innovation-related information. This is an important consideration in the biotechnology context, because the costs associated with both aspects of open-source-style free revealing are likely to be higher than in software—an issue to which we return in Chapters 6 and 7.

First, open source licensors are not prohibited from selling the licensed technology; they may do so at any price the market will bear. Of course, the fact that an open source license must guarantee licensees' freedom to imitate the licensed technology and distribute it to others means that that price tends to be driven down to the marginal cost of reproduction and distribution. In the software context, this is close to zero. But the low selling price of open source software is not an inherent feature of the open source model; rather, it is the consequence of market forces that could be expected to operate somewhat differently in other settings. In the case of biotechnologies that are less highly codified than software (such as those that revolve around a particular laboratory technique) or are embedded in tangible objects (such as vectors) whose production costs are sensitive to economies of scale, the marginal cost of reproduction and distribution may be quite high. The higher these costs, the more licensees may be willing to pay the licensor for extra "copies" or tangible embodiments of the technology, and the fewer distributors may come forward to compete with the original licensor. Second, there is nothing to stop an open source licensor from offering the same technology on both proprietary and open source license terms at the same time. As we saw in the last chapter, this "dual licensing" approach generates a surprising amount of income for many open source programmers—some of whose customers prefer, and are willing to pay for, a more conventional licensing arrangement. Finally, it is arguably compatible with open source principles for licensors to charge a one-off license fee, as distinct from royalties, which by their nature as ongoing payments create the danger of lock-in. This possibility is discussed further below.

If open source licensing is neither inherently anti-intellectual-property nor anticommercial, what is its central purpose? The simple answer is that it facilitates collaborative technology development. However, as this is also the primary purpose of many proprietary licenses, some further explanation is required.

Proprietary software licenses of the kind that every computer user has encountered on splash screens or shrink-wrap packages are designed to facilitate a product market in software code. They exploit the copyright owner's ability to sell the executable version of a software program separately from the associated intellectual property—in particular, the right to make and distribute copies—in support of an "invent one, sell many" (proprietary manufacturing) business model.

Other commercial software licenses have a somewhat different purpose. In Eric Raymond's estimate, perhaps 5 percent of all programming effort is directed toward writing code for sale. The rest consists of writing and maintaining enterprise software—customized financial and database software of the kind that is used by every medium- and large-size firm—and "back end" software like device drivers or embedded code for airplanes, cars, mobile telephones, household appliances, and industrial equipment.[19] Much of this work is outsourced by prospective users, and the license contracts that govern the relationship between programmer and user bear only a passing resemblance to mass shrink-wrap or click-wrap software licenses.[20] It is here, where the freedom to use and reuse effective solutions to common programming problems makes obvious business sense, that open source licensing has come into its own.

In biotechnology, we have seen that product development is usually too large and complex a task for even the best-resourced industry participants to undertake in isolation. Different entities colonize different phases in the product value chain, from basic research to regulatory approval and marketing. Technology is often licensed at an early stage of development, before the precise nature and utility of the product is known.[21] In this context, the function of a license agreement is not usually to facilitate the sale of a product directly to the end user—though some biotechnology licenses, such as seed-label contracts, do perform this function—but to facilitate the integration of valuable information from a range of sources. In other words, most proprietary biotechnology licenses are designed, like open source licenses, to promote collaborative development of early-stage technologies.

The *difference* between open source licenses and these proprietary licenses relates to the structure of collaboration they seek to engender. The purpose of a collaborative proprietary license is to provide the licensee with just enough access and freedom to operate to move the technology another step along the value chain. The licensor wants the technology to be improved and integrated into downstream products, but this goal is in tension with the need to preserve the characteristics of excludability and rivalry-in-use that make it possible to treat the technology as a private good—the use of which may be conditioned upon payment of licensing fees, exchange of cross-licensing rights, or other concessions. In this model, license provisions dealing with exclusivity and sublicensing determine who is entitled to exercise the licensed rights, while field-of-use and territorial restrictions set limits on the conditions on which they may be exercised.

By contrast, an open source licensor chooses to forgo the value of the technology as a private good in order to establish—or reestablish—it as a public good. Some proprietary licensing—such as non-exclusive licensing of genetic technologies and research tools for a small up-front fee as recommended by bodies such as the NIH and OECD—goes almost as far; adopting a fully open source approach effects the subtle but significant shift from network to bazaar-style governance.[22] An open source license allows anyone, anywhere, for any purpose, to use, copy, modify, and distribute the licensed software—for free or for a fee—without having to pay royalties to the licensor.[23] In other words, it tips the scales all the way toward maximizing external contributions to technology development, enabling the licensor to invite the broadest possible range of participants to help realize the technology's full potential.

In practice, of course, most open source software development takes place within the context of a project such as Linux, Apache, or any of the many other open source projects now in existence.[24] Each project is an innovation system sustained by social interactions directed toward a common set of technical goals. In this context, as Steven Weber has pointed out, the role of the license agreement promulgated by the initial developers of the technology is akin to that of a written constitution.[25] As such, it serves a dual purpose in facilitating collaborative development: articulating the terms of the collaborative effort, and demonstrating credible commitment to those terms on the part of each contributor. Let us consider these two functions in turn, starting with credible commitment.

The role of credible commitment in engendering cooperation in the face of uncertain outcomes was explained by economic historians Douglass North and Barry Weingast in their well-known study of the seventeenth-century monarchies of France and England.[26] In France the monarch was above the law, in the sense that no legal obligation could be enforced against him. In England, though the modern doctrine of the "rule of law" did not yet exist, the king was understood to rule "through" the law—meaning that obligations entered into by the king were enforceable in the courts. When the two monarchs went to war and needed to borrow money, investors charged the French king much higher interest rates because they could not be certain of ever recovering the principal of their loans. Thus, an apparent weakness became a strength: by accepting limits on his own power, the English king increased his borrowing capacity, and hence his capacity to wage war. The general point is that without some means of tying your own hands—so that others can be sure you won't stab them in the back—you may have to make much bigger concessions in order to secure cooperation.

How does this apply to open source licensing? To encourage others to invest their own resources in improving and building on a particular technology, the owner of that technology must be able to reassure potential collaborators that they will have ongoing access and freedom to operate. Without such reassurance, potential collaborators will be wary of becoming dependent on the technology and reluctant to expend time, money, and effort on improvements in case they find themselves ambushed by demands for royalties or other restrictions somewhere down the track. Open source licenses enable technology owners to overcome these obstacles to collaboration by demonstrating a credible commitment to transparent, unencumbered technology transfer. They do this by shifting the emphasis from protecting the rights of the intellectual property owner—the primary focus of conventional proprietary licensing—to protecting *users'* rights.

What of the second function of an open source license, that of articulating the terms of collaboration? The specific terms of collaboration established by open source licensing schemes vary from one license to another, but all open source licenses fall into one of two classes.

"Academic" or "permissive" open source licenses protect users' rights only with respect to the particular version of a given technology that is owned and distributed by the licensor. Empirically, these licenses tend to be short and to the point; their terms reflect the traditional academic bargain in which an innovator donates his or her work for use by all comers, with few or no restrictions.[27] The only substantive obligation imposed by many academic licenses is the obligation to acknowledge the licensor's contribution in any derivative works—thus giving effect to the author's right of attribution, one of the "moral rights" recognized by copyright law in many jurisdictions. The earliest and best-known example of an academic open source license is the Berkeley Software Distribution (BSD) License, first employed by the University of California to license a package of software tools and utilities assembled by Bill Joy as a graduate student in 1978.[28]

Licenses of the second class are known as "copyleft" or "reciprocal" licenses. As we saw in Chapter 1, the earliest and best-known example of a copyleft license is the GPL. Copyleft licenses seek to protect the rights not only of the current generation of users, but also of future generations. I return to a discussion of the copyleft mechanism and the challenge of devising a functional equivalent in biotechnology licensing below. For the time being, simply note that the fundamental set of user rights protected by copyleft licenses is the same as that protected by the simpler academic licenses. The distinguishing feature of this class of licenses—an obligation imposed on licensees, if and when they choose to distribute derivative works, to do so under the same terms as the original license—is an extension of the basic function shared by all open source licenses, to protect users' rights.

To summarize: The purpose of an open source license is to facilitate collaborative technology development. Like a conventional biotechnology license, it does this by articulating the terms of collaboration and enabling the licensor to make a credible commitment to those terms. Unlike a conventional proprietary license, however, the terms of the collaboration uncompromisingly favor the rights of technology users and developers over those of intellectual property owners.

The obvious next question is, what *are* the rights guaranteed to all users under an open source license? Among proponents of free and open source software, they are referred to collectively as "software freedom."

Philosophers and legal theorists distinguish among different types of rights and freedoms. Technically, the rights of a licensee under an open source license are a mix of liberty rights (legally protected freedoms to engage in certain behavior) and entitlement rights (legal obligations on the part of someone other than the right bearer to provide something to the right bearer).

Rights of both sorts are found in the various attempts that have been made to define software freedom. For example, consider Rosen's five principles of open source licensing, themselves a synthesis of the key provisions of the OSD:[29]

1. Licensees are free to use open source software for any purpose whatsoever.
2. Licensees are free to make copies of open source software and to distribute them without payment of royalties to a licensor.
3. Licensees are free to create derivative works of open source

software and to distribute them without payment of royalties to a licensor.
4. Licensees are free to access and use the source code of open source software.
5. Licensees are free to combine open source and other software.

Most of these principles relate to liberty rights, but the fourth deals with an entitlement—specifically, the entitlement to receive a copy of an open source software program's source code.

The FSD also enumerates the elements of software freedom so as to highlight both liberty and entitlement rights:

> Free software refers to four kinds of freedom, for the users of the software:
>
> - The freedom to run the program, for any purpose (freedom 0).
> - The freedom to study how the program works, and adapt it to your needs (freedom 1). Access to the source code is a precondition for this.
> - The freedom to redistribute copies so you can help your neighbor (freedom 2).
> - The freedom to improve the program, and release your improvements to the public, so that the whole community benefits (freedom 3). Access to the source code is a precondition for this.[30]

In his book *The Success of Open Source,* Weber offers yet another formulation of the essential features of open source licensing that is particularly useful as a starting point for translating open source principles into biotechnology because it is expressed in very general terms. According to Weber,

> open source licensing schemes generally try to create a social structure that:
>
> - Empowers users by ensuring access to source code.
> - Passes a large proportion of the rights regarding the use of the code to the user rather than reserving them for the author. In

fact, the major right the author as copyright holder keeps is enforcement of the license terms. The user gains the rights to copy and redistribute, use, modify for personal use, and redistribute modified versions of the software.

- Constrains users from putting restrictions on other users (present and future) in ways that would defeat the original goals.[31]

Here again we see an entitlement to source code, combined with freedom from legal encumbrance. The only privilege of ownership retained by the licensor is that of enforcing software freedom; the only permissible constraints are those deemed necessary to promote that primary goal.

We may conclude that software freedom has two critical components: (1) the liberty to use, develop, and commercialize the licensed software without proprietary restrictions, and (2) the entitlement to source code. How might these freedoms be realized in the biotechnology context?

Open Source Biotechnology Versus Simple Free Revealing

As a licensing strategy, open source is predicated on the existence of a proprietary right. But we have seen that a nonproprietary technology transfer strategy need not be open source. Instead, it could rely on straightforward free revealing. What advantages, then, might an open source strategy have over straightforward publication as a way of encouraging widespread adoption and ongoing technology development?

In software, copyright arises automatically when the program is first fixed in a tangible medium—the making of the first "copy."[32] Essentially, so long as an author has not copied a work, he or she will be able to claim copyright in it. Under United States copyright law, there is no mechanism for waiving a copyright that merely subsists, and no accepted way to dedicate an original work of authorship to the public domain before the copyright term for that work expires. Thus, a license is the only recognized way to authorize others to undertake the authors' exclusive copyright rights.[33]

By contrast, patent protection—prevalent in biotechnology—is neither automatic nor cost-free. Obtaining a patent requires specialist skills, may take several years and entail substantial uncertainty, and typically costs on the order of tens of thousands of dollars for each international filing. Maintenance and enforcement can also be very expensive, especially if the licensor becomes involved in litigation. Moreover, the threshold for patent protection is higher than for copyright: patent rights depend on an inventor clearing the key hurdles of novelty, utility and nonobviousness, or inventive step.

These differences have two implications for open source biotechnology licensing. One is that for biological innovations, unlike software, open source is not the default version of a free-revealing strategy. Instead, we must ask why anyone would go to the trouble of prosecuting a patent only to license it on open source terms. Why not simply refrain from obtaining the patent in the first place?[34]

The other implication is that in biotechnology, anyone who chooses an open source strategy over straightforward free revealing runs the risk of alienating some of those who would otherwise support a nonproprietary approach. Even an intellectual property right that is licensed on open source terms has the drawback of adding to the complexity of the intellectual property landscape. It is true that open source licensing generally creates fewer transaction costs and is inherently more transparent than a proprietary licensing strategy. Nevertheless, in some contexts, claiming ownership over an innovation may generate a negative rather than a positive signal to potential collaborators, creating ill will among prospective users and decreasing the chances of a technology being widely adopted or built upon. This is especially likely where the ownership claim is particularly broad (as in the case of the notorious "junk DNA" patents) or where user-developers have a strong belief that the technology ought to be in the public domain. For example, public-sector human genome scientists at one time considered releasing sequence data under copyleft-style open source licenses, but ultimately rejected the idea because open source was not open *enough* to allay their collaborators' fears of excessive central control of the data stream.[35] Of course, these are also circumstances where an open source, as distinct from a *proprietary* approach, is most likely to be worth considering: a broad patent is the best basis for a copyleft-style commons, and open source collaboration is boosted by the contributions of user-developers who respond to nonproprietary incentives. Nevertheless, there is a danger that open source could be seen not just as the seizure of enemy weapons but as the escalation of an unwanted arms race.

The upshot is that in biotechnology, those who choose to adopt an open source license will likely do so as the result of a deliberate calculation that takes into account not only the pros and cons of a nonproprietary versus proprietary approach, but also the pros and cons of different nonproprietary strategies. In that case, what are some of the factors that could be expected to weigh in favor of open source over straightforward free revealing?

First, ownership rights do arise automatically with respect to some components of many biotechnologies. Some biological innovations incorporate software code, data, written protocols, or other elements that may be subject to copyright protection. Similarly, many biological innovations have tangible material components such as cell lines or germplasm that constitute personal property, irrespective of any *intellectual* property rights that may be associated with the technology. In such cases, a license may help reduce the transaction costs of transferring the technology to other prospective users by clarifying the owner's intention to make the technology available on nonproprietary terms.[36]

Second, in some cases a patent owner may wish to open source the relevant technology toward the end of its life, after its value as a generator of proprietary licensing revenue begins to decline. Granting an open source license is then an alternative to abandoning the patent, which may be undesirable for a variety of reasons. In that case, the choice whether to adopt an open source approach comes after the decision to obtain a patent, not before, and the higher cost of patent protection compared with copyright protection will have little bearing on the decision whether to adopt an open source license. Even though a substantial investment may have been made in order to obtain and maintain intellectual property protection, these costs are properly regarded as sunk costs rather than costs associated with adopting an open source approach.

Suppose, however, that the technology is to be protected by a patent and that protection has not yet been secured. Why might an open source approach be preferable to straightforward free revealing in that case?

Where the technology owner intends to adopt a copyleft-style open source license, the rationale is clear. As we saw in Chapter 1, copyleft uses copyright as a sort of hook, baited with software freedom, to compel follow-on innovators to contribute to a protected commons. Any similar mechanism in the biotechnology context would have to rely on the existence of a prior intellectual property right: one cannot impose conditions on the distribution of a technology one does not own. For academic- or permissive-style open source licenses, though, what would be the purpose of pursuing patent protection in the absence of an intention to impose a copyleft-style license restriction?

One answer is that failure to assert ownership over a technology before making it available for public use leaves open the possibility that someone else will patent the technology and pursue a proprietary exploitation strategy, to the detriment of the original inventor as well as other users. Straightforward free revealing is especially risky where there is a proliferation of overlapping intellectual property rights or the field of innovation is especially competitive or litigious. In practice, relatively minor improvements—such as "me too" drugs in the pharmaceutical industry—can get the benefit of patent protection; thus, straightforward free revealing can help downstream patentees, leaving the upstream innovator to find a way around patents for which he or she laid the foundations.

It might be thought that free revealing would in itself be enough to defeat subsequent claims because of the requirement that patentable inventions be novel and inventive, but that is not always the case. The name given to free revealing in this context is "defensive disclosure" or "defensive publishing." To have a chance of defeating subsequent patent claims, a defensive disclosure must qualify as prior art for the purposes of assessing novelty under patent legislation in the relevant jurisdiction or jurisdictions. It must also be easily accessible to patent examiners, who are frequently too overworked and underresourced to conduct a comprehensive search of all possible sources of prior art references. Various methods of publishing prior art information in order to make it more likely to come to the attention of patent examiners do exist, but each has its own drawbacks with regard to cost, accessibility to inventors or patent examiners, or time delays between submission and publication.[37]

Thus, defensive disclosure is not always easy or cheap and does not always give reliable protection against subsequent accusations of infringement. In some circumstances, obtaining a patent before freely revealing one's technology may be a better guarantee against downstream appropriation—at the same time, enhancing the credibility of the inventor's own commitment to biotechnology freedom.

A second reason why a biotechnology innovator might choose to obtain patent protection despite intending to license the technology on "academic" or "permissive" open source terms is to retain the option of making the technology available under more than one license—the approach known as "dual licensing." While everyone must be free to access open source technology under an open source license, some licensees prefer more tailored terms. One obvious reason for such a preference is that the licensee wishes to avoid incurring any copyleft obligation with respect to modified or improved versions of the technology. However, even in the absence of a copyleft obligation, a licensee may also be motivated by any of a variety of considerations having to do with the complex web of business relationships in which biotechnology research and development takes place. In any case, only an intellectual property owner has the ability to offer users a choice between nonproprietary and proprietary alternatives.

A third reason to choose open source licensing over straightforward free revealing is that patent ownership gives an innovator the ability to set terms of use and exclude anyone who will not abide by those terms, thereby opening up a wider range of strategic options than mere publication or defensive disclosure. The most obvious example is again that of copyleft, but this is not the only possibility. Another is that the innovator may wish to make use of a litigation deterrent or "yank" clause. Found in many licenses, both open source and proprietary, this is a clause that terminates (yanks) the licensee's rights under the license if he or she sues the licensor (for example, for infringement of one of the licensee's patents).[38] Such a strategy is not available to an innovator who has no property right in his or her innovation.

Finally, certain pathways to development may be facilitated by the existence of an intellectual property right even if the owner has no intention of pursuing either a proprietary or a copyleft-style licensing strategy. Bringing a mere invention to fruition as a useful innovation typically involves bringing together disparate resources according to an unpredictable timetable. In this highly contingent process, intellectual property rights are a signal to potential investors, private and public, that an innovation has substance and is worth supporting. Patents in particular are commonly regarded as a proxy for the innovative capacity of organizations and individual employees. Being able to "tick the box" regarding patent rights may help an innovator to access resources—including, but not limited to, monetary support—that are critical to the success of an innovation at an especially vulnerable stage in its development. As long as intellectual property protection is regarded as desirable by those who invest in biomedical or agricultural innovation, this will be the case even where development of the relevant technology has less to do with proprietary exclusivity than with establishing networks of users, eliminating design flaws, and demonstrating the value of the technology in different environments.

Implementing Open-Source-Style "Biotechnology Freedom"

Where an open source biotechnology licensing strategy is judged preferable to simple free revealing, how should it be implemented? Earlier I argued that modeling open source licensing in biotechnology is not a simple matter of editing existing open source licenses; nor is it a question of drafting a new license according to established open source standards. Even apart from the difficulty of choosing the best or most authoritative starting text, both of these sources contain terminology specific to software and copyright, and concepts for which there are no straightforward substitutes in the biotechnology context. To choose wisely among possible implementations, it is necessary to model function rather than form. Recall that software freedom has two critical components: liberty rights and entitlement rights for users of the software. Addressing these components in order, what are the issues that arise in translating the "liberty" component of software freedom into biotechnology?

As noted earlier, fundamental to any kind of licensing is the existence of a property right. Every law student learns that property rights are like bundles of sticks, each of which represents access to or control over a distinct stream of benefits. Important "sticks" in the case of real property include the rights to sell, lease, mortgage, donate, subdivide, or grant easements over a particular parcel of land. Different technologies—even the different components that make up a single technology—are associated with different bundles of sticks.

In the United States, the Copyright Act of 1976 gives the owner of copyright in a piece of software the exclusive right to do or authorize certain activities.[39] In practice, the most important of these are reproduction of the original work in copies, preparation of derivative works based on the original work, and distribution of copies of the work to the public by sale or other transfer of ownership or by rental, lease, or lending.

By contrast, the United States Patents Act does not give a right to use, but grants the owner of a patent the right to exclude *others* from making, using, offering for sale, or selling the invention throughout the United States or importing the invention into the United States, and, if the invention is a process, the right to exclude others from using, offering for sale, or selling throughout the United States, or importing into the United States, products made by that process.[40] Copyright and patent statutes in other jurisdictions confer similar (though not identical) bundles of rights. Plant variety rights and personal property in biological materials are also associated with unique bundles of rights.

One approach to translating software freedom into biotechnology would be to try to map open source licensees' freedom to exercise rights that would otherwise be exclusive to the copyright owner onto the various rights associated with biotechnology patent ownership. For example, "reproducing an original work in copies" might be regarded as roughly analogous to "making" a patented invention; "selling" an invention or a product made using an invention might appear to correspond to one of the ways a copyright owner might choose to "distribute" a copyrighted work. But identifying a functional equivalent for each of the rights granted to users by an open source copyright license in every one of the bundles of rights commonly used to protect biotechnology innovations would be a difficult and perhaps impossible task.

Aside from the basic difference between an exclusive right and a right to exclude, one reason is that the copyright holder's statutory rights to "copy," "prepare derivative works," and "distribute" are not consistently incorporated into either (1) the statements of open source principles referred to earlier in this chapter or (2) the actual language of copyright grants in open source licenses. Thus, before one could begin the mapping process it would be necessary to translate the actual terms of open source software licenses into legally meaningful concepts.

More fundamentally, though, such an atomistic approach to defining biotechnology freedom could never be completely successful because intellectual property rights are just too different from one another. Each has its own history—its own context-dependent accretions of meaning and practice. Even the overarching concept of "intellectual property" is of relatively recent origin. There is no a priori reason to suppose that any given stick in one intellectual property bundle will have a functional counterpart in every other. In consequence, any attempt to build a concept of biotechnology freedom by reference to particular legally defined aspects of software freedom would run the risk of letting important freedoms fall through the gaps between different regimes.

Even if that outcome could be avoided, a cobbled-together set of principles could not be expected to achieve the same recognition and acceptance as a notion of freedom that arises organically within the relevant technology community. The way open source programmers talk about software freedom is very telling. While they may disagree vehemently about how to articulate open source principles—even about what is and is not open source—each would say of software freedom, "I know it when I see it." Similarly, the ultimate test of any definition of biotechnology freedom is whether it can be internalized to the point of becoming a sort of gut reaction among industry participants—just as proprietary thinking is today.

Fortunately, there is an alternative to the reductionist approach that offers a much simpler and more robust conceptualization of biotechnology freedom. The key to this alternative is recognizing that although software freedom—like any other freedom—is inherently valuable, it can also be seen as a means to an end. That end is free competition. A corollary of the fundamental liberties and entitlements that make up the ideal of software freedom is that open source licenses are essentially procompetitive: they promote low barriers to entry and dismantle the monopoly powers associated with intellectual property rights. On this view, "biotechnology freedom" should incorporate whichever specific freedoms are necessary to eliminate the competitive advantage enjoyed by intellectual property owners *per se* over other technology users and distributors.

How do open source software licenses achieve this goal? An open source licensor makes his or her software available to "anyone, anywhere" on royalty-free, nondiscriminatory terms. Anyone who obtains a copy of open source software is entitled to a license that allows him or her to compete on a level playing field with the copyright owner as both (1) user and (2) distributor of the licensed technology.

As a user, each open source licensee is free from field-of-use and territorial restrictions commonly used in proprietary licensing to protect the intellectual property owner or other licensees from competition in a particular market segment.[41] Following this model, a biotechnology license granted "for research and noncommercial use only" could not be open source. Similarly, an open source license may not impose a requirement to report to the licensor, or to disclose the means and manner of any internal use of the licensed technology.[42] (In copyleft licenses, it is distribution, not internal use, that triggers the copyleft obligation to disclose source code.)

As a distributor, an open source licensee is also free from restrictions commonly found in proprietary licenses. An open software source license may not restrict the number of copies a licensee is allowed to distribute, the identity or geographic location of the recipients, or the price the licensee asks them to pay (anywhere from zero to the highest price the market will bear).[43] The same goes for derivative works, with the qualification that under a copyleft license, the licensee may be constrained to deal with others as he or she has been dealt with by the licensor. Under any open source license, any person to whom the software is distributed may in turn become a licensee and exercise the same rights of distribution.[44] As Steven Weber has remarked, open source licensing is based on intellectual property—but it is a concept of property configured around the right to distribute, not to exclude.[45]

Seeing software freedom as a way to promote competition makes it easy to comprehend aspects of open source licensing that might otherwise seem confusing. Take as an example the requirement that open source licenses be royalty-free: that is, open source licensees must be free to make copies or derivative works of the licensed software and to distribute them without payment of royalties to a licensor.[46] As we saw earlier, this does not mean that a licensor cannot charge for copies of an open source software program: commercial distributors of open source software routinely sell copies of their own software, together with open source software they have licensed from others, to paying customers. A common explanation is that these licensors are charging for the service of distributing the software, not for the rights to copy, modify, and distribute it. In practice, however, the distinction is immaterial—and so is the actual price. Provided it is a one-off, even a substantial up-front charge may be compatible with open source principles. What *matters* is that the fee structure must not create a continuing obligation that could, even in theory, give the licensor indirect control over the licensee's subsequent use or distribution of the technology. As Lawrence Rosen explains,

> When you go to a computer store and Red Hat Linux sells for $49.95 and it is a box with a CD in it that cost them $5, is that a royalty? Is that the cost of their marketing? I don't know, and I don't care. I don't have to go into their books and see how they account for that revenue stream. The point is, whatever they charge you for, you only have to buy once.[47]

Looking at the ideal of biotechnology freedom through the lens of competition also makes it clear why the freedom to fork code is regarded by some as the defining characteristic of open source licensing. Under the terms of an open source license, anyone who is dissatisfied with the conduct of a project leader—on technical, administrative, political, or even purely personal grounds—is free to take the collaborative effort in a new direction.[48] In practice, forking is rare, largely because the benefits are usually not worth the hassle and uncertainty of persuading others to join a new branch of the project. For example, where the differences are technical ones it is often easier to continue contributing to the main project and simply devote a few extra resources to tweaking the results to meet one's own needs. But the ever-present possibility of a code fork makes project leaders responsible to their co-developers and ensures that no individual or group unduly dominates the process of technology development. Conversely, there is no danger that a potentially useful tool will be left on the shelf simply because of the waning interest or incapacity of an initial innovator: intellectual property owners who license their technology on open source principles are thereby precluded from acting as technological "dogs in the manger." As Eric S. Raymond, the author of *The Cathedral and the Bazaar,* wrote in an open letter to Sun in 2004, "Anyone who doesn't like Linus [Torvalds'] decisions about Linux can fork the codebase, start his own effort, and compete for developer and user attention on a legally equal footing. *That* is the essence of the bazaar."[49]

By this analysis, the point of software freedom—and by extension, biotechnology freedom—is that none of the market-ordering techniques traditionally employed by intellectual property owners are available to open source licensors. This commitment to free competition reflects a fundamentally different understanding of what motivates innovation from that embodied in standard economic justifications for intellectual property rights. Proponents of open source licensing perceive the greatest danger to innovation not in a lack of incentive to innovate or in poor coordination of complementary efforts, but in granting initial innovators too much control over downstream development.

The competition-oriented approach to defining biotechnology freedom sets some broad parameters, but leaves plenty of room for debate within the biotechnology community as to exactly how much control on the part of the initial innovator should be regarded as too much. This is no bad thing; it is through such deliberation that a true open source movement may take shape in biotechnology.

Up to this point we have seen that by analogy with software freedom, biotechnology freedom depends on two things: first, the licensed technology must be legally unencumbered, and second, it must be technically transparent. The foregoing discussion focused on the first requirement—the *liberty* rights that guarantee freedom to operate with any technology that is licensed on open source terms. In software, open source licensees also have an *entitlement* to access the source code of the licensed program. How does this entitlement translate to biotechnology?

In Chapter 1, I described source code as the form of a software program that can be read and understood by human beings. A computer program is a sequence of instructions to be executed by a computer; but the vast majority of computers can only execute instructions encoded as strings of binary numbers, so a computer program is first written in source code, and then compiled by another program into binary or "executable" form.

Clause 2 of the OSD articulates the open source software entitlement to source code as follows:

> [In order to be open source, the] program must include source code, and [the license] must allow distribution in source code as well as compiled form. Where some form of a product is not distributed with source code, there must be a well-publicized means of obtaining the source code for no more than a reasonable reproduction cost—preferably, downloading via the Internet without charge. The source code must be the preferred form in which a programmer would modify the program. Deliberately obfuscated source code is not allowed. Intermediate forms such as the output of a pre-processor or translator are not allowed.[50]

According to the OSI website, the rationale for requiring open source licensors to give users access to un-obfuscated source code is that "you can't evolve programs without modifying them. Since our purpose is to make evolution easy, we require that modification be made easy."[51] The FSD says access to source code is necessary to allow users to study how a program works, adapt it to their needs, improve it, and release improvements to the public.[52] On either view, access to source code is required in order to give practical effect to the legal freedoms granted by an open source license. Without source code, those freedoms are largely meaningless because the software is nothing but a black box.[53]

What is the equivalent of software source code in biotechnology? In a purely technical sense, there is no real equivalent. Source code is simultaneously functional and descriptive: anyone who has access to the source can compile it and run the program. In other words, source code not only explains how to use or modify the technology, it *is* the technology. Even the most highly codified of nonsoftware biotechnology tools do not share this peculiarity. To take a robust example, consider a simple DNA construct. The base pair sequence of a DNA molecule is often likened to source code; yet substantial extrinsic information is required to make sense of a DNA sequence. As one of the leaders of the human genome project put it, "the raw, unannotated genome is not a usable tool in the hands of the average biologist."[54]

Does this difference between software and biotechnology tools imply that the open source licensee's source code entitlement cannot be translated into biotechnology? It is true that the diversity of biotechnology tools and their relative lack of codification make it impossible to frame a disclosure requirement in simple, universally applicable technical terms like the software requirement to provide access to source code. But this is not necessary in order to give practical effect to biotechnology freedom. Practitioners of any given class of biotechnology tools share a working understanding of what information and materials are required to get up to speed on new developments. Further, biotechnologies are exchanged every day under licenses and material transfer agreements (MTAs) in which those requirements are articulated to the satisfaction of both parties. Returning to the DNA example, a sample annex to an MTA for a simple DNA construct illustrates the point:

Provide the following information in electronic format

1. Name of sequence (plus accession number if available)
2. Sequence length
3. DNA sequence—in FASTA format (include "¿" and indicate start and stop codon with underline)
4. Protein sequence—in one-letter code
5. Organism of origin
6. How sequence obtained—e.g., Two Hybrid, Sequencing, Database

7. Nature of sequence—please stipulate if it is a genomic sequence, a predicted cDNA, or a reverse-transcribed mRNA (real cDNA); full length or partial (if partial provide also predicted full length); what motifs are present and where; what and where are the exons.
8. Homology analysis—alignments, BLAST search results (only if can do securely) indicating date of search and databases searched.[55]

It is therefore sufficient to express the biotechnology equivalent of the open source software licensee's entitlement to source code in general functional terms, leaving implementation to licensors and enforcement to licensees. But what should those terms be?

Fortunately, there already exists a standard of technical transparency that is familiar and has proved workable in a biotechnology setting. To obtain a valid patent in the United States, an applicant must make a full and clear disclosure of the invention as prescribed by 35 U.S.C. 112, first paragraph, which provides:

> The specification shall contain a written description of the invention, and of the manner and process of making and using it, in such full, clear, concise, and exact terms as to enable any person skilled in the art to which it pertains, or with which it is most nearly connected, to make and use the same, and shall set forth the best mode contemplated by the inventor of carrying out his invention.

In the case of biological materials, it is not always possible to fulfill every aspect of the disclosure requirement without making a sample of the material available to the public, so the law permits the deposit of biological materials in a public repository as part of the patent application process.[56]

The policy behind the patent disclosure requirement is to ensure that inventors live up to their side of the so-called patent bargain, in which the state grants an inventor exclusionary rights for a limited period in exchange for making the invention available to the public for further research and development. The source code requirement in the OSD serves the same basic purpose of putting the public in full possession of the technology, though in a different overall context. It is therefore no surprise that clause 2 of the OSD corresponds closely with the three patent law requirements of written description, enablement, and best mode of practice. Source code is both written description and enablement, and open source licensors are obliged to provide source code in the preferred form for modification—the best mode.

Thus, at least in theory, disclosure in the form of a written description that meets the standard set out in 35 U.S.C. 112, paragraph 1—supplemented where necessary by a deposit or transfer of biological material—should be adequate to give effect to the legal freedoms guaranteed under an open source biotechnology license. For biotechnologies that are protected by a valid patent, this standard will already have been met; for open source licensing purposes, the same standard of disclosure could be applied to technologies or components of technologies that are protected by some other form of intellectual or personal property right.

It might be argued that in practice, patent applicants have devised numerous ways to minimize the quality and quantity of information they must reveal to the public in patent specifications.[57] As a result, the public's right to access information about a patented invention may not be as strong as the open source licensee's right to access source code—certainly, it is possible to obtain a software patent without disclosing the relevant source code. But does this mean that the standard of disclosure set by the patent statute in fact falls short of that required to implement open source principles in biotechnology?

It does not, because the motivations of a patent applicant are not the same as those of an open source licensor. Even a proprietary licensor has a practical incentive to disclose enough information to enable the licensee to practice the technology; licensors frequently provide hands-on training to licensees at their own expense in order to ensure effective technology transfer, over and above the disclosure requirements of a patent application. In an open source context, where free revealing is a deliberate strategy, most licensors have an even stronger motivation to make it not just *possible* but *easy* for others to use, understand, and improve their technology. This is especially true in biotechnology, where the expertise available for collaborative technology development is more limited than in software and the costs of acquiring information are higher.

Of course, the cost of acquiring biotechnology-related information correlates with its relative lack of codification, which also makes the information more expensive for licensors to provide. Recall that some formulations of software freedom specify that licensees must have "free" access to source code, meaning both zero price and freedom from control by the licensor. The OSD also expresses a preference for source code to be available for downloading from the Internet without charge. The actual *requirement,* however, is that licensors provide access to source code for "no more than a reasonable reproduction cost."[58]

Clearly, a reasonable cost of reproduction is likely to be higher with respect to many biotechnology tools than it is for software. In particular, it would almost certainly be impractical to insist on free-as-in-beer access to hardware and wetware components as a requirement of open source biotechnology licensing. A strict analogy with open source software principles would require biotechnology licensors to distinguish between the costs of preparing written documentation or obtaining supplementary materials (not recoverable) and the cost of printing or mailing such information and materials (recoverable). In practice, however, the same market logic applies as in the case of up-front license fees. Provided any payment is a one-off and is accompanied by the freedom to redistribute information or materials at any price, there is no reason to prohibit the imposition of even a reasonably substantial fee for access to the biotechnology equivalent of source code.

Note that the patent "best mode" requirement applies to the information held by the applicant at the time of filing;[59] by analogy, an open source biotechnology licensor should be permitted to impose further charges for providing new information about the best mode of practicing an innovation. On the other hand, the licensee should be free to decline the new information and refuse to pay such fees. By this analysis, any arrangement under which a licensee remained beholden to the licensor beyond the initial transfer, such as through a compulsory subscription fee, would conflict with the goal of protecting biotechnology freedom.

Copyleft: Freeing Whole Data Streams

So far our discussion of open source licensing in biotechnology has been limited to those aspects of biotechnology freedom that would be guaranteed by all open source biotechnology licenses. To appreciate the added dimension introduced by copyleft or "reciprocal" license terms, it is helpful to consider the licensed technology as an element in a continuously evolving data stream. Copyleft-style licenses aim to secure the freedom of all elements downstream of the licensed technology, not just the technology itself. In software, they do this by requiring that any distribution of derivative works be licensed on the same terms as the original program.

The problem that copyleft-style licenses seek to address may be likened to the diversion of water from a river to irrigate private lands. When a follow-on innovator improves or extends a free technology and then pursues a proprietary exploitation strategy with regard to that improvement or extension, those who contributed to the development of the free technology are excluded from access to and freedom to operate with new technologies that could not have been developed without their input. They are like upstream landowners who limit their water use in order to maintain the overall viability of a river system, only to find that downstream users refuse to show the same restraint. While downstream users who incorporate open source innovations into proprietary technologies are sometimes seen by some as taking unfair advantage of others' public-spirited conduct, the main objection from a pragmatic perspective is that upstream innovators are also frequently downstream users or end consumers. As such, they have an interest in maintaining robust innovation and competitive quality and pricing all the way to the end point of technology development.

In biotechnology, this type of problem is most evident in controversies over benefit-sharing. When germplasm developed over many generations of cultivation is incorporated into a lucrative new crop variety, or a plant identified as having medicinal properties is used to develop a new pharmaceutical drug, it is a matter of justice to reward contributions made by poor farmers or keepers of traditional knowledge. But while such contributions have often been made in the absence of any expectation of a profitable innovation emerging from the data stream, in other settings an expectation of sharing in the benefits enjoyed by follow-on innovators may be central to the decision to commit resources to innovative activity in the first place. In that case, equitable benefit-sharing is not the only issue: in addition, some means of protecting the initial innovator's investment is needed in order to provide the necessary incentive to innovate.

Of course, this is the very logic that is used to justify intellectual property protection in fields, like biotechnology, where the economic rewards of innovation depend largely on the unpredictable outcome of a cumulative process of technology development. By giving upstream innovators a right that is enforceable against downstream innovators and end users, intellectual property prevents the whole cumulative effort from unraveling backward from the point at which something useful or profitable is developed. In fact, both copyleft licenses and conventional proprietary licenses use intellectual property in this way; but they protect different stakes in the final outcome.

Most proprietary licenses aim to secure the licensor's stake in any monetary rewards generated by commercially successful downstream innovations. As we saw in Chapter 2, valuation of intellectual property inputs in circumstances where the ultimate outcome of research and development is highly uncertain leads to high transaction costs because it is risky for both licensor and licensee. Some licensors attempt to sidestep the danger of undervaluing their technology against future returns on downstream innovations by imposing reach-through royalties. A reach-through royalty provision conditions royalty payments on the use or sale of follow-on innovations instead of (or as well as) tying them directly to the licensed technology. This approach helps overcome immediate obstacles to further innovation, but it can lead to the problem of "royalty stacking," in which the final innovator in a chain of sequential innovation is faced with multiple claims on a finite revenue stream.[60]

Copyleft terms are essentially reach-through terms that protect a licensor's stake in continuing access to and freedom to operate with downstream innovations instead of in the monetary profits such innovations may generate. Substituting technology freedom for monetary rewards as the quid pro quo for contributing to cumulative innovation removes the need to quantitatively predict the value of a particular technology to future research and development, thereby reducing transaction costs. At the same time, it eliminates the problem of royalty stacking, because—unlike money—technology freedom can be enjoyed by an unlimited number of contributors without consuming any finite resource. For these reasons, technology freedom can be a more efficient form of incentive to innovate than the rewards associated with exclusive proprietary control.

Some proprietary licenses also contain reach-through terms aimed at securing ongoing access and freedom to operate with downstream innovations for the licensor. The difference between these licenses and copyleft-style licenses is that technology freedom applies to everyone who wishes to become a user or distributor of the downstream technology, not just the initial licensor. The copyleft-style obligation to distribute derivative works on the same terms as the original license has been characterized as a kind of "grant-back"—a provision that requires the licensee to grant the licensor rights to use or own improvements or discoveries created using the licensed technology.[61] But the term *grant-back* implies a tit-for-tat type of interaction in which the licensor gains a special privilege in exchange for his or her contribution. In reality, the generosity mandated by copyleft-style license terms is of a different kind, known colloquially as "passing it forward." The licensor may (and usually expects to) benefit from the performance of a copyleft obligation, but only as a member of the wider community. Similarly, characterizing copyleft licenses as "reciprocal" licenses can be misleading if reciprocity is taken to imply a closed loop between licensor and licensee. In fact, reciprocity in this context refers to the fact that, in cumulative technology development, licensors and licensees may find themselves at different times on different sides of the same license terms.[62]

This distinction between proprietary reach-through and grant-back terms and copyleft terms is important because conventional reach-throughs, grant-backs, and other similar terms have sometimes been regarded as potentially anticompetitive. It has been suggested that copyleft-style terms may raise the same concerns, especially in a patent context.[63] But this anxiety is unfounded, because the analogy between conventional and copyleft terms breaks down at exactly the point where conventional terms become suspect. Unlike conventional terms, in which a licensor restricts licensees' freedom in order to maintain a competitive advantage, copyleft-style terms restrict licensees' freedom in order to create a competitive market in which the licensor retains no advantage relative to other prospective users or distributors of downstream technologies.

As we have seen, not all open source software licenses are copyleft licenses. Is an analogous mechanism likely to prove attractive to biotechnology licensors? Certainly, the private appropriation of formerly free data streams generates the same concerns in biotechnology as it does in software. As in the software case, how much such appropriation actually matters to the overall viability of the data stream will vary from one commercial and technology setting to another. Some data streams are more robust than others: like well-supplied river systems, they can sustain numerous proprietary diversions without drying up. Protecting these data streams with the (loose) licensing equivalent of irrigation quotas may be unnecessary and even counterproductive.

For individual licensors, the decision whether or not to employ a copyleft-style open source biotechnology license will depend on the specific constellation of incentives operating on potential contributors in a given sphere. Sometimes technology freedom with respect to the licensed technology itself will be enough to sustain the collaborative effort; sometimes a little "GPL judo" may be required to direct innovative momentum toward a common goal. Depending on the circumstances, it may well be feasible to trigger a virtuous cycle of cumulative innovation without achieving 100 percent compliance with the copyleft ideal. Only if the incentives of potential contributors are likely to be seriously undermined by the existence of free riders is there a need to enshrine a copyleft-style obligation in the license terms.

In this connection, it is important to realize that a number of the economic benefits of participation in open source technology development mentioned in the previous chapter are independent of either licensor or licensees having access to any follow-on innovation apart from that which they generate themselves. This suggests that an academic or permissive open source license can achieve at least some of the commercial goals of open source as effectively as a copyleft license. Indeed, where the licensor's primary goal is to encourage widespread adoption of the initial innovation, an academic-style license may be more effective because imposition of a copyleft obligation can be expected to deter at least some potential adopters who would like to be able to commercialize their own improvements on a proprietary basis.

The translation of copyleft-style provisions into biotechnology licenses creates two further dimensions of choice in defining the portion of the data stream that is to be made "free." The first relates to the act that triggers the copyleft-style obligation to provide access to and freedom to operate with downstream technologies. In software, no such obligation arises unless and until the downstream technology is distributed: internal uses of downstream technologies are not caught by the copyleft "hook." Some copyleft licenses rely on the definition of distribution in the copyright statute; others employ a broader definition.[64] Lawrence Rosen's Open Software License refers to "external deployment" of the technology;[65] from a modeling perspective, this is a useful expression because it captures the policy objective of copyleft licensing without invoking a body of law that is specific to copyright.

The other dimension of choice in drafting a copyleft-style biotechnology license is the definition of downstream technologies to which the copyleft obligation is to apply. In software, copyleft applies to the distribution of derivative works, but not all copyleft licenses seek to encompass every downstream technology that would fall within the statutory definition of that term.[66] Some licensors allow licensees to choose their own terms for the distribution of a subset of technologies that would normally be counted as derivative works; the same latitude should presumably apply to the equivalent concept in an open source biotechnology license.

Of course, the design of a copyleft-style license in biotechnology would be subject to the limits of ownership. Like a restriction enzyme that cuts DNA, a given intellectual property right can cut a data stream only at specific points and in specific ways. The portion of a data stream that is "freed" by operation of a "patentleft" license provision would not be bounded in precisely the same way as one defined by copyleft. Personal property rights in biological materials would create yet another set of constraints and opportunities.

To illustrate the point: Consider how the differences between copyright and patent law might impact on the design of reciprocal license terms. In one sense, patentleft would be even stronger—that is, more restrictive and more protective of technology freedom—than copyleft. Copyright law gives the owner an exclusive right to authorize the preparation of derivative works.[67] Improvements that are not derivative—that is, independent creations not based on the original work—are not subject to copyright and therefore cannot be made subject to a copyleft license provision. By contrast, a patent owner has the right to exclude others from making, using, selling, offering for sale, or importing the claimed invention, including in the process of developing or exploiting an improvement; if an improvement is otherwise infringing, the fact that it was not *derived* from the patented invention is irrelevant. Permission to practice an independently developed improvement can therefore be made subject to a patentleft provision.[68]

In other ways, though, patentleft may be weaker than copyleft. An improvement may be patentable even if it infringes another's patent. Both the owner of the initial patent and the owner of the improvement patent may therefore require each other's permission to practice the improvement. Depending on the value of the improvement, this may give the developer of an infringing improvement on a patented technology more bargaining power with which to resist the imposition of copyleft-style license terms than is possessed by the author of an unauthorized derivative work.[69]

Whether such differences represent distortions or mere variations on the copyleft theme is ultimately a matter for the specific technological communities whose activities shape and are shaped by the data streams in question. Each variation creates its own balance of incentives to innovate—its own compromise between allowing private appropriation of downstream technologies and guaranteeing ongoing technology freedom. As in the software context, there will be some failures as well as multiple successful equilibria.

An Institutional Design Perspective on License Enforcement

So far in this chapter I have highlighted certain key features of open source licensing—specifically, credible commitment, competition, and optional copyleft-style reciprocity. Clearly, however, these principles underdetermine the choice of provisions to be included in an actual open source biotechnology license. Even within the relatively strict interpretation of what it means for a license to be "open source" outlined in this chapter, would-be licensors have at their disposal a wide range of design options, including whether to adopt an academic or copyleft-style license, the particular size and shape of a copyleft-style "hook," and the breadth of specific disclosure requirements.

In identifying the principles that should guide these lower-level design choices, as well as decisions about other important aspects of licensing practice, such as enforcement, it is helpful to think in terms of institutional design. Steven Weber has emphasized the role of open source licenses as formal expressions of the social structure that surrounds each open source project and constitutes a community from what would otherwise be a mere collection of individuals.[70] Bruce Perens, author of the OSD, has remarked that open source licensing is essentially a form of social engineering.[71] Theoretical insights about institutional design are thus directly relevant to the construction of open source biotechnology licenses that can actually help licensors achieve what they set out to achieve.

Open source licensing is an institution whose primary goal is to promote a benefit. In the case of open source licensing generally, the benefit is collaborative, nonproprietary technology development; in the case of copyleft-style licenses, it is the continuous growth of a universally accessible technology commons. Institutional design principles based on rational choice theory suggest that where the primary goal of an institution is to promote a benefit rather than to prevent harm, complier-centered strategies are more effective than deviant-centered strategies.[72]

A deviant-centered strategy assumes (in the words of philosopher David Hume) that "every man ought to be supposed a knave, and to have no other end in all his actions than private interest."[73] Deviant-centered strategies generally rely on heavy sanctions (which may be either positive or negative—that is, they may be rewards or penalties). Complier-centered strategies, on the other hand, aim to strengthen the initial orientation toward compliance shared by all who, for their own varied and self-interested reasons, wish to see an institution's goals realized. Complier-centered strategies make preferential use of "screens" or "filters" over sanctions as a front-line response to the threat of noncompliance with institutional rules and standards. Screening techniques can be used to maximize the participation of agents who are disposed to comply with institutional standards relative to those who are inclined to act as "knaves." They can also be used to maximize and reinforce the range of compliant options available to all agents operating within the institutional framework.[74]

Open source software licensing is a veritable showcase of complier-centered institutional design. The freedom to fork is an example of a screen applied to agents; as Weber and others have pointed out, open source licenses are explicitly designed to empower "exit" rather than "loyalty" in the case of irreconcilable differences between project leaders and followers.[75] Examples of screens applied to agents' *options* include allowing licensees to use derivative works internally without incurring any copyleft obligation, or employing a relatively narrow definition of derivative works in a copyleft license. In biotechnology, screening to maximize compliant options might include permitting a reasonable level of cost recovery for the transfer of biological materials or uncodified techniques.

Where sanctions are found to be necessary as part of a complier-centered strategy, the principles of institutional design suggest that they should be imposed in a "deliberatively supportive" way—that is, in such a way as to avoid triggering a switch from cooperative to self-interested deliberation on the part of the target agent and all others within the system.[76] It is unsurprising, then, that open source sanctions (for example, enforcement of license obligations or OSD conformity assessment) tend not to be employed aggressively. Instead, they are used to reinforce community values through collective deliberation, and recipients are offered every opportunity along the way to act and be seen to act as "good citizens." The threat of legal action is kept in reserve as far as possible.[77]

To litigation-hardened biotechnology industry participants, this complier-centered approach may seem naïve. In fact, it is both sophisticated and successful, and should be preserved as far as possible in open source biotechnology licensing.

Consider as an example the design of an open source biotechnology license provision setting out the licensee's entitlement to information equivalent to software source code. In an open source setting, the entitlement to source code or its equivalent is a safety net, needed to deal with the minority of licensors who are not genuinely seeking to engender collaborative development. The design of an effective safety net depends on the motivations of those "knaves." In biotechnology, one possible disingenuous motivation is to exploit the goodwill and funding opportunities that may be available to those who call themselves open source licensors. In such a case, the success of collaborative technology development may not matter to the licensor—but reputation does matter, and the threat of community censure may be enough to induce more effective disclosure. Another possible reason for a licensor to test the limits of the open source entitlement is to avoid disclosing the equivalent of source code for modifications to or improvements on innovations licensed under copyleft-style terms. In that case, an overly strong disclosure requirement would exacerbate the licensor's unwillingness to share and encourage noncompliance; meanwhile, other potential users of the licensed technology might be put off altogether. In both situations a complier-centered institutional design is more likely to achieve the overall goal than a deviant-centered design.

In this chapter I have argued that it is possible to conceptualize open source licensing independently of the specifics of the software industry, in terms that might be applicable in other fields including biotechnology.

Specifically, I have suggested that open source licenses are not inherently anti-intellectual-property; rather, they are a legitimate, if unconventional, form of intellectual property management. Nor are open source licenses inherently anticommercial; on the contrary, they enable an economically significant class of commercialization strategies, known as "free-revealing" or "nonproprietary" strategies. By choosing an open source license, a licensor demonstrates a credible commitment to allowing his or her technology to be treated as a contribution to bazaar-style production, whether on academic/permissive or copyleft/reciprocal terms. Both classes of open source license guarantee users "software freedom," which in biotechnology might be thought of as "biotechnology freedom" or, generically, "technology freedom." From an instrumental perspective, the essential point of technology freedom is to put licensor and licensee on an equal competitive footing, giving licensees full power to use and develop the technology and distribute it to others. Distribution occurs on licensees' own terms, in the case of a permissive license, or on the terms on which the technology was granted to them, in the case of a copyleft-style license. Both classes of open source software license guarantee the user an entitlement to the source code; while there is no technical equivalent in biotechnology, the disclosure and enablement requirements of patent law represent an acceptable conceptual equivalent.

Besides satisfying these general open source criteria, however, open source biotechnology licenses must reflect the specifics of biotechnology research and development. This chapter has done no more than scratch the surface of the work required to formulate workable open source biotechnology licenses for specific technologies or classes of technology. Many practical questions remain to be resolved, ideally by prospective licensors and licensees engaging together in a process of iterative learning, guided by sound principles of institutional design. In the next chapter, we explore the role of community in supporting the evolution and regulation of a fully fledged open source licensing practice in biotechnology. Meanwhile, the generic model of open source licensing outlined in this chapter is offered as grist to the experimental mill.

6

Foundations of the Biobazaar

So far in this book I have (1) made a case for the desirability of open source research and development in biotechnology and (2) described the open source approach in general terms, as distinct from terms specific to the software context in which it emerged. While this discussion went some way toward showing how open source principles might be applied in the biotechnology context, it left a number of questions about the feasibility of the biobazaar unresolved. The time has now come to confront those questions directly. Could open source succeed in biotechnology?

Despite the strong parallels highlighted in Chapter 1, there are many technical, legal, commercial, and cultural differences between software and biotechnology. For example, we saw in the previous chapter that innovations in these two fields are subject to different forms of intellectual property protection, and that the prevalent form in biotechnology—patent protection—is more difficult and costly to obtain than copyright. More generally, the amount of capital needed for biotechnology research and development is larger than for software development, which, it has been suggested, requires nothing more than "a laptop, an Internet connection, and a packet of Doritos."[1] Unlike software programs, some biotechnology products cannot be marketed without undergoing expensive, time-consuming regulatory approval.

Open source software development also proceeds quite differently from biotechnology research and development. In general, the latter takes place in an institutional setting, whereas at least some of those who contribute to open source software development are individual hobbyists with no financial backing or formal credentials. Innovation-related information is not as highly codified in biotechnology as in software; coordination of biotechnology project contributions via the Internet therefore cannot be as cheap or convenient. It might also be argued that biotechnology lacks an open source community equivalent to that found in software; that open-source-style business opportunities do not exist in biotechnology and related industries; and that the culture of these industries is too strongly proprietary to sustain open-source-style collaboration.

How important are these differences between software and biotechnology? The answer depends on which features of the open source approach are essential to its success and which are not. A difference may be real, yet immaterial to the implementation of open source principles in a biotechnology setting. The point of constructing a generic model of open source—the focus of the previous two chapters—has been to provide a conceptual basis for distinguishing between genuine potential obstacles and red herrings.

Taking that model as a guide, I argue in this chapter and the next that none of the differences between software and biotechnology constitutes an insurmountable obstacle to implementing an open source "biobazaar." Whether the open source approach will actually be widely adopted in this new setting is, of course, another question. I engage with it in the final chapter.

Biobazaar, Old and New

In Chapter 4, we saw that open source software development is an Internet-era manifestation of a long-established mode of production—the bazaar—that is only now being recognized as distinct from firm, market, and network governance structures. Earlier examples of production systems dominated by bazaar governance include publicly funded academic research of the kind that gave rise to the computer software and biotechnology industries.

Perhaps the single most important point to recognize in connection with the feasibility of open source biotechnology is that the bazaar governance structure that underpins open source production is already present in this context. This preexisting version of the biobazaar is not, however, identical to open source. Rather, it takes the form of strong academic and other nonprofit involvement in many crucial aspects of biotechnology research and development. Even the most cursory overview of contemporary biotechnology-related industries, such as that provided in Chapter 2, highlights the continuing importance of this type of bazaar production—not just at a precommercial stage, but throughout the research and development process.

In the process of drug development, for example, basic biomedical research and target validation (stage 1 of the development process) takes place overwhelmingly in universities or government laboratories, in the United States and elsewhere. Most of it is funded by the NIH and its overseas counterparts. Traditionally, preclinical development (including the search for new drug candidates—stages 2 and 3) was the province of pharmaceutical companies, which specialized in producing small molecule drugs using the techniques of analytical and synthetic chemistry. Since the 1980s, biotechnology companies have also entered the field, producing new drugs primarily by making or modifying very large molecules like proteins or hormones using the techniques of molecular biology. However, public and private nonprofit players also have the capacity to engage in this aspect of biomedical research and development.

The diversity of entities potentially involved in the development of a new drug is illustrated by the story of the first HIV/AIDS medication, AZT. By the account of former editor in chief of the *New England Journal of Medicine* Marcia Angell, the AZT molecule was originally synthesized in a nonprofit setting as a possible cancer treatment. German researchers subsequently discovered it to be effective against viral infections in mice, and it was acquired by the pharmaceutical company Burroughs Wellcome for possible use against herpes. Soon after the AIDS disease mechanism was discovered in a race between researchers at the NIH and the Pasteur Institute in Paris, a team established by the NIH screened the AZT molecule as a possible treatment along with many other existing antiviral agents. Working closely with researchers at Duke University, this team carried out preclinical tests and established the efficacy of AZT in early clinical trials. At this relatively late stage, Burroughs Wellcome patented the molecule to treat AIDS and became involved in conducting further clinical trials. The company subsequently obtained FDA approval to market AZT and became its first manufacturer and distributor.[2]

Thus, although pharmaceutical and, to a lesser extent, biotechnology companies are often thought of as the only entities capable of preclinical drug development, in fact this capacity also exists in the nonprofit sector. The same is true at later stages of the process. Clinical trials (stage 4) involve coordinating data gathered by doctors in teaching hospitals and private offices. Once upon a time, this task fell to the pharmaceutical companies themselves. However, a relatively new type of for-profit firm, the contract research organization, now does most of the actual work of conducting clinical trials. Most clinical trials are still ultimately paid for by the drug companies; but the NIH itself sponsors about 10 percent of clinical trials, carried out largely in academic medical centers.[3]

Traditional bazaar production also persists in agricultural biotechnology. For example, we saw in Chapter 2 that plant biotechnology research and development is an aspect of plant breeding, the first link in the value chain of seed and agricultural input production. Although such research and development activity is increasingly dominated by a handful of large companies, public-sector and nonprofit institutions continue to participate. Internationally, members of the Consultative Group of International Agricultural Research (CGIAR) carry the burden of developing crops for which there is no significant commercial market, as well as traits that might benefit poor farmers and consumers. At the national level, national agricultural research systems (NARS) consist of government laboratories and government-funded research institutions, especially universities. In the United States, for example, the United States Department of Agriculture and the land-grant universities support minor and horticultural crop interests as well as contribute to research and development for developing countries. Public-sector and nonprofit institutions also play a role in delivering the benefits of biotechnology research and development to farmers and consumers.

Of course, it is in the interests of industry participants who are heavily committed to the proprietary approach to downplay the significance of nonproprietary contributions in both red and green biotechnology. But we must be careful not to confuse rhetoric with reality: participants whose missions are predominantly nonproprietary are engaged in research and development activities all the way from target validation through to clinical trials, from plant breeding through to extension services. It is true that their present activities emphasize early-stage research and aspects of downstream technology production that tend to be neglected by commercial entities. However, the point remains that the *capacity* for bazaar-style production in biotechnology exists all along the value chain.

As we shall see, the continuing existence of this earlier version of the biobazaar goes a long way toward answering objections to the feasibility of a version modeled on the more recent phenomenon of open source software. Nevertheless, an open-source-inspired biobazaar would differ from traditional academic-style bazaar production in biotechnology in three important respects.

First, compared with traditional academic research, open source technology development is more strongly integrated with the Internet as a tool for coordinating every stage of knowledge production, from content generation through quality control to integration. As Yochai Benkler points out, computer networks are bringing about a change in the scope, scale, and efficacy of peer production such that it can be applied to larger and more complex tasks.[4] Second, open source licenses introduce a new mechanism for establishing and maintaining connections among participants. Third, traditional forms of bazaar or commons-based peer production in biotechnology were confined to the nonprofit sector. As commercial players became involved in biotechnology research and development, technology transfer was mediated either by proprietary mechanisms or by straightforward free revealing. By contrast—due largely to the legal and social functions performed by open source licenses—an open-source-style biobazaar would offer a means of incorporating contributions from both noncommercial and commercial sources on a nonproprietary basis.

These three differences between the form of the bazaar found in traditional academic bioscience and open-source-style bazaar production provide the focus for the rest of the discussion in this chapter and the next. They imply that to establish the feasibility of open source biotechnology, it is necessary to show how existing practices could be extended, if need be, (1) to harness the full power of Internet-enabled peer production; (2) to incorporate open source licensing of intellectual and personal property in biotechnology innovations; and (3) to capture contributions to nonproprietary technology development from commercial as well as noncommercial sources. The first two factors are dealt with in this chapter. The third is discussed in Chapter 7.

Harnessing the Power of Peer Production

Recall from Chapter 4 that—perhaps ironically—the efficiency of the bazaar compared with firm, market, and network governance structures derives from the weakness of its control and incentive mechanisms. In particular, self-selection of contributors to bazaar production, combined with the removal of proprietary barriers, permits larger clusters of potential contributors to interact with larger clusters of information resources in search of new projects and opportunities for collaboration.[5]

Perhaps the most spectacular example of this aspect of bazaar governance in the context of open source software is the phenomenon of Linux, released by a single programmer in a usable but unpolished state and adopted by thousands of other user-developers within a few short years. In his book *The Wealth of Networks*, Yochai Benkler gives examples of distributed production outside the field of software development: NASA Click-workers, Wikipedia, Project Gutenberg, and SETI@home, among others.[6] Like Linux, these projects involve large numbers of participants whose contributions to information production are coordinated via the Internet.

It might be supposed that open-source-style production could be replicated in biotechnology only if open source biotechnology projects have a pool of potential contributors as large as the pools for these other production systems. But how big a pool of contributors is actually required for open source development to succeed?

A preliminary point is that, depending on the nature of the project, the pool of potential contributors to distributed knowledge production is not necessarily smaller in biotechnology than in other fields. In some existing large-scale Internet-based distributed computing projects, the object of production is essentially the same type of knowledge as that generated in biotechnology research and development. One example is Folding@home, a project that makes use of contributors' spare CPU cycles to model protein folding, misfolding, aggregation, and related diseases, including Alzheimer's disease, Huntington's disease, Parkinson's disease, and cancer. Another is Fightaids@home, which screens existing compounds for possible use against HIV/AIDS. The Folding@home project is run by a nonprofit academic institution that freely reveals the results in the form of journal publications and, subsequently, raw data. As of January 2007 the project had registered nearly 597,000 individual contributors.[7]

Although some high-profile instances of open source and bazaar production do involve very large numbers of participants in single projects, it does not follow that the number of potential recruits to an open source biotechnology project needs to be in the tens or hundreds of thousands for it to have any prospect of success. In the software context, there is no hard evidence available concerning the ratio of potential to actual contributors with respect to any particular project. There is also no reason to suppose that the ratio is constant from one project to another, or that any formula one might arrive at in relation to software projects could be carried over into biotechnology. What we *do* know is that even within the software field there are wide variations in the size of developer groups, and many projects are worked on by a mere handful of programmers.

Empirically, the question whether there is an *optimal* size for open source development remains open. Some research in the software context suggests that a developer community reaches critical mass when it contains approximately thirty to forty people,[8] but anecdotal evidence suggests that a smaller team—six to twenty people—is optimal. The reason is that in a larger team it is too difficult for contributors to keep track of all the interactions between different aspects of the project. Even larger open source projects are often composed of small subteams, with overall goals being defined by programmers who are technically proficient but not deeply involved in the actual coding.[9]

If the size of dedicated biotechnology companies is taken as an indication of the minimum number of active contributors needed in order to achieve productive goals in biotechnology research and development, it would seem that there is a rough correspondence between this number and that observed in relation to many successful open source software projects. Most biotechnology companies—as distinct from pharmaceutical or chemical companies—are small: only a handful employ as many as a thousand scientists. As a rough estimate, the majority have a technical staff of between 15 and 150 researchers. As one open source software entrepreneur points out, the open source world is full of fifteen-person projects, and reasonably full of hundred-person projects.[10]

In any case, what matters to the success of an open source development effort is arguably not so much the sheer number of potential contributors, but the participation of a core of highly innovative contributors who are able to "start the ball rolling."[11] Bonaccorsi and Rossi describe a small but efficacious subgroup that establishes a critical mass of other participants;[12] in user innovation terms, these are "lead users."[13] Thus, it is not the case that an enormous pool of actual or potential contributors is essential for the success of every open source project. Nevertheless, it is worth considering some of the factors that may affect the availability of sufficient numbers of core participants in a biotechnology setting.

One such factor is the cost of engaging in biotechnology research and development. Biotechnology is often perceived as highly capital-intensive, especially compared with software development. Is this perception accurate? If so, what are the implications for the feasibility of open source biotechnology?

According to Benkler, the reason why the bazaar, or "commons-based peer production," is coming into its own in the networked information age is that the declining price of physical capital involved in information production and the declining price of communications make human capital a salient component of production costs. In the capital-intensive model of information and cultural production that prevailed in the century and a half before the advent of the Internet, the bazaar's superior ability to match human capital with information resources represented less of an advantage over other governance structures than it does now that the physical capital required for both fixation and communication of information is lower and more widely distributed.[14]

If that is so, we should ask what kind of physical capital is required for fixation and communication of biotechnology-related information. The answer varies from one context to another, because, as we saw earlier, *biotechnology* is an umbrella term for a range of tools used in a variety of industry settings. For some types of research, a "dry" laboratory—one that relies entirely on computational tools—is sufficient. Some biotechnology research is conducted in the field (on farms or in hospitals), requiring few or no specialized facilities. Where a wet laboratory is needed, the cost of establishing such a facility depends to a large extent on the cost of mundane data stream elements such as building materials, electricity, and water supply. These costs vary enormously from one location to another: in developing countries the cost of labor—for example, to build benches and fume hoods—is generally much lower than in developed countries, while the cost of materials is often higher because of lower-quality transport infrastructure.

Even within a given field of research and development, costs vary significantly from project to project, not only according to project goals but also according to the strategy adopted to achieve those goals. Often the same scientific questions can be approached in a number of different ways. Some methods may be highly capital-intensive, requiring expensive specialized machinery and large quantities of proprietary reagents, while others require a much smaller budget and rely more heavily on the brainpower of individual investigators.

Despite these variations, it is clear that the capital costs of innovating are generally higher in biotechnology than in software. Does it follow from Benkler's analysis that biotechnology is inherently unsuited to Internet-enabled peer production?

As a preliminary point, it is important to recognize that although biotechnology is an inherently more expensive technology than computer software, it is still remarkably cheap. Nearly every other type of technology—including computer hardware—is more costly to reproduce and distribute. This is one reason why biotechnology is regarded in development circles as a promising solution to problems of poverty and food insecurity.

Moreover, the physical capital costs of biotechnology research and development are falling rapidly, just as the physical capital costs of computer-based information production fell rapidly over the past several decades. In an open letter dated October 2000, engineer Rob Carlson and biologist Roger Brent argue that if current trends continue, the basic tools needed for molecular biology research may soon be within the reach of individual hobbyists in developed countries and farmer collectives in developing countries on a scale comparable to the distribution of Internet-connected PCs:

> Considerable information is already available on how to manipulate and analyze DNA in the kitchen. A recent *Scientific American* "Amateur Scientist" column provided instructions for amplifying DNA through the polymerase chain reaction (PCR), and a previous column concerned analyzing DNA samples using home-made electrophoresis equipment. The PCR discussion was immediately picked up in a Slashdot.org thread where participants provided tips for improving the yield of PCR. Detailed, technical information can be found in methods manuals, such as *Current Protocols in Molecular*

Biology, which contain instructions on how to perform almost every task needed to perform modern molecular biology, and which are available in most university libraries. More of this information is becoming available online. Many techniques that once required Ph.D.-level knowledge and experience to execute correctly are now performed by undergraduates using kits. . . . DNA synthesis [is] becoming faster, cheaper, and longer, and it is possible that in ten years specified large stretches of sequence will be generated by dedicated machines. Should this capability be realized, it will move from academic laboratories and large companies to smaller laboratories and businesses, perhaps even ultimately to the home garage and kitchen.[15]

More importantly, the cost of physical capital is not the central organizing principle of production in biotechnology research and development. Biotechnology, like software, is a knowledge industry—by definition, an industry in which human knowledge and creativity are more important inputs than physical goods. My discussions with leaders of large and small projects in biomedical and agricultural biotechnology and with funders of research in developed and developing countries indicate that despite substantial variations in overall costs, in biotechnology the capital costs of both fixation and communication almost always account for a significantly smaller proportion of the total ongoing project budget than labor costs. Thus, human creativity is very much a salient resource in biotechnology research and development. According to Benkler's analysis, biotechnology is therefore a suitable candidate for bazaar production.

Of course, even though biotechnology research and development *can* be conducted very cheaply, the trend in developed countries over the past several decades has been toward ever larger and more complex projects that require ever more substantial capital investment. In biotechnology, "big science" is typified by large-scale genome sequencing efforts such as the Human Genome Project. In its early days—especially before it became clear that this approach could yield significant results—this project was widely criticized on grounds that intersect with the concerns over commercialization of academic research described in Chapter 2. Most relevant to the present discussion was the charge of elitism: it was thought that the trend toward big science in biology would increase the gap between "haves"—laboratories, institutes, and companies with vast resources, facilities, and funding—and "have-nots," with the result that small laboratories lacking the resources to use high-throughput methods would no longer be able to participate in research and development.[16]

In the postgenomic era, this controversy has largely died down, replaced by the realization that individual investigators rely on the outputs of big science, while big science relies on the drive and creativity of individual investigators.[17] The challenge for contemporary scientific institutions, then, is to reconcile the need for individual creativity and scientific entrepreneurship with the increasingly sophisticated and costly resource requirements of leading-edge biological research.

One response is to rely more heavily on computational methods, which permit researchers to simulate high-throughput analyses and test specific hypotheses without the need to conduct "wet" laboratory experiments. Clearly, this approach is fully compatible with open source knowledge production. Indeed, the increasing involvement of researchers with "hard science" backgrounds—engineers, computer scientists, and physicists—in biological research is one of the driving forces behind the nascent open source biotechnology movement. The genomic revolution created a demand for people with new skills to manage and interpret large data sets; these new biologists have brought with them an appreciation of the advantages of open source development, just as the physicists who helped found the discipline of molecular biology in the 1930s brought with them a revolutionary commitment to methodological reductionism.

Another response is to try to make the most sophisticated laboratory resources and equipment more widely accessible. One way to approach this goal is to establish central user facilities that are made available to scientists from smaller laboratories.[18] However, this approach could still be regarded as elitist because the required investments are likely to be made only in rich countries. Globally, a more equitable approach would be to build collaborative links between large and small laboratories in which larger players would make their data and technology available to smaller players along with the freedom to use, adapt, and disseminate it for their own purposes. If scientists at the periphery of the global system of biotechnology research and development were free to engage with scientific problems and make use of others' contributions without waiting for permission to join a formal network, their relative lack of power and resources would not prevent them from influencing the trajectory of technology development. Such a collaboration would look a lot like open source biotechnology; in other words, open source development might actually be part of the *solution* to creeping capital costs in biotechnology research and development.

Such an outcome cannot be taken for granted, however. According to Benkler, projects will succeed in harnessing the power of Internet-enabled bazaar production only if they possess certain characteristics. These are modularity, granularity, and low-cost integration.[19]

A *modular* project is divisible into components that can be independently produced so that production can be incremental and asynchronous, pooling the efforts of different people with different capacities who are available at different times. *Granularity* implies that modules are predominantly fine-grained or small in size, yet heterogeneous so that the project can accommodate variously sized contributions. *Integration* entails both quality control and the bringing together of contributions to make a finished product. If the cost of integration is too high, says Benkler, either it will fail or the integrator will seek to appropriate the residual value of the common project, leading to dissipation of motivations to contribute *ex ante*.[20]

To what extent are these characteristics either already present or able to be incorporated into biotechnology research and development?

Regarding modularity, the answer is straightforward. As we have seen, distributed production—and hence, modularity—is already a characteristic of biotechnology research and development in both noncommercial and commercial settings. In a noncommercial setting, modularity is essential to academic peer production. The collaborative sequencing of DNA among academic laboratories provides a concrete example:

> We divided up the job by each starting at the same place on a chromosome and sequencing away from one another in opposite directions. That way we had only one overlap between the labs to worry about per chromosome. If it seemed like one lab had a particular problem covered, then the other left it to them.[21]

Modularity, also known as "task partitioning," is equally necessary to permit collaboration among commercial players. We have seen that the knowledge base of biotechnology research and development is generally too broad for the full range of necessary skills and resources to be integrated within a single firm. The solution is essentially for entities that are able to capture the sale value of end products to outsource various aspects of upstream research and development. This is achieved not only through service contracts, as in conventional outsourcing, but via a complex web of joint ventures, research partnerships, strategic alliances, minority equity investments, and licensing arrangements. Even those aspects of research and development that one might expect to require a fully integrated approach—for example, late-stage pharmaceutical development—are often modularized to permit outsourcing to service providers such as contract clinical research organizations and firms that specialize in packaging information for regulatory approval. Indeed, with research and development laboratories scattered around the globe, large pharmaceutical and agribusiness companies themselves must find ways to modularize their operations. If biotechnology projects were inherently nonmodular, all these forms of intra- and interorganizational collaboration would be impossible.

The key insight with respect to both modularity and granularity is that these are not inherent qualities of a particular technology or subject matter. It might be assumed that modularity is an inherent characteristic of software architecture; however, it is quite possible to write software programs in a nonmodular fashion. In the context of open source software, modularity is a question of best practice—a good programming habit—because it keeps the costs of contribution and coordination of contributions to a minimum and thus maximizes incentives to participate in open source development.[22]

In both software and biotechnology, therefore, modularity and granularity are a matter of construction. To grasp this point more fully, recall that the underlying reality of scientific research is that of a continuous and evolving flow of production—a "data stream"—on which scientists and others impose artificial boundaries for a range of purposes, one of which is to facilitate the exchange of data across organizational boundaries.[23] The packaging of portions of a data stream proceeds according to conventions that vary from one subfield to another. Importantly, these conventions are not immutable. In fact, the concept of a data stream is itself derived from sociological studies of scientists' early efforts to modularize their research in new ways so as to enable collaboration within and among firms in the emerging biotechnology industry.[24]

Thus, even if a particular biotechnology project *as presently constituted* is insufficiently modular and granular to accommodate an open source development methodology, it is possible to redraw the boundaries—in other words, to "refactor" the data stream. (To refactor a body of software code is to make changes that improve its internal structure without changing its function. The point of refactoring is to make the design easier to change in the future—in this case, by permitting an open source approach to further research and development.) This is not to say that there are no constraints on refactoring. The range of possible ways to package the elements of a data stream is limited not only by convention but also by the nature of the data itself. In the biotechnology context, for example, the complexity of living systems means that an apparently small change to one part of the system may generate unintended effects that are inherently unpredictable and often also delayed, making them difficult to detect even after the fact. Nevertheless, at present the modularity and granularity of biotechnology projects are effectively limited not by the inherent nonmodularity of biological systems but by the prevailing structure of data streams in biological research.

An interesting illustration of the possibility of restructuring data streams in biotechnology is the introduction of abstraction barriers that allow the behavior of living systems to be predicted within a specified range of conditions. This is the goal of a relatively new subdiscipline within biotechnology known as "synthetic biology." Synthetic biology is about decoupling the design of biotechnologies from their fabrication. According to its proponents, in a refactored biology the bottom level of design would involve the engineering of parts—for example, consisting of circumscribed protein–DNA interactions. At the next level, those parts would be used to engineer systems; at the next level, the systems would be used to engineer cells, then ensembles of cells, and, at the highest level of design, multicellular systems. This separation of biological engineering layers would enable engineering specialization: with a sufficiently detailed description of the components at each level—the behavior of parts or the inputs and outputs of systems—designers could choose to work at that level without needing to know in detail about lower-level interactions. The change would not be in the physical properties of biological molecules, but in the way the system is *described*. Nevertheless, synthetic biology would impose a new set of constraints and introduce new technological possibilities.[25]

Up to this point we have seen that biotechnology research and development is already modular. We have also seen that modularity and granularity are not fixed properties; rather, they are aspects of project design that are nevertheless ultimately constrained by the nature of the subject matter. The existence of the academic biobazaar implies that it is possible to construct biotechnology project modules such that they are sufficiently granular to permit the operation of bazaar-style incentives, at least in that particular institutional context. However, Benkler's analysis suggests that in the open source version of the bazaar, projects are especially fine-grained—even to the extent that single individuals with no formal credentials or any ongoing commitment to a project are able to make a worthwhile contribution. Could biotechnology research and development be made this granular? And is such a high degree of granularity actually needed for open source to succeed?

Clearly, the degree of granularity that can be achieved with respect to a given biotechnology project depends partly on the fixed costs of engaging in biotechnology research and development. The participation of casual amateurs will obviously be discouraged if the smallest worthwhile contribution to a project depends on access to a wet laboratory. We have already covered the issue of capital costs, but there is another, related, issue that is also relevant to the question of granularity—namely, the level of skill and commitment required to make a contribution in the biotechnology setting.

One frequently raised objection to the feasibility of open source biotechnology is that the level of skill and commitment needed to conduct biotechnology research and development is much higher than that required to produce software code. According to the executive director of one nonprofit biotechnology research institute, "It's harder to build things in biology than it is to write code. . . . You really can't do it part-time. . . . For fluency in nucleic acid manipulations, I'd say the typical person here has had eight years of post-undergraduate education."[26] By contrast, "every 16-year-old in the developed world today has a PC on their desk hooked up to the Internet. . . . It's the pet hobby of the masses."[27]

One problem with this argument is that it tends to play down the amount of skill and training needed to make a real contribution in the software field. In fact, as management scholar Georg von Krogh and his co-authors point out, software development is a knowledge-intensive activity that often requires very high levels of domain knowledge, experience, and intensive learning on the part of contributors.[28] On the other hand, it is true that to some degree the necessary skills in software can be acquired "on the job." Indeed, learning through feedback from other developers is commonly cited as one of the intrinsic or process-oriented benefits motivating contributions to open source software projects. To what extent might this be the case in biotechnology?

One answer is that even if most biotechnology researchers *have* in fact had substantial amounts of formal training, it is not clear either that this training needs to be in a biology-related discipline, or that any kind of formal training is actually necessary for most research-related tasks. On the first point, in MIT's first course on synthetic biology, in 2003, only about half of the students had biology backgrounds. The rest were trained in mechanical or electrical engineering or media arts and sciences.[29] On the second point, Boston's Whitehead Institute, in its work for the Human Genome Project, reportedly made substantial use of researchers trained through a six-month course at a local community college where they were taught how to synthesize DNA, make plasmids, transform bacteria, and extract the inserted DNA.[30] An even lower level of formal accreditation was required to join the Human Genome Project at the United Kingdom's Sanger Centre:

> We would recruit unskilled people, who would . . . have no need of academic qualifications. We judged them on school achievements, interview and something by which I set great store: the pipetting test. I showed the candidates how to use a pipette—a hand-held tool for manipulating small volumes of liquid—and invited them to have a go [as] an indication of their manual dexterity.[31]

The structure of this project did leave room for researchers to progress to higher levels of skill and commitment, yet maintained scope for more casual contributions from experienced researchers:

> [As director of the Sanger Centre] I got used to the idea that people would . . . come in at the level of routine tasks and learn what they could and then move up as high as they could, but there were also people who were coming for a short period who would pass through, even though they were highly qualified, and be happy to contribute something temporarily.[32]

These examples suggest that it would indeed be possible for contributors to acquire at least some of the necessary skills through the process of participating in open source biotechnology projects, with or without formal training in biology or related disciplines. But even if formal training is deemed essential, there is no reason why participation in open source projects could not be incorporated into the training process.

For example, many heads of university research laboratories who also have teaching responsibilities maintain grant money and an ongoing set of project goals for the purpose of having interesting and worthwhile projects to offer prospective research students. Attracting good students at the undergraduate or internship level is an important investment in future productivity because these students may turn out to have promise as Ph.D. students or laboratory managers: between them, these workers are often responsible for the day-to-day running of a lab and also for a large proportion of its intellectual output, as more senior investigators tend to be preoccupied with university and grants administration. Good students given good projects often generate useful data, though not always in sufficient quantities to warrant publication in a high-ranking journal. The cumulative effect of such projects, if coordinated through the bazaar, might not be scientifically glamorous; but it could well be economically valuable.

From the trainee's perspective, the incentive to contribute to open source development in this situation would be a process benefit akin to those observed in the software context—an opportunity to acquire new skills and to connect with others in the field. From the trainer's perspective, any investment in innovative activity would be made in pursuit of a goal other than exploiting the innovation itself—namely, educating the trainee. In neither case is proprietary exploitation of the trainee's contribution an essential part of the incentive to innovate. Thus, in principle there need be no barrier to freely revealing a research student's results in the context of an open source project. As a corollary, the pool of potential contributors to open source biotechnology may be considered at least as large as the population of life sciences students engaged in any kind of laboratory research at universities and other educational institutions around the world.

Why the qualification, "in principle"? Because, as noted above, where individual innovators depend on institutions for access to the resources needed to generate contributions to open source technology development, there is always the possibility that restrictive institutional policies may hinder individuals' engagement with open source development. In a university setting, for example, employment or service-provision contracts could stipulate that intellectual property is to remain the property of the university, which might limit participation in open source collaboration, as might the terms of commercial sponsorship or grants from funding agencies.[33] Similarly, commercial companies normally keep a tight rein on intellectual assets generated by employees.

Despite the stereotype of the hobbyist hacker, this kind of problem is not limited to biotechnology; it also arises in the software context. As we saw in Chapter 4, some participants in open source software development projects are independent programmers—amateur or professional—but many are company or university employees whose participation is supported by employers on the basis that it serves some broader institutional purpose.

In the next chapter, I will argue that the same logic should apply for both nonprofit and for-profit institutions in relation to open source biotechnology. For now, simply note that although some contributions to open source biotechnology development generated using institutional facilities might be made on behalf of the institution itself, others would more properly be regarded as a harnessing of institutional overcapacity. As Benkler points out in *The Wealth of Networks,* laboratory funding is silo-based; machines that are redundantly provisioned in laboratories have downtime that, coupled with a postdoctoral fellow in the lab—a figure that is perhaps rather too readily assumed to have plenty of time on his or her hands—is "an experiment waiting to happen."[34]

Overcapacity is not restricted to academic laboratories. In the pharmaceutical and agricultural industries, research and development capacity must be matched with fluctuating operational requirements. From time to time, commercial research and development facilities find themselves operating below capacity; with too few projects, such facilities are in danger of becoming inefficient, having to drop or interrupt long-term projects and losing experienced employees. The problem is serious enough that otherwise conservative firms have been known to experiment with new business models for the sake of retaining access to capability and expertise at a particular site.[35] One solution for a firm in this situation might be to encourage underemployed scientists to contribute to open source projects. The potential payoffs of such a strategy for the firm itself are discussed in the next chapter. However, we have already seen (in Chapter 4) one advantage of open source over other possible solutions to the problem of temporary overcapacity: unlike other modes of production, bazaar production does not require a long-term or substantial commitment by contributors. Hence, the governance structure of open source would permit a firm to devote resources to open source development only when other projects were not available.

The plausibility of this scenario is enhanced by the observation that many for-profit firms in the biotechnology and related industries already have a practice of allowing employee scientists to spend 10 to 20 percent of their time on discretionary projects—unrelated to the firm's own strategic goals. Some also have programs equivalent to the academic institution of the sabbatical.[36] Employee scientists, whose career success depends to some extent on maintaining a professional reputation outside the firm, might well choose to devote this time to open source projects, especially if the design of those projects emphasizes the visibility of individual contributions. Given that many scientifically and socially worthwhile projects fall short of the extremely high threshold of profitability necessary to induce firms to commit resources to development, such projects might still have substantial value even if they were confined to the development of products of no commercial interest to these scientists' employers.

The foregoing discussion suggests that even if biotechnology research and development cannot be made as fine-grained as software development, this need not constitute an insurmountable barrier to the implementation of open source principles in biotechnology. Instead, we might simply expect that the profile of contributors to open source biotechnology projects would differ somewhat from that of contributors to open source software projects, in that there would be relatively fewer contributions from hobbyists and more from students, postdoctoral fellows, and other individuals employed in noncommercial or commercial institutions.

Benkler's final criterion for successful Internet-enabled bazaar production is low-cost integration. Here again we must expect some differences between open source software and open source biotechnology. In the open source model, peer-based quality control mechanisms and integration of contributions to make a useful technology require the exchange of innovation-related information among participants. Where the relevant information can be fully codified in the form of computer files, the Internet renders exchange cheap and virtually instantaneous, irrespective of the number of recipients. In biotechnology, however, innovation-related information is often uncodified and may be embedded in tangible objects, including biological materials. This information is more costly to transfer than digital information, raising the costs of integration.

Of course, Internet communications are not the exclusive province of software developers. As management scholar Michael Porter has observed, the Internet is an enabling technology that can be used in almost any industry and as part of almost any strategy.[37] The same kinds of tools used by open source software developers to achieve cheap asynchronous communication and to track, archive, and search project-related information are also available to biotechnology researchers. Open source biotechnology projects might also make use of low-cost Internet video conferencing to simulate face-to-face communications. Internet-based conferencing is already in routine use by members of large-scale collaborations in biotechnology. To these researchers, as to open source software developers, "the Internet is very, very important. We couldn't function without the Internet!"[38]

The continuing existence of the noncommercial or academic

biobazaar constitutes an even stronger argument that bazaar production in biotechnology is feasible despite the need to exchange nondigital information. Before the advent of the Internet, collaborative projects in biotechnology (and in software) made use of a range of low-cost mechanisms for exchanging information. In the early days of genome sequencing, for example, mapping data from the *C. elegans* genome was made available electronically over the predecessor of the Internet; researchers developed a system of incremental updating to avoid having to send all the data on tape each time more information was added. The map—an integrated graphical representation of all contributions to the overall effort—was on display in a variety of forums:

> We had regular updates in the *Worm Breeders' Gazette,* the informal newsletter of the worm community; we showed it at conferences; and anyone could request clones at any time, free, immediately, whatever they wanted, so that they could look for genes. . . .
>
> Being thousands of miles apart wasn't really a problem. We used e-mail a lot, and [talked] on the phone. . . . Individual members of the . . . labs visited each other regularly. The highlight of the year was the annual lab meeting, when we took it in turns to host a visit from all the members of the other lab . . . to see at first-hand how the other group was working.[39]

This description illustrates an insight of the innovation management literature on horizontally networked user innovation (bazaar production): information that is costly to transfer in small increments can be stored and transferred in batches at a time when the incremental cost of transferring each separate "byte" is very low (in the above example, during scientific meetings or staff exchanges). In fact, open source software developers also supplement electronic communications with rare but highly effective face-to-face meetings at conferences and "hackathons."[40] Such episodically low-cost methods of transferring information have clear advantages over consistently low-cost methods such as Internet and email for the transfer of some types of information.

A slightly more costly but evidently still viable option for the exchange of tangible objects such as biological samples is to send them by post. A number of companies around the world specialize in shipping biological materials (mainly, though not exclusively, for clinical trials, which involve bringing materials from many locations to be analyzed in a central laboratory).[41] Same-day domestic delivery is standard, with international shipping taking two or three days. For a specialist courier service, the cost is two or three times higher than ordinary postage; on the other hand, specialist services offer the assurance that samples, including live animals, will reach their destination in good condition and in compliance with regulations such as customs and quarantine regulations or clinical trial protocols. An intermediate option, suitable for many transfers, is to use a mainstream courier service.

It is true that transferring information through the post or at annual conferences is slower and more expensive than simply pressing the "return" key to contribute code to an open source software project. Nevertheless, experience in relation to existing biotechnology projects suggests that this difference between software and biotechnology is of little practical consequence. For example, Nobel laureate Alfred Gilman, director of the Alliance for Cellular Signaling—a project supported by the first of several NIH "glue grants" designed specifically to support large-scale collaborations—notes that although the project involves "a fair bit" of exchange that cannot occur via the Internet, such exchange "certainly isn't any kind of hassle. . . . I've never even thought about that, it's so far down the list."[42]

Another factor that could be expected to make bazaar-style integration more costly in biotechnology than in software is the need for regulatory approval for some biotechnology products. Clinical testing, which is required in order to obtain marketing approval for new drugs, is the most expensive—though the least creative—aspect of drug development. Obtaining regulatory approval for field-testing and release of genetically modified organisms can also be onerous—indeed, the burden of securing regulatory approval is regarded by some industry participants as an even greater impediment to innovation in agricultural biotechnology than either structural barriers or problems accessing intellectual property.

How do these comparatively high costs of integration in some areas of biotechnology research and development affect the feasibility of the biobazaar? To answer this question, it is helpful to recall the reasoning behind Benkler's conclusion that low-cost integration is essential to the success of Internet-enabled bazaar production. Benkler argues that if the integration step is too costly, then either the integrator will seek to recover those costs by appropriating the resulting technology, or integration will fail.[43] If potential contributors have reason to fear at the outset that an integrator will attempt to appropriate the technology, this is assumed to erode the incentive to contribute in the first place.

By this reasoning, the issue is not, in fact, the absolute cost of integrating contributions to bazaar production. Rather, it is the cost of integration relative to the size of nonproprietary incentives or available collective action subsidies. Provided that even one actor, whether in the public or the private sector, is prepared to meet the costs of integration without seeking to exploit the investment according to a proprietary strategy, it does not matter whether integration is cheap or expensive on an objective scale.

As an example, consider the problem of integrating contributions to the development of a new drug. We have already seen that a publicly funded government agency—the NIH—sponsors a significant proportion of clinical trials for new drugs. Private nonprofit organizations such as the Rockefeller Foundation or the Gates Foundation might also choose to sponsor the integration of contributions to bazaar-style development in agriculture or biomedicine. These possibilities represent subsidies to collective action made by nonprofit entities in the public interest—but could we also envisage nonproprietary integration on the part of commercial actors?

One possibility is that generic manufacturers might become involved in integrating contributions to open source drug development. Generic manufacturers are for-profit firms that focus on the manufacture and distribution of existing drugs for which any patent rights have already expired. As we have seen, manufacturers of generic drugs face a less onerous regulatory burden than manufacturers of new drugs: generic drugs need not undergo clinical trials, provided the manufacturer can show that they are equivalent to an already approved drug. To date, the difficulty of establishing equivalence in relation to biologics—large-molecule drugs produced using biotechnology—has deterred the growth of a generic industry outside the field of traditional small-molecule drugs. However, if open source biotechnology research and development were to prove capable of generating a range of entirely new, nonproprietary drug candidates, generic manufacturers might well find it worthwhile to shepherd these drugs through the regulatory process—perhaps with the assistance of specialist contract service-providers, since established generic firms have not traditionally needed this kind of expertise. Relevant to the question whether there would be a sufficient commercial incentive for them to do so is the fact that patents are not the only form of exclusive rights in the pharmaceutical context. The manufacturer of a new drug approved by the FDA also obtains exclusive marketing rights for a period of several years. Such exclusivity, which is not dependent on patent ownership, might be sufficient to induce commercial actors to engage in the more costly aspects of integration while still not being a strong enough proprietary or quasi-proprietary right to deter upstream contributions by other bazaar participants.

This last speculation raises an important point: even if potential contributors to open source biotechnology development did perceive a risk of the technology being appropriated at the integration stage, this would not *necessarily* extinguish the drive to contribute. We saw in Chapter 5 that not all contributors to open source software projects are concerned about maintaining technology freedom with respect to *follow-on* innovations. If they were, there would be no place for "academic" (or "permissive" or "BSD-style") open source licenses. In the biotechnology context, one might imagine that at least some contributors to socially valuable technology development may be more concerned to ensure that the technology reaches its end users—patients or farmers—than to ensure that it is available on nonproprietary terms. Of course, this is more likely to be the case for, say, a malaria vaccine than for a new equivalent of PCR. But in general, the kinds of technologies to which scientists will be most anxious to preserve ongoing access because of their usefulness as research tools are not the ones that require the most substantial investments at the integration stage.

How plausible is the suggestion that a commercial drug manufacturer might be prepared to invest in the relatively costly later stages of drug development in the absence of proprietary exclusivity? In fact, even large research-based pharmaceutical firms—key players of the knowledge game—have historically shown themselves willing to do so under appropriate conditions.

A United States precedent is the development of the Salk polio vaccine.[44] Developed by Jonas Salk at the University of Pittsburgh in 1953, this vaccine was unprotected by intellectual property rights because it was insufficiently scientifically novel to meet patentability requirements. The university did not have the capacity to generate enough vaccine for the necessary large-scale field trial to ascertain both the safety and effectiveness of the vaccine and the best protocol for large-scale manufacture, so it sought the help of several large pharmaceutical companies. The companies knew that getting the vaccine through the next stage of development would require substantial investment, would be complex and technically difficult, and entailed substantial risk (it was not guaranteed that the vaccine would ever be approved). On the other hand, if the vaccine could get past the field-trial stage, it would be highly profitable: public fear of polio was then at its height, and all parents wanted their children vaccinated. The pharmaceutical companies in this case were willing to develop a field-trial vaccine without patents and even without an exclusive contract. Given the size of the "pie," the advantage of being ready to move the moment the vaccine was approved was considered worth the investment risk.[45]

Another possible solution to the problem of integration lies in combining public funding with the benefits of commercial competition. For example, activist Jamie Love and Human Genome Project bioinformaticist Tim Hubbard suggest a role for different types of organizations that could operate as private businesses selling the service of allocating contributions according to the wishes of contributors.[46] They suggest that funds to support the production of public goods could be raised by taxation,[47] with taxpayers having the right to designate which competitive intermediaries would distribute "their" taxes. Similarly, for voluntary contributions, "matching funds organizations" could perform due diligence on specific projects and then invite donations (with the proviso that if not enough is donated to make the project viable, the donations will be returned). Such organizations would be competitors in the sense that they would publish their distribution policies and track records and contributors would choose among the distributing organizations.

In a discussion of potential solutions to the problem of integration, examples from the pharmaceutical industry are useful because this is the context in which the cost of integrating contributions to open source biotechnology development—and therefore the risk of appropriation—is potentially highest. However, the same logic applies in other contexts. For example, we have seen that noncommercial entities and commercial entities with a nonproprietary business model are engaged at all stages in the development of agricultural biotechnologies: hence, the capacity for nonproprietary integration of open source contributions does exist.

At the start of the last chapter, I highlighted three related but distinct aspects of the open source model: technology development, licensing, and commercialization. So far this chapter has focused on technology development. A number of frequently raised objections to the feasibility of open source biotechnology relate to this aspect of the open source model. For example, it might be supposed that the pool of potential contributors to biotechnology projects is too small to sustain open source development; that biotechnology research and development requires capital investment, skill, and commitment beyond that which is available in a bazaar setting; and that the exchange of codified information and tangible materials required in biotechnology research and development is too costly and cumbersome to permit an open source approach. In answering these objections, I have argued that although there are certainly many technical differences between software and biotechnology, none represents a serious obstacle to the implementation of open-source-style technology development.

Before we leave this aspect of the open source biobazaar, one more point is worth making regarding the technical differences between software development and biotechnology research and development. Many of these differences boil down ultimately to cost: while both technologies are inherently cheap, software is cheaper than biotechnology. When considering which governance structure is best suited to coordinating biotechnology research and development, it is a fallacy to compare the costs of bazaar production in biotechnology with the cost of open source software development. Instead, the relevant comparison is between the costs of bazaar production and the costs of other modes of production in the same technological sphere. If the sum of transaction and production costs in the bazaar is lower than in the firm, the market, or the network, then economic agents will be best served by transacting through the bazaar—even if the investments required are higher in biotechnology than they would be in software.

As a brief illustration, consider the cost of quality control—an aspect of Benkler's integration step—in an open source biotechnology setting. For all of the reasons discussed in this section, rigorous peer review of contributions is more costly in biotechnology than it is in software. But this cost exists no matter which governance structure is adopted. In this example and as a general principle, *the higher the cost, the more sense it makes to spread the burden.*

Bazaar production offers a way to achieve this. But as we saw in Chapter 5, the spreading of cost and risk in a specifically open source bazaar depends on the existence of open source licenses, both as an articulation of fair terms of collaboration and as a legal protection against appropriation (that is, defection from the collaborative effort). To determine the feasibility of open source biotechnology, then, we must ask whether any generally accepted open source biotechnology licenses yet exist. If not, is there any plausible mechanism by which they might be developed?

Open Source Licensing

The last chapter demonstrated that it is possible to articulate a generic model of open source licensing that makes sense in the biotechnology context. However, something still needs to be said about the means by which this model might be implemented in actual license clauses relating to specific biotechnologies. How will the work of formulating open source biotechnology licenses and ensuring that they are acceptable to the relevant community actually get done?

In the discussion that follows, I seek to convey a sense of the issues that would-be open source biotechnology licensors and their supporters might expect to encounter in the early stages of the open source biobazaar. Of necessity, this discussion is speculative; I do not claim to offer either an exhaustive catalogue of problems or a comprehensive program of solutions. Rather, the point is that for each obstacle that is envisaged, it is possible to imagine a way forward—principally by drawing on networks and resources in what begins to look very much like an open source biotechnology community.

One important set of potential obstacles to implementing the generic open source principles articulated in Chapter 5 in biotechnology relates to the legal technicalities of biotechnology licensing. In biotechnology, a typical license agreement is a highly technical document, carefully drafted by specialists and incorporating a range of "boilerplate" provisions to deal with various contingencies. By contrast, many open source software licenses omit formal provisions that most lawyers would consider important.[48] While careless or sloppy drafting of license agreements is clearly undesirable, the informality of open source software licenses serves an important purpose. Technical legal language and clauses dealing with issues that are not central to the transaction generally make a license more difficult to read and understand and less widely applicable. The absence of such technicality has certainly contributed to the widespread adoption of open source software licenses.

By facilitating the direct involvement of technology users in formulating license terms, the simplicity of key open source software licenses has also contributed to their fine-tuning as instruments that accurately reflect software authors' collective understanding of the terms of open source collaboration. Recall Steven Weber's insight that open source licenses are de facto constitutions as well as contracts. It follows that an open source licensor is effectively engaged in a process of institutional design. But designing new institutions is risky, and most lawyers see their primary role as protecting their clients from legal risk. Thus, although the involvement of lawyers as "norm entrepreneurs" has been critical at various stages in the evolution of open source software licensing, the necessity of involving lawyers or other licensing professionals in the everyday execution of license agreements would have been a considerable hindrance to the overall evolution of the open source paradigm.[49]

This is not to say that the development of open source biotechnology licensing can or should proceed without the help of legal experts. My goal in Chapter 5 was to articulate in general terms those aspects of open source licensing that contribute to the success of open source; subjecting particular licenses to critical scrutiny was not relevant to that purpose. Nevertheless, open source software licenses do have weaknesses that have been documented and debated in the legal academic literature and the blogosphere (the virtual world of web-logs, or "blogs"). Naturally, anyone who sets out to draft an open source biotechnology license will seek to reproduce the strengths of open source software licenses and leave the weaknesses behind.

Further, drafting a workable biotechnology license in accordance with open source principles poses numerous technical challenges. As noted in the previous chapter, these include finding ways to disseminate license terms and ensure proper offer and acceptance in contexts where license documentation cannot easily be packaged with the technology itself, as well as complying with the requirements of biosafety and security regulations, consumer protection laws, local laws such as the Bayh-Dole Act in the United States, and so on. Although it is beyond the scope of this book to engage with such technicalities, they ought to be competently addressed in the drafting process.

Given the inherent complexity of biotechnology licensing, these considerations point to the need for enlisting lawyers and other licensing experts in the development of open source biotechnology licenses. But can such expertise be made available to ordinary licensors, and in a way that permits them to express their unconventional intentions?

One frequently raised objection to the feasibility of open source licensing in biotechnology is that cultural differences between tribes of lawyers specializing in different fields of intellectual property law would result in a dearth of expert assistance for patent owners wishing to license their inventions in an unconventional manner. Patent lawyers have been described as "guys in green eyeshades: very, very technical."[50] Meanwhile, the copyright bar is seen as "a beast of a different color"—particularly on the West Coast of the United States, where the proximity of sun, sand, and sea and the company of hackers and other free spirits may combine to produce a lessening of lawyerly inhibitions and a greater willingness to participate in the development of new licensing norms.

One way to overcome this problem is to decouple the roles of lawyers as norm entrepreneurs and as legal technicians. In the copyright setting, this has been the approach of the Creative Commons licensing initiative, led by academic lawyers in the United States and elsewhere. In this model, legal academics work together with nonlawyer team members with close links to the community of prospective license users, helping users to develop innovative licensing models. Technical drafting work is carried out largely by practicing lawyers either on a pro bono basis or paid by academic grants. Such work is interesting and exciting compared with most run-of-the-mill billable matters, but it is nevertheless perfectly compatible with lawyers' professional conservatism. At the time I visited the Creative Commons' Stanford headquarters in 2003, there was an oversupply of volunteers.

Thus, even though certain radical legal personalities may have figured prominently in the history of open source software licensing, what is needed in open source biotechnology is actors dedicated to promoting open source licensing, together with patent lawyers who like a technical challenge. There is no reason to think either is in short supply.

Would this "decoupling" approach be viable in the absence of such a close association between technology owners and legal practitioners? The question of payment need not be a stumbling block. Assuming that a greater proportion of prospective open source licensors in the biotechnology context than in the software context will be affiliated with existing institutions, we might expect that many such licensors will be able to command the services of legal professionals through ordinary business channels. Such in-house or fee-for-service efforts might generate some useful spillover; for example, a license that is made available on the licensor's website on terms that permit copying and preparation of derivative works is a resource that may be accessed by any would-be licensor. Prospective licensors who cannot afford to pay for professional legal services may qualify for assistance from organizations such as Public Interest Intellectual Property Advisors (PIIPA), an international nonprofit volunteer service that makes intellectual property counsel available to developing countries and public interest organizations who seek to promote health, agriculture, biodiversity, science, culture, and the environment.[51]

A more serious challenge in the absence of an intermediary dedicated to promoting the open source model is to ensure that the institutional design intentions of the would-be licensor are not overridden by the professional draftsperson's commitment to established licensing practice. To avoid this result, prospective licensors need a way to engage with licensing experts that permits them to retain some control over the drafting process.

One possibility would be for public or private funding bodies to develop and encourage the use of a suite of model licenses, with reusable precedents made available in a publicly accessible database. The use of model licenses is generally regarded as both desirable and feasible in biotechnology licensing as a way of reducing transaction costs; while there may be some impediment in the United States to making federal funding contingent on a particular mode of commercialization, institutions such as the NIH could still play a role in facilitating open source as a licensing option.[52]

Again, this is an approach that has been employed in relation to copyright content licenses by the Creative Commons initiative. A copyright owner can go to the initiative's home page and compose a suitable license by clicking on a menu of options relating to specific license provisions. For example, a license may or may not require attribution, payment for commercial use, or reciprocity with respect to derivative works—a "share-alike" option analogous to copyleft. Each option is represented by a unique symbol known as a "commons deed," so that prospective licensees who are familiar with the system can ascertain at a glance the key terms of use associated with any given content.[53]

One criticism of the Creative Commons approach is that this system gives prospective licensors insufficient guidance as to the choice of licensing terms. However, such "module libraries"[54] could be incorporated into a larger toolkit for open source licensing that, like other toolkits for user innovation from semiconductors to food preparation, would aim to codify all the expertise (in this case, legal expertise) necessary to customize an off-the-shelf product (in this case, a license).

In fact, such a system has been proposed from time to time as a way of reducing the costs associated with creating a new open source software license. However, such proposals have not been met with much enthusiasm. Part of the reason is that developing a foolproof system—one that rendered the choice of terms entirely user-friendly while maintaining the overall coherence of the license agreement as an enforceable contract—would actually be quite difficult.

In the Creative Commons context, the solution to the problem of user-unfriendly license terms has been to introduce two new layers to the license agreement. A copyright owner can make any content available under a Creative Commons license agreement simply by applying one of a small number of commons deeds alongside the usual copyright notice. The Creative Commons website displays each commons deed as an icon hyperlinked to a plain language version of the corresponding license—the "human readable" layer of the agreement. Behind the plain language version is the actual license, designed to be enforceable in a court of law. This is the "lawyer readable" form of the agreement. The use of Creative Commons licenses on the Internet can also be tracked by means of metadata tags—a "machine-readable" layer.

The obvious downside of this approach is that it may generate uncertainty as to which version of the license agreement is authoritative in case of conflict. Arguably, a "lawyer readable" license that is drafted in plain language in accordance with international best practice would be a superior solution. But in any case, in the biotechnology context a certain degree of technicality is probably inevitable because of the diversity of subject matter and the difficulties of defining rights in living materials. For example, the NIH's Uniform Biological Material Transfer Agreement (UBMTA)—a standard contract that allows signatory institutions to transfer materials using a boilerplate "Implementing Letter" executed by provider and recipient scientists—is widely regarded as too legalistic and cumbersome, despite genuine efforts to make it user-friendly.[55]

Another part of the reason why mix-and-match menu options for creating new licenses have not been embraced in the open source software context is that encouraging would-be licensors to draft new licenses that are precisely tailored to their own needs is regarded as inherently undesirable. License proliferation is a serious issue for both open source software licensing and open content licensing schemes, for two reasons. First, the larger the number of licenses, the higher the information cost. (A related concern is that seeking to address problems of technology freedom through licensing leads to unnecessary legal incursions into what might otherwise be a law-free zone.) Second, unless the solution space open to users of a mix-and-match system is very carefully defined, a larger number of licenses means a higher chance that licenses will turn out to be incompatible with one another in a cumulative innovation setting.

These are genuine concerns, but it is important to keep them in perspective. Conventional proprietary licensing practice places no effective restraints on license proliferation: hence the high transaction costs that give rise to anxiety about a tragedy of the anticommons in biotechnology. The charge of over-lawyering may carry some weight in relation to owners of copyright in cultural goods, who, it is suggested, may be encouraged by open licensing initiatives to assert legal rights where they might not otherwise think to do so.[56] This concern is far less salient in the biotechnology context because the anticipated disaster has already happened. Thus, Rebecca Eisenberg describes the emergence of a two-tiered pattern of exchange in biotechnology: technology transfer officials preside over formal legal agreements that are constantly undermined by informal exchanges among researchers unwilling to tolerate the delays and restrictions of the formal process, while researchers may find their research ultimately cannot be commercialized because of third-party rights that they had chosen to ignore.[57] In this situation, provided steps are taken to keep the associated transaction costs as low as possible, some means of preserving technology freedom that is both legally sophisticated and user-friendly would be an unambiguous net gain.

The fact is that once a decision has been made to work within the existing framework of intellectual property rights, the tension between minimizing license proliferation and freedom of choice for would-be licensors is inescapable. It might be thought that this tension, described by legal scholars Thomas Merrill and Henry Smith as the problem of "optimal standardization," is a reason to avoid defining property rights via contract. In other words, it might be seen as an argument against the open source licensing approach and in favor of pursuing technology freedom through the alternative strategy of law reform. However, while it is not my intention to promote either approach at the expense of the other—in any case, they are not mutually exclusive—we should recognize that similar problems arise at higher levels of lawmaking.[58]

The approach of the Open Source Initiative (OSI) to the problem of license proliferation has been typically deliberative and complier-centered. In late 2005 the OSI chartered an advisory committee to look into the phenomenon, its consequences, and possible solutions. The committee set out to classify existing licenses into three tiers: "preferred," "recommended but not preferred," and "not recommended."[59] Undertaking to consult with license stewards—authors or primary users of particular licenses with standing to influence others' license choices—before making any recommendation with respect to particular licenses, it pledged never to take action that would require licensors to relicense existing projects. Rather, license stewards would be encouraged to consider relicensing and to publicly "deprecate" existing licenses to make it clear that they should not be used for future projects or project contributions. Meanwhile, the committee flagged the possibility of providing educational materials to help prospective licensors choose an appropriate license, perhaps in the form of a matrix describing the traits of existing approved licenses. More broadly, it has signaled the intention of soliciting input from the community (individual and corporate) on all aspects of license proliferation. Two publicly archived mailing lists dedicated to license proliferation were open to anyone to join, though in one of them only committee members might post. The committee's draft report has now been published; the OSI is taking public comments on the draft report and will make decisions based on those comments "in due course."[60]

Does the OSI's approach to the problem of license proliferation offer any guidance to would-be open source biotechnology licensors and their supporters seeking to achieve optimal standardization? Obviously, the problem of license proliferation is one that has yet to arise in relation to open source biotechnology: what is needed is not a retrospective solution, such as that adopted by the OSI, but a means of forestalling unnecessary proliferation. If it were possible to go back in time to the establishment of the OSI, one might imagine that its founders would choose to define their license certification criteria in such a way that new licenses would be eligible only if (1) they conformed to the relevant standard *and* (2) they were substantially different from any existing license. Such an approach could easily be adopted in the biotechnology context, *assuming* the existence of a standard-setting and conformity assessment body for open source biotechnology licenses. At present, there is no counterpart to the OSI in the biotechnology context. Could there be? Should there be?

We saw in Chapter 5 that the translation of open source licensing principles from the software to the biotechnology context leaves room for a range of different interpretations. Indeed, the word *translation* itself, as used by sociologists, implies that as an idea passes from one person or institutional setting to another, it will inevitably be modified and in some sense distorted in response to local conditions, including the needs and priorities of the translator.[61] If the adoption of an open source approach to biotechnology research and development is regarded as desirable, then it is important to preserve the resulting diversity. In other words, given that it is not yet apparent how best to implement the open source paradigm outside the software context, would-be sponsors of open source biotechnology projects should not be constrained to follow any particular interpretation. Rather, they should be encouraged to try it their own way—in public—so that everyone can see and learn from the results.

Assuming that such an evidence-based, consequentialist approach to implementing open source in biotechnology is appropriate, supporters of the open source biobazaar ought to think carefully before taking action that would legitimize some implementations and delegitimize others. On the other hand, a voluntary licensing standard or set of best-practice guidelines would be useful in two ways. First, it would help prospective licensees to decide whether a biotechnology license that purports to be "open source" in fact offers the technology freedom implied by that term. Second, it would enable prospective licensors to take full advantage of the economic and social insights to be garnered from the experience of open source software licensing, not just the name.

The first point is important because open source technology development is a form of collaboration in which shared principles play a key role in establishing trust. In network-based collaborations, collaborators are able to establish trust by checking each other's credentials and to build on this initial trust relationship through an extended series of mutually beneficial transactions. Because bazaar governance relies on weaker ties, potential contributors may not be in a position to identify one another and may not make any long-term commitment to the project. In the open source version of the bazaar, potential contributors depend to some extent on each other's stated commitment to a particular set of principles, embodied in open source license agreements, to overcome mutual suspicion. In the software context, the fact that a technology owner describes the terms of a proffered agreement as "open source" may be part of what induces potential developers to invest their own resources in improving the technology. If there is no way of knowing what the initial developer means by "open source," then the value of that concept as a way of building trust is eroded.

Of course, this problem is exacerbated if the term is not used in good faith. This issue arises from time to time in the software context, but it may pose more of a threat in biotechnology because the implementation of open source principles in biotechnology must be worked out in the shadow of the open source software phenomenon. The term *open source* is now a widely recognized brand; as such, it is open to potentially destructive exploitation that may be either opportunistic or strategically motivated. A widely accepted licensing standard—which need be no more elaborate than an articulation of broad principles—could help shelter the growth of a body of licensing practice that supports the goal of biotechnology freedom.

Thus, it appears that some type of voluntary standard would be useful in relation to open source biotechnology licensing, provided the chosen standard is not too inflexible. In the software context, flexibility is achieved partly through the use of complier-centered sanctions and partly by leaving room for competition among standard-setting bodies. For example, the OSI has no proprietary or quasi-proprietary right to the term *open source* itself; anyone is free to apply the term to any license, provided it is not inaccurately described as "OSI-approved." When the OSI is alerted to what it considers a misuse of the term *open source,* its first-line response is to attempt to persuade the user to either drop the term or submit the relevant license for approval.

Would such a "soft" approach be viable in the biotechnology setting? Arguably, this depends on the degree to which the various aspects of community that exist in the open source software context could be replicated in biotechnology. Indeed, one argument against the feasibility of the open source biobazaar is that there is no such community in biotechnology.[62] To address this objection fully, it is necessary to step back—though only temporarily—from the specific issue of licensing standards.

As we saw in Chapter 4, the concept of an open source "community" is problematic. In that chapter I argued that direct relationships between participants are not essential to ideal-type bazaar governance, but that such ties may enhance productivity in real-life open source production systems in several ways: as aspects of network governance; by reinforcing private incentives in the face of residual collective action problems; by providing contributors with sociability, support, information, a sense of belonging, and a social identity; and by eliciting broad support for the open source brand.

We have already seen that biotechnology research and development is characterized by a significant degree of network governance; indeed, the idea of the network as a distinctive governance structure arose out of empirical studies of the biotechnology industry.[63] Logically, it is likely that increasing bazaar governance in biotechnology would reinforce the strength of existing networks (and hence, the relevant aspects of community). As innovation management scholars Chris DeBresson and Fernand Amesse point out, this is because, although close-knit relationships between partners in a research and development network create resilience in the collaboration, networks formed entirely of such close ties are resistant to change: cliques are typical of cartels and stable oligopolies, not of dynamic networks of innovators.[64] It is now well recognized that innovative networks are sustained not so much by the density of internal relationships as by the existence of "weak ties"—distant, unstable relationships—and openness to outside linkages.[65] Free revealing of innovations by self-selected contributors to bazaar production generates just this type of weak tie between participants, thereby enhancing both bazaar and network governance in the relevant production system. Thus, empirical research has shown that the most successful knowledge production networks in biotechnology exist in close proximity to centers of traditional academic bazaar production—for example, major universities—and are characterized by a mixture of closed information "conduits" (formed of the strong ties mediated by exclusive proprietary rights) and open "channels" (which permit knowledge spillover).[66]

What of the collective action aspects of community—do they exist in biotechnology? Throughout this chapter I have emphasized the fact that biotechnology evolved out of and remains rooted in academic research, a mode of knowledge production in which it is easy to see elements not only of the bazaar, but also of a collective action model in which private incentives are supplemented by both public subsidies and community governance. Unsurprisingly, collective-action–style community governance in biotechnology research and development is easiest to see in the context of large-scale collaborations, where project leaders take on many of the same responsibilities that have been identified as key aspects of the open source leadership role: external advocacy, enforcing shared conventions of behavior, motivating contributions, and maintaining momentum by ensuring that individual contributors' goals are sufficiently well met to minimize the risk of defection.[67] Such collaborations may incorporate contributions from both noncommercial and commercial participants. Thus, in biotechnology, "community" in the collective action sense need not be confined to the noncommercial sphere.

In addition to aspects of community related to networks and collective action, communities also exist in biotechnology that provide their members with sociability, support, information, a sense of belonging, and a social identity. Biotechnology research and development is full of technical communities that scientists in both commercial and noncommercial settings regard as their professional base, frequently using the term *community* to describe not just the colleagues with whom they personally collaborate, but the whole "invisible college" of scientists focused on a particular area of research (for example, the "worm community"). A scientist's technical community includes past and present college advisers and professional supervisors, students and workplace subordinates, academic rivals or commercial competitors, likely reviewers of publications and grant applications, and the intended readership of his or her latest paper.

Smaller user communities also grow up around specific biotechnology research tools; such communities swap tips about how to use the technology effectively in different settings, organize workshops to teach others the relevant skills, and put pressure on manufacturers to bring out improved versions.[68] Some of these user communities are effectively brand communities, and their presence signals the possibility of a broader open source brand community emerging in biotechnology.

Thus, all of the elements of community that contribute to the productivity of open source software development are also present in biotechnology. While they have yet to crystallize around open source, the fledgling open source biotechnology efforts described in Chapter 8 may provide the necessary nucleation point.

Returning now to the specific question of the role of an open source biotechnology community in setting and enforcing licensing standards, it appears that those aspects of open source "community" that support the existence of standard-setting and conformity assessment bodies such as the OSI *could* be replicated in biotechnology. I have already argued that some such effort *should* ultimately be undertaken. However, it does not follow that the OSI itself is a perfect model.

Interestingly, the OSI's announcement of its intention to address the problem of license proliferation sparked some controversy as to the scope and legitimacy of its authority; critics pointed out that the OSI board is not formally accountable to any defined membership. Further, it was argued that the OSI's views are not representative of all elements of the "community," especially programmers based outside the United States.[69] In general, discussion of governance issues in the software context is relatively unsophisticated, being regarded as a distraction from more immediate goals. Thus, proponents of open source in both software and biotechnology may have something to learn from the experiences of standard-setting and conformity assessment bodies in other fields.[70]

One possible model is the International Social and Environmental Accreditation and Labelling (ISEAL) Alliance, a collaboration of standard-setting and conformity assessment bodies whose common purpose is to promote voluntary social and economic standards as policy instruments in global trade and development. Like the OSI, many of these bodies administer certification schemes based on trademarks or service marks. If "open source" were to become a voluntary standard for intellectual property management in biotechnology, it might bear a close resemblance to standards established and maintained by ISEAL members, such as the international standard-setting body Fairtrade Labelling Organizations (FLO) and the organic producers' federation, the International Federation of Organic Agriculture Movements (IFOAM).

Why is the ISEAL example of interest in relation to open source biotechnology licensing standards? Because members of the ISEAL Alliance have had years—in some cases, many decades—of experience dealing with the same challenge that would face any licensing-standards body that may emerge in the context of the biobazaar. That challenge is to establish and preserve the legitimacy and mainstream credibility of voluntary standards even where there is substantial disagreement among stakeholders as to what the content of the standards themselves should be.

Recently, the ISEAL Alliance formulated a Code of Good Practice for voluntary standard-setting.[71] The code embodies a conviction, grounded in practical experience, that the key to the legitimacy of any voluntary standard is to incorporate democratic values into the process of stakeholder consultation at every stage. Note that the word "democratic" in this context does not refer to electoral democracy; rather, it refers to the ways in which a standards body can remain accountable to an active constituency and facilitate participation and deliberation. Thus, the code sets out to explain what any open source biotechnology licensing standards body would need to know: how to operationalize the values of participation, deliberation, and accountability in formulating, marketing, and defending an open source biotechnology "brand."

The point of this example is not to advocate a particular approach to standard-setting in open source biotechnology. Rather, it is to encourage proponents of open source biotechnology and related initiatives to look beyond their immediate environment—and beyond the software model of open source—for ideas and principles that could help them design institutions that most effectively support the fundamental goal of protecting technology freedom. An eclectic approach to modeling is important, because even if would-be open source biotechnology licensors and their supporters accept the generic open source licensing principles articulated in Chapter 5, these principles clearly underdetermine the content of workable open source biotechnology licenses and licensing standards.

There is more work to be done in developing open source biotechnology licenses. However, there is nothing to stop it from being done—provided interested parties are prepared to engage in open, constructive discussion. This book is intended as one contribution to that discussion. Another important contribution would be the establishment of a credible, independent forum designed to optimize deliberation among diverse participants. However, as with open source technology development, the *sine qua non* of an evolving body of open source biotechnology licensing practice is the existence of motivated contributors—an issue that is addressed in the next chapter.

In this chapter I have argued that bazaar-style production in biotechnology is inherently feasible. The best demonstration of its feasibility is the continuing existence of a version of the bazaar that predates open source—namely, traditional academic research. However, to establish the feasibility of an *open-source*-style biobazaar, it has been necessary to consider the differences between the traditional academic mode of production in biotechnology and the open source model. One such difference is the degree to which each of these two versions of the biobazaar makes use of full-scale Internet-enabled peer production. Another is the use, in the open source model, of intellectual property licenses formulated to promote non-proprietary technology development and commercialization.

Are these differences material or immaterial to the implementation of open source biotechnology? In the course of this chapter we have encountered a number of perceived obstacles arising from these differences, but none seems insurmountable. One important issue remains to be addressed, however. Apart from public funds, where would the money come from for open source biotechnology development? Would commercial entities have a role to play in the new biobazaar?

7

Financing Open Source Biotechnology

In Chapter 6 we saw that bazaar governance in biotechnology research and development has hitherto been mostly confined to the noncommercial sphere. In the traditional academic biobazaar, technology transfer to commercial entities is mediated either by straightforward free revealing or (since the 1980s) by a variety of proprietary mechanisms. The first alternative permits private commercial actors to unfairly lock up the benefits of publicly funded bazaar production; the second creates proprietary incursions into what was once nonproprietary territory, shrinking the scope of bazaar production and threatening ongoing innovation.

The open source model described in this book differs from the traditional biobazaar in that it stresses the participation of commercial actors motivated by private incentives. But it also differs from mainstream commercial practice in that open source methods of commercializing a new technology do not rely on proprietary exclusivity. This is the promise of open source biotechnology: that by incorporating both noncommercial and commercial contributions to biotechnology research and development on a nonproprietary basis, it is possible to preserve *both* the benefits of commercial technology transfer *and* a robust science and technology commons.

Could a nonproprietary mode of production really have any

commercial application in the intensely proprietary world of biotechnology research and development? Or is open source biotechnology essentially just a form of nostalgia for the (arguably nonexistent) "good old days" of publicly funded, nonproprietary science?

The first step in arguing for the feasibility of open source biotechnology is to extend the discussion in Chapter 4 of bazaar incentives and nonproprietary strategies for exploiting innovation to the biotechnology, pharmaceutical, and agricultural industries.[1]

Having considered some of the potential commercial benefits of open source, we then turn to the matter of costs. From the perspective of a potential contributor, the costs of open source production include both the opportunity cost of *not* pursuing a proprietary exploitation strategy and the actual costs of implementing a nonproprietary strategy. (Established firms would also need to consider the transition cost of changing from one strategy to another.)

In this chapter, a systematic survey of the opportunity costs for different types of institutions, commercial and noncommercial, suggests that industry participants may not always have much to lose by adopting a nonproprietary approach. Further, I show that the actual costs of open source production could be met by contributions from entities that would benefit from nonproprietary research and development but lack the capacity to engage directly in innovative activity. Such indirect contributions could either supplement or take the place of traditional sources of investment capital, which often depend (or are perceived to depend) upon maintaining proprietary exclusivity.

It might be objected that even if a nonproprietary strategy appears rational on the basis of the type of analysis presented here, the culture of the biotechnology industry is such that industry participants would still be unwilling to depart from the prevailing proprietary approach. I defer the answer to this objection to the final chapter.

Bazaar Incentives and Business Strategies in Biotechnology

In Chapter 3, I described a fundamental tension at the heart of intellectual property policy. Intellectual property rights permit owners to impose restrictions on access to and freedom to operate with innovation-related information. Such restrictions have both static and dynamic costs: they allow producers to charge a nonzero price for goods (information) that have a zero marginal cost, and they raise the price of information inputs into future rounds of innovation.[2] Nevertheless, the conventional assumption is that proprietary exclusivity is essential to innovation, either as an incentive to private investment or as a means of coordinating the exchange of information needed to enable cooperative and cumulative development.

This assumption is central to the knowledge game. Strategically deployed, intellectual property rights allow powerful multinational players to discipline markets and even states, extracting substantial economic rents. As long as proprietary rights are regarded as essential to innovation in biotechnology, any threat to proprietary privileges can be met with the argument that if consumers are not prepared to pay monopoly or oligopoly prices—justified as reflecting the high costs of research and development—they will simply have to do without life-saving drugs or life-sustaining new crops.

The success of open source demonstrates the falsehood of this core proprietary assumption, at least as a general proposition. Knowledge production coordinated through the bazaar does not rely on proprietary exclusivity; nonproprietary incentives *can* be sufficient to drive socially and economically useful innovation. Further, while open source software is probably the best-known example of bazaar production in an industry setting, we have seen that other examples exist—not just outside the software context, but also outside the broader sphere of digital information production.

Perhaps the earliest documented example of a real-life bazaar involving commercial competitors was the development of the Cornish pumping engine, a key technology of the British industrial revolution.[3] Other examples include early techniques for mass-producing steel in the United States, the first personal computers, and some types of extreme sports equipment.[4]

The wide range of settings in which it has been observed that innovators (1) respond to a range of incentives other than proprietary exclusivity and (2) sometimes find it more profitable to freely reveal their innovations than to pursue a proprietary strategy suggests that there is nothing unique about software in this respect. The obvious next question, left unanswered at the end of Chapter 4, is whether there is something special about biotechnology that makes bazaar incentives and nonproprietary business strategies uniquely *in*effective in that context.

In fact, as the examples below will indicate, bazaar-style incentives do operate in biotechnology, and opportunities do exist for commercial firms in biotechnology to employ nonproprietary exploitation strategies of the kind that drive commercial involvement in open source software development. The discussion that follows revisits the various incentives and strategies canvassed in Chapter 4, this time with specific reference to biotechnology.

Recall that in the user innovation lexicon, a manufacturer is a commercial entity whose primary incentive to innovate is the prospect of profits obtained by treating the resulting technology as a saleable product. A proprietary strategy permits a manufacturer-innovator to exclude competitors in markets for products embodying the technology—this is the archetypal approach of the fully integrated pharmaceutical company business model. A proprietary strategy also enables the innovation itself to be "sold" (or, more frequently, licensed) in information markets—the archetypal approach of the dedicated biotechnology firm.

Conventional business models in knowledge industries like software and biotechnology are examples of this proprietary manufacturing strategy. But we saw in Chapter 4 that self-interested, profit-maximizing innovators are also sometimes driven by other incentives that are independent of the expected sale value of a new technology and therefore have a much weaker affinity with proprietary exploitation strategies.

For example, some research and development activity is motivated, not by the expected outcome of the innovative process, but by the process itself. In the software context, process benefits include fun, learning, and a sense of belonging to a community. At first glance, this would appear unlikely to be a powerful class of incentives in biotechnology because most biotechnology research and development takes place within institutions, where process benefits that accrue to individuals might be expected to carry less weight than they do in the software context. But there are two reasons why we should not dismiss the possibility of process-driven investments in biotechnology.

First, some types of free revealing that are commonplace in commercial as well as noncommercial biotechnology settings may best be explained by reference to process benefits that accrue to institutions indirectly, by heightening the effectiveness or job satisfaction of individual innovators operating within an institutional framework. As noted in Chapter 6, many commercial firms in biotechnology and related industries have a practice of allocating some staff time to discretionary projects. Most for-profit companies also allow and even encourage staff to submit publications to academic journals and to attend academic and industry conferences to give talks, present posters, and maintain ties with others working in the same scientific discipline. These practices suggest that companies recognize that their best employees are motivated by the same kinds of rewards as individual contributors to open source software projects—learning and maintaining skills, building a professional reputation (either for its own sake or to enhance future employment prospects), and connecting with a community.

Second, even at the corporate level there are benefits to be gained from engaging in the innovative process, irrespective of the value of the resulting technology. To see why this is so, consider an analogy with the individual learning a software programmer acquires through the process of writing code. Sociologist Woody Powell points out that in technologically intensive fields like biotechnology, where there are large gains from innovation and steep losses from obsolescence, competition is best regarded as a learning race. The ability to learn about new opportunities requires participation in them; hence, for biotechnology firms and pharmaceutical and chemical companies, as well as for nonprofit institutions in biotechnology and related industries, a wide range of interorganizational linkages is critical to learning and technology development. This is so even though it must be expected that some of these collaborations will fail to bear fruit in the form of potentially lucrative new technologies. According to Powell, biotechnology firms that focus too closely on the details of individual transactions risk missing the boat as the larger field rides the waves of rapid technological change. In such a setting, organizational learning is as important to the firm as maintaining an up-to-date skill set is to individual software programmers. Says Powell, "Process matters."[5]

Of course, the fact that some research and development activity is motivated by expected process benefits does not necessarily mean that the innovator will choose to exploit the resulting technology in a nonproprietary fashion. Nevertheless, a corporation that invests in research and development activities for the sake of organizational learning may well benefit from a nonproprietary exploitation strategy, because such a strategy lowers the transaction costs associated with forming interorganizational linkages, thereby indirectly enhancing a firm's absorptive capacity and ability to act on new knowledge.

Thus, process incentives—an important driver of contributions to open source technology development—do operate in biotechnology. What about incentives relating to the creation of a useful technology that has economic value as a tool or intermediate good, irrespective of its sale value?

User innovation features prominently in open source development; indeed, by some accounts, it is the very stuff of which open source is made. The very first of Eric Raymond's "lessons" in *The Cathedral and the Bazaar* is that every good work of (open source) software starts by scratching a developer's personal itch.[6] Similarly, the innovation management literature characterizes open source software development as an instance of "horizontally networked user innovation."

To what extent should we expect to see user innovation in the biotechnology context? Theory predicts that the incentive for users to innovate for themselves will be particularly strong when (1) the information required to generate new technological developments is "sticky"—that is, costly to transfer from one person to another—and (2) different users need or want different versions of the technology.[7] The reason is that most innovation takes place through an iterative process of "learning by doing." To be effective, at least some of the "doing" must take place in the setting where the innovation will eventually be used. A user already has information about that setting, whereas a manufacturer has to acquire it—not just once, but at each iteration of designing and testing the prototype innovation. The stickier the information, the higher the cost of involving a manufacturer in the process. If user need is heterogeneous, this cost is exacerbated because the manufacturer has to acquire information from a range of sources. But heterogeneity also means the market for any particular version of the technology will be smaller, which means that the costs of innovating are likely to outweigh the benefits for manufacturers. By contrast, users will still be able to profit by using the innovation internally to bring about cost savings or process or quality improvements.[8] Thus, I argued in Chapter 3 that whereas proprietary manufacturers of biotechnology-related goods such as drugs and seeds concentrate on serving mass markets at the expense of smaller and niche markets, open source production would help solve the problem of unserved need by permitting users to adapt technologies for diverse uses.

Users are often innovators with respect to software code (both proprietary and open source) because most code written for in-house use is integrated with its environment in ways that make re-using or copying it very difficult. For this reason, a manufacturer will not generally find it worthwhile to develop and sell a one-size-fits-all version, whereas many users are still prepared to invest in a tailor-made solution. But, as noted above, the effectiveness of this type of incentive is not confined to the software context. Internal use value has also been shown to suffice as an incentive for innovation in diverse industrial settings. In addition to the examples mentioned earlier in connection with bazaar production outside the software industry, these include semiconductor and printed circuit board design and assembly and the development of library information systems, pultrusion (a process for manufacturing plastic products), wind turbines, and alternative currency systems.[9] Further examples of user innovation in settings closely related to biotechnology research and development include plant breeding and the development of scientific instruments, agricultural equipment, and automated clinical chemistry analyzers.[10]

Conditions conducive to user innovation exist in many areas of biotechnology. First, biotechnology-related information is often sticky. The stickiness of any given piece of innovation-related information depends partly on the qualities of the information itself. *Codification, tacitness, generalizability,* and *embeddedness* are all terms that refer to these inherent qualities.[11] We have seen that most of the information associated with biotechnology research and development is relatively uncodified. Scientific data are not stable, well-defined entities; rather, they are elements of an evolving data stream that are often embedded in tangible objects whose form affects the purposes for which the data can be used. Information stickiness is also affected by extraneous factors such as the amount of information to be transferred and the attributes or choices of information providers or information recipients.[12] For example, information providers may decide to charge for access to information, while information recipients may lack relevant tools or complementary information or may be less efficient for some other reason at acquiring particular information. In both scenarios, the stickiness of the information is increased. In biotechnology, this kind of information stickiness is implicit in the conceptualization of data as a continuous stream that cannot easily be transferred without being artificially partitioned into separate "bytes." The proliferation of intellectual property rights in biotechnology also contributes to information stickiness.

To perceive the multiple dimensions of information stickiness at play in biotechnology research and development, consider the apparently straightforward process of turning a published DNA sequence into a usable input for further innovation. A scientist who sees a published sequence and decides to conduct his or her own follow-on experiment would normally ask the author of the publication to mail a sample of the actual DNA molecule, usually incorporated into some kind of vector—a method known as "clone by phone."[13] If the request is refused, he or she might try to isolate the DNA from a sample of the organism from which it was initially derived. This takes some effort, both in the laboratory and beforehand in locating and obtaining the sample. There may be other problems, depending on the nature of the organism: it may be dangerous, like smallpox, or difficult to work with, like some humans. A third option is to place an order for custom-made synthetic DNA based on the published sequence. But DNA synthesis is quite costly, and it can be done only in short sequences. A long piece of DNA is both fragile and sticky in a literal sense: as you are working on one end, the end you have already made starts sticking to itself, goes bioactive, and does things you don't want it to do.[14]

The second condition identified by theorists as being conducive to user innovation is heterogeneity of user need. Depending on the particular technology, user need in biotechnology is often heterogeneous. One reason is the coexistence of "big" and "small" science in this field. The scale of a project has a direct impact on users' ability and willingness to pay for off-the-shelf innovations developed by manufacturers. Clearly, many smaller projects have budget constraints that may lead them to prefer makeshift solutions over expensive manufacturer-developed tools; but size cuts both ways. Ready-made cDNA arrays may be quite reasonably priced for some purposes, but if you are doing an experiment that requires several hundred arrays at thousands of dollars apiece, the sheer scale of the enterprise may make it more affordable to "roll your own."[15]

Another reason why user demand is often heterogeneous in the biotechnology context is that molecular biotechnologies exploit characteristics of living organisms that are shared at a molecular level by a huge variety of plants, animals, and microbes, many of which are, in turn, used by humans for a variety of different purposes. This means that the same basic molecular biological technique may have applications in animal husbandry, crop selection, and medical diagnostics; however, it may be necessary to "port" the technology to each new biological "operating system" in order to take account of higher-level biological differences and varied use environments. Even if a tool is designed for use with a single type of organism, there may be a need for customization to take into account external factors such as growth conditions, user sophistication, availability of other inputs, and so forth.

So, for example, farmers in developing countries may grow different varieties of the same crop plant on the east and west sides or upper and lower slopes of a single hillside because of variations in soil type, rainfall, and other cultivation requirements.[16] Similarly, a diagnostic kit intended to be carried in the field by military or emergency personnel has design requirements that are different from those for a kit intended to be kept in a temperature-controlled hospital room for long-term, large-scale use by trained doctors and nurses. Some would regard pharmaceutical drugs as the ultimate off-the-shelf technology; but one of the hottest emerging fields in biotechnology is pharmacogenomics—the development of techniques to identify a priori which patients will respond well or poorly to a particular drug, with the ultimate goal of allowing drug treatments to be tailored to individual patients.[17]

Thus, both of the conditions that logically favor user innovation (information stickiness and heterogeneity of user need) exist in biotechnology. In fact, user innovation is widespread in this setting. For example, many large projects incorporate technology development into their experimental schemes, investing a percentage of project funds to create tools that help researchers to meet project objectives. The Alliance for Cellular Signaling initially allocated a portion of its funding to developing tools to measure concentrations of very large numbers of cellular lipids.[18] Similarly, John Sulston in his book *The Common Thread* describes user innovation with respect to DNA-sequencing machines in the public-sector Human Genome Project.[19] The scale of the sequencing job meant that existing tools were highly inefficient, at least in their off-the-shelf format. While funders hoped for "some magic new technology," the researchers got to work making their own changes:

> Running gels might seem fiddly and labour-intensive, but we were constantly finding better ways to use them. ABI was promising machines with more and more lanes—already we had modified their machine to run forty-eight lanes, and sixty or even ninety-six lanes were talked of. . . . There were people working on automating all the tedious and time-consuming jobs such as picking clones. We had ex-

> cellent support from our bioinformatics teams in developing innovative software to track the samples and make sense of the results.[20]

In conversation, Sulston notes that such activities were not isolated:

> We weren't the only ones. There were a number of people . . . ripping the ABI machines apart and drawing the data out. . . . People decided to stick wires inside the machine to tap it off in a different way, and so they were using this as a real breadboard.[21]

Unsurprisingly, user innovation is part of the history of many key inventions in biotechnology. When hybridoma technology was first invented, it was regarded as a means of investigating the genetic control of antibody diversity, not as a way to produce monoclonal antibodies as reagents. Only through deployment in a variety of user environments was the technology's broader potential—that is, its full use value—realized.[22] PCR was also the product of an inventor "scratching his own itch."[23] In molecular diagnostics, one reason gene patents are seen as harmful is that so many practitioners rely on "home brew" diagnostic tests that are subsequently blocked by patent claims.[24] "Home brew," "roll your own," "breadboard"—all these expressions signify innovation for users, by users.

The foregoing discussion establishes the existence in biotechnology of another key incentive driving commercial participation in open source technology development—namely, the use value of an innovation. But many of the specific examples given above are of technologies that first emerged in a noncommercial setting. Do commercial players in the biotechnology and related industries also innovate for themselves, as distinct from innovating in order to sell or license the resulting technology to others? And when they do, what reasons might they have for making those innovations available on open source terms?

Let us consider the second question first. As noted earlier with respect to process incentives, the fact that innovative activity is primarily motivated by the expected use value rather than sale value of the resulting technology does not necessarily mean that the innovator will choose to exploit the resulting technology in a nonproprietary fashion. However, we saw in Chapter 4 that when the usefulness of a technology is more important to the innovator than its value as an end product, the innovator may actually *increase* his or her returns by following a nonproprietary strategy. The reason is that free revealing can enhance the usefulness of a technology in a variety of ways.

One example given in Chapter 4 was that the value of a tool to its user is often higher if the user is able to understand fully how the tool works. This factor is relevant in biotechnology because many biotechnologies make use of complex living systems as components of the technology. Providing access to the technology in as transparent a form as possible allows users to interpret unpredictable results with greater confidence, as well as imagine new uses that might not have occurred to the initial innovator.

A nonproprietary strategy may also enhance the use value of a technology by making it accessible to a larger number of users. Even in the absence of follow-on innovation, an expanded user base may enhance a tool's overall value through positive network externalities (network effects). It might be objected that network effects are less important in biotechnology than in software, because the demand for technical interoperability of biotechnology tools is not as strong. But there are two answers to this objection.

First, even apart from the fact that many important research tools in biotechnology *are* software programs, there are many situations in which technical compatibility of biotechnology tools would enhance their usefulness. For example, when microarray manufacturers use different proprietary chip formats, data sets obtained using microarrays sold by different manufacturers are not easily compared or integrated. Further, the importance of technical interoperability is likely to increase as biotechnology matures into a true engineering discipline. Efforts currently under way in the field of synthetic biology to generate a suite of "biobricks"—short sequences of synthesized DNA with different functions that can be joined together in biological circuits by means of standardized connector sequences—are a case in point.[25] The value, commercial and otherwise, of such a technology would clearly be enhanced by open standards.

Second, recall from Chapter 4 that technical interoperability is not the only potential driver of network effects. In biotechnology, government regulation of various aspects of product development provides self-interested innovators with a different reason to favor nonproprietary strategies as a means of establishing their own technologies as industry standards. An example of a biotechnology standard designed to boost commercialization by reducing the costs of regulatory compliance is the recently issued (2006) Publicly Available Specification "PAS 83: Guidance on Codes of Practice, Standardised Methods and Regulations for Cell-Based Therapeutics—From Basic Research to Clinical Application," developed by the British Standards Institution at the behest of the United Kingdom's Department of Trade and Industry. The purpose of this standard is to ensure that lack of clarity over best practice and provenance does not act as an obstacle to the commercial exploitation of stem cell science (specifically, its translation into therapeutic and biomedical research applications), an area in which the UK hopes to take a global lead.[26]

More generally, as we saw in Chapter 4, network effects include "certification signal" or "peer review" effects that enable users to treat a technology as reliable on the basis that it has been tested or checked by other users.[27] This is one of the recognized advantages of open source technology development: as Raymond puts it, "Given enough eyeballs, all bugs are shallow."[28] The description in Chapter 2 of problems associated with restrictive licensing of diagnostic tests provides a concrete negative example. With only a handful of test providers, regulators may find it is not cost-effective to develop adequate proficiency testing and laboratory inspection regimes, and a lack of comparative data may allow systematic errors to go undetected. Thus, any field of biotechnology research or product development where quality control is important could, in principle, be subject to network effects that increase the returns to free revealing.

Of course, the most obvious reason why a nonproprietary exploitation strategy may enhance the usefulness of a technology is that those to whom a technology is freely revealed may contribute to its further development. There is no question that this logic also applies in biotechnology; if it did not, anxieties concerning the impact of restrictive proprietary practices on incremental innovation in biotechnology would never have arisen. Nevertheless, it is interesting to consider a concrete example.

First developed in 1975 at the UK Medical Research Council's Laboratory of Molecular Biology in Cambridge, Cesar Milstein and Georg Kohler's technique for producing monoclonal antibodies from hybridomas soon proved to be of enormous commercial significance, permitting hundreds of start-up biotechnology companies in the United States and elsewhere to generate cash income by selling the antibodies as research reagents and as part of diagnostic kits. But the broad use value of monoclonal antibodies as a powerful new tool in biology and medicine had to be established by progressively "domesticating" and stabilizing the technique in a range of fields, a process that required the innovative input of many different researchers at a range of institutions. Initially, this process was aided by the fact that the inventors themselves did not pursue proprietary rights over their invention, although various aspects of hybridoma technology were subsequently patented.[29]

The evolution of hybridoma technology also provides an illustration of my previous point that a technology is often more useful the more widely it is disseminated. The success of hybridoma technology was linked to its standardizing power: it permitted the production of monoclonal antibodies with stable and uniform properties, allowing comparison of results obtained by laboratories around the world.[30] Interestingly, worldwide standardization of antibodies as reagents was achieved only through the participation of commercial firms. In this case, for-profit innovators were more strongly in favor of broad dissemination than were their academic colleagues; from a commercial perspective, it made sense to make the antibodies available to any qualified investigator who requested them in order to grow the market for a generic tool, whereas the academics preferred more restrictive terms that would have allowed them to preserve an edge in scientific competition.

So far we have seen that in biotechnology, as in other fields, when an innovator makes a new technology available on nonproprietary terms, that technology is likely to become more readily available, cheaper, better, and/or more transparent—in short, more useful. Although the conclusion itself seems obvious, it is helpful to have considered some biotechnology-related examples for the sake of illustrating the next step in any nonproprietary business strategy, which is to convert the specific enhancements in use value that may be achieved through free revealing into private rewards.

Commercial firms that are user-innovators can capture a return on their research and development investment in a very straightforward manner, by using the improved technology in a research program or as a component of a production process. Depending on the nature of the improvement, this may result in cost savings, risk sharing, or efficiency gains. Thus, a firm that conducts in-house gene sequencing as part of its overall research and development effort would benefit from improvements to sequencing technology. Pharmaceutical firms would benefit from improvements in chemical libraries, molecular synthesis techniques, protein assays, instrumentation for high-throughput measurement of biological assays, or analysis methods for predicting toxicological and other properties of new molecular entities.[31]

The SNP Consortium was a concrete example of this kind of business logic at work in the pharmaceutical industry. A high-quality, high-density SNP map was perceived to be a research tool that would benefit everyone involved in genomic research. By collaborating, the for-profit members of the consortium—AstraZeneca, Bayer, Bristol-Myers Squibb, Glaxo Wellcome, Hoechst Marion Roussel, Hoffman–La Roche, Novartis, Pfizer, Searle, and SmithKline Beecham—expected to create a commonly accepted SNP map more quickly, with shared financial risk and less duplication of effort, than if each company had proceeded on its own. Further, they anticipated that a collectively constructed map would be of greater density and therefore potentially of greater utility to the pharmaceutical industry.[32]

Clearly, prospective technology users are likely to pool their resources to develop technologies only if the technologies will not be a source of competitive advantage for the participants. One would not expect pharmaceutical companies to license their latest blockbusters on nonproprietary terms. However, the SNP example suggests that there might be a place for open source development even in this most proprietary of industry settings, at least to the extent that it would complement proprietary business strategies by reducing costs, risks, and product development time. In effect, open source would be a means of facilitating precompetitive collaboration. (Note that the incentive for pharmaceutical companies to participate in the SNP effort had another important dimension, discussed below.)

It might be supposed from this example that nonproprietary precompetitive collaboration on the part of major agricultural or pharmaceutical firms would be confined to upstream research and development. That is one obvious application. But history shows that if the opportunity is lucrative enough, pharmaceutical companies may also be induced to invest in nonproprietary development farther along the value chain, as they did for the Salk polio vaccine. Today, the key to identifying opportunities for open-source-style precompetitive collaboration in the pharmaceutical context is to focus on high-cost, high-risk aspects of the drug development process that do not generate any significant competitive advantage for the manufacturer by differentiating its end products from those of its competitors. Two such areas are predictive toxicology and the management of clinical trial data.

What is predictive toxicology? Once upon a time, the major roadblock to drug development was a scarcity of promising drug targets. In the postgenomics era this is less of a problem; nowadays, a key concern is that many drug candidates fail too late, after substantial investments have already been made. A tool that could help identify unpromising drug candidates, either predictively or by early analysis of the toxicology profile, before they reach the second or third phase of clinical trials, would save pharmaceutical companies a lot of money in research and development costs. Even more lucrative would be a reduction in the time taken in clinical trials, which occur *after* the patent clock has started ticking. Such rewards might be large enough to justify joint investment in developing, or supporting the development of, an open source toxicology tool.

Similarly, the management of clinical trial data is an area in which joint investment in the development of nonproprietary tools could make commercial sense. At present, data management is often inefficient and error-prone, in circumstances where errors have important regulatory and product consequences. At the same time, data management tools are not a major source of competitive advantage to prospective users. Here again, the drive to share costs and risks could motivate users to contribute to open source development.

One sphere in which pharmaceutical companies have separately invested significant resources over the past decade, only to be disappointed in the results, is bioinformatics.[33] At one point in the late 1990s, bioinformatics companies were coming into existence seemingly at the rate of several new companies each day, offering to solve the problem of managing, correlating, and mining the enormous data sets generated by novel DNA- and protein-sequencing techniques. Most of these companies did not succeed, partly because the big pharmaceutical firms ultimately decided to build their own bioinformatics capabilities. Even with excellent employees, these in-house bioinformatics divisions could not be optimally efficient because their activities were clearly outside the companies' core competencies. Without a vehicle for sharing the expense of developing bioinformatics software, individual companies' bioinformatics teams arguably represented an overhead rather than an asset.

Consistent with this observation, bioinformatics is one area where open source biotechnology has, in fact, gained a substantial foothold. The first real-world examples of open source biotechnology businesses are bioinformatics companies that have adopted business models along the lines of those described in Chapter 4.[34] Widely used open source bioinformatics tools include the NCBI toolkit (home of BLAST), scripting language projects such as BioPerl, and advanced projects such as BioLisp and hidden Markov model libraries. Other well-known open source bioinformatics software projects include BioJava, BioPipe, BioRuby and BioPython, DAS, MOBY, OBDA, and EMBOSS. The open source nature of these platforms saves duplication of effort with respect to basic programming features, but permits customization in response to heterogeneous user need. This combination of characteristics is important because the complexity of biological data sets is such that there are few off-the-shelf solutions to bioinformatics problems. Instead, they must be solved on the spot by people who understand the local system, ideally using tools that are freely available so as to avoid having to reinvent the wheel.[35]

So far, our discussion of actual and potential nonproprietary biotechnology business strategies has been limited to innovators who can expect to benefit directly from the existence of open source technologies. But in the software context, contributors to open source development also include suppliers of complementary goods and services and technology distributors. Are these strategies viable in biotechnology?

As we saw in Chapter 4, while nonproprietary strategies relying on technology distribution are well known in the software context, they can be difficult to implement successfully because customers have the option of downloading the same software, minus value-adding, from the Internet for free. This type of strategy may actually be more promising in the biotechnology context because the uncodified nature of much biotechnological information means that users may well be prepared to pay a premium in order to avoid some of the trouble and expense of optimizing a freely revealed protocol.

One service-oriented biotechnology strategy would be to carry out the type of precompetitive research and development described above on behalf of end users of the technology in exchange for a fee. We saw in our discussion of open source software strategies that end users contribute to open source development both by devoting their own staff and resources to open source projects and as the ultimate customers for other commercial contributors. At present, many dedicated biotechnology firms are heavily committed to proprietary licensing because they hope to strike an exclusive deal with large pharmaceutical or chemical companies. But an open source biotechnology start-up that could boast the support of one or more large pharmaceutical companies for even a part of its research program would be in a good position to attract additional funding, including venture capital.

How likely is it that such a nonexclusive approach would work in the real world, where large agricultural and pharmaceutical firms tend to make jealous partners? While it can be challenging for a smaller technology firm to maintain an "open relationship" with multiple larger firms, it is not unprecedented. For example, early in the 1990s, before the public-sector Human Genome Project made sequence data publicly available, at least one dedicated biotechnology firm employed a database license that required any customer who used the database to discover and characterize a full length gene to "grant back" nonexclusive freedom to operate with respect to the gene sequence, not just to the biotechnology company itself but also to all of its other customers. Initially viewed by licensees with a degree of suspicion, the purpose of such provisions was to allow the firm's customers to use the data without fear of infringement suits. But these nonexclusive grant-backs ultimately came to be seen as an additional source of value and a competitive advantage for the database provider: instead of gaining access only to the biotechnology firm's own data, customers were able to access information generated by all of their own competitors.[36] Such a strategy is, of course, still proprietary; but the difference between this type of "club" arrangement and open source is a matter of R&D governance, not commercial logic.

This example is one of a broader class of open source business models in which nonuser contributors to open source development link their freely revealed contributions to complementary markets in goods or services. The software version of this strategy is already in use in the biotechnology context in relation to bioinformatics software: if the software itself is open source, a bioinformatics company's competitive advantage might derive from the user interface rather than from the underlying algorithms that analyze DNA chips, protein chips, or sequencing gels. In the race between Celera and the public-sector initiative to sequence the human genome, it has been suggested that Celera could have given away its genome sequence data and made money by selling genome annotations.[37] Had Celera won the race, the strategy it in fact adopted would have been far more lucrative; but as it turned out, Celera lost and the company collapsed—an outcome that might have been avoided had it adopted an open source value-adding approach. Other examples of complementary marketing would include an assay kit manufacturer making assay protocols freely available in order to boost demand for the kits, or the owner of a cell line or other biological material making the material itself available on open source terms, but charging a consulting fee to provide advice on how to use it.

Is it realistic to suggest that a for-profit biotechnology company might choose to exploit an innovative technology by using it to leverage the appeal of related products or services? Both service provision and the supply of products such as reagents, laboratory instruments, and other research and development inputs are mainstream business models in both the agricultural and health sectors. Not all such companies are potential contributors to open source biotechnology development (at least directly—see below), because many do not have the capacity to engage in research and development activities in their own right. However, it is actually quite common for biotechnology companies that do have this capacity to adopt a mixed-revenue model that includes the provision of goods or services complementary to its own intellectual property rights and technological know-how. For example, a company might generate revenue by granting licenses to manufacturers of laboratory kits incorporating its technology but also by performing contract research and data analysis services or selling software.

Such mixed business models have always been more attractive to biotechnology entrepreneurs outside the pharmaceutical sector because of the difficulty of attracting venture capital investment in fields where an initial public offering of shares is unlikely to yield spectacular returns. A shortage of investment capital in these areas of biotechnology has led start-ups to try to establish an income stream as early as possible instead of following the classical biomedical model of burning capital through several investment rounds before launching an actual product. However, in recent years an unfavorable investment climate and a dearth of opportunities to collaborate on technology development with larger firms has prompted a number of former biotechnology platform companies in the health sector to alter their business models to incorporate either a product pipeline or the provision of contract services. Under appropriate circumstances, such firms might find a nonproprietary approach attractive, because although proprietary licensing of platform technologies can be very lucrative in the short term, the broader and more successful the platform, the greater the incentive for others to invent around it. In other words, a proprietary platform strategy is vulnerable to ongoing innovation in the field. Freely revealing the platform or enabling technology in order to encourage widespread adoption while concentrating on building services or proprietary products that sit on top of the platform may be a better strategy in the longer term. Of course, in biotechnology, conservation of biological functions at the molecular level means that some enabling technologies are more or less invulnerable to inventing around. The owner of such a technology has little commercial incentive to make it available on open source terms. Nevertheless, a free-revealing strategy may be attractive to other industry participants as a means of *preempting* the imminent establishment of a proprietary standard. I return to this point below.

Perhaps the most interesting case of a nonproprietary service-based business model in biotechnology is that of Diversity Arrays Technology, a privately held Australian company built around a patented molecular marker technology invented by its founder and director, Andrzej Kilian. The company's position with respect to intellectual property in its core technology is unique: the patent owner is Kilian's former employer, CAMBIA, a nonprofit research institute that has since pioneered the use of open-source-like licensing in biology for a number of its patented technologies, including diversity arrays technology (DArT). A description of CAMBIA's

"Biological Open Source" initiative is provided in Chapter 8; here the point is that for the commercial company, a conventional proprietary strategy is out of the question because the same technology is available at a relatively low price from another source.

Thus, instead of relying substantially on licensing revenue as in the conventional model for a dedicated biotechnology firm, the company offers genotyping services for a range of crops, development of DArT arrays for new species, application of the technology to the genome-wide study of DNA methylation, sales of dedicated software and a technology-specific laboratory information management system, and sales of technology packages that can be used by others to "port" the DArT technique to new crops or use environments. Its business model is designed to exploit the synergy between service provision and ongoing development of the technology: as in the case of hybridoma technology, progressive deployment of the technology in a range of settings is seen as the key to growing the market and enhancing the technology's use value through a process of iterative learning. Technology-specific know-how and a strong scientific reputation are crucial to maintaining a competitive edge.

It might be argued that this company's choice of a nonproprietary business model is not a free choice but instead makes a virtue of necessity. Even if that were true, it would not undermine the case for open source commercialization: in business, no strategic choice is unconstrained, and there will always be situations in which, for one reason or another, a proprietary exploitation strategy is not viable. But in fact, in this case the choice of a nonproprietary strategy reflects the same commercial logic that drives service-oriented open source software business models.

Specifically: (1) As a company that obtains a fraction of its total revenue from licensing fees and another fraction from complementary services, Diversity Arrays Technology has something to gain by increasing the number of users of its technology, even if those users did not pay license fees, because a larger user base expands the market for related offerings (in this case, genotyping and technology development services). (2) As a company that devotes significant resources to high-risk research, Diversity Arrays Technology benefits by sharing costs and risks with a group of external user-developers. Hence, the firm has established a "DArT network" for the exchange of innovation-related information and materials among developers and service providers working in diverse locations on diverse crops; its stated aim is to use any intellectual property rights arising from ongoing research and development to attract partners to the network rather than to restrict access to the relevant technology. (3) Building a user community helps build a reputation for cutting-edge innovation that makes it easier to attract investors, collaborators, and quality staff. (4) As a company that uses its own technology in the provision of goods or services, Diversity Arrays Technology benefits directly from any improvements in the use value of the technology. (5) The longer the firm stays in the field that it has helped create through its own innovation, the more it builds on its initial competitive edge while still capturing what in the software context would be called "developer mindshare" and growing overall demand for the technology.

Clearly, even though its executives have not sought to publicize this fact, Diversity Arrays Technology is very close to being an open source biotechnology company. In fact, the company's business model has been directly inspired by that of the open source software enterprise Red Hat. And on its own terms, this company is a success. The agricultural services market in which Diversity Arrays Technology operates is particularly challenging because of the low profit margins associated with breeding and growing conventional crops. (While DArT is a molecular biotechnology, it is not directed toward developing genetically engineered crops.) Yet the company, established in 2001, is a thriving small business with a growing income stream and international customer base. Kilian jokes that whereas most biotechnology companies have a sound business plan and fail, Diversity Arrays Technology has a questionable business plan and succeeds.[38]

Thus, there are commercial firms in biotechnology that possess both the capacity to innovate—that is, to contribute to technology development—and the ability to generate revenue in complementary markets using a nonproprietary strategy. Another way for nonusers to capture the enhanced use value of a freely revealed technology is as an investment in the firm's overall brand and reputation. Free revealing of patented technologies is, in fact, sometimes used as a deliberate strategy to enhance the reputation of companies in the biotechnology and related industries. For example, some agribusiness firms have made substantial technology donations to developing countries in an attempt to counter the negative public relations effects of opposition to genetically modified crops in developed countries.[39] Merck's funding of research on ESTs (expressed sequence tags) might have had a broadly similar motivation, along the lines of "We're going to have to spend this money anyway, so we might as well make the information public and get some kudos."[40] These examples suggest that contribution to open source biotechnology development on the part of companies that already enjoy strong name recognition could be a way to achieve specific public relations goals.

Skeptics might dismiss the idea that firms in biotechnology and related industries might find it worthwhile to freely reveal what would otherwise be proprietary information in order to boost their reputation in the marketplace, on the basis that a firm's reputation is economically insignificant next to cold, hard cash. But we saw earlier that in the rapidly developing field of biotechnology, the locus of innovation is found in "learning networks" rather than in individual firms. Importantly, firms differ in their ability to form the collaborative links necessary to participate in these networks. Commercial success depends on being perceived both as highly innovative and as a good collaborator.[41] Becoming the leader of a successful open source development project would be an extremely effective way to signal these qualities to potential partners. It would also help to create trust and goodwill, increasing the firm's chances of securing favorable outcomes in repeat bargaining situations with other industry participants.

The foregoing examples demonstrate that in biotechnology, as elsewhere, free revealing on the part of a technology owner may enhance the value of the technology to users. It also appears that there exist opportunities in biotechnology and related industries to convert this enhanced use value into private economic rewards, whether directly (as in the case of user-innovators) or indirectly (as in the case of distributors or suppliers of complementary goods and services).

Thus, either process-oriented or use-value-oriented open source business strategies might be employed by commercial participants in biotechnology research and development. Together, these cover a sufficiently broad range of commercial opportunities to confirm that, as foreshadowed earlier, there is nothing special about biotechnology that precludes the possibility of open source commercialization. But before we leave the subject of bazaar incentives and nonproprietary business models in biotechnology, it is worth considering two further potential motivations for commercial contributions to the biobazaar.

In Chapter 4 we saw, of those whose primary incentive to innovate is the prospect of profits obtained by treating the resulting technology as a saleable product—that is, "manufacturers"—not all adopt a proprietary approach. In the pharmaceutical industry, generic drug manufacturers are not usually manufacturer-innovators in the technical user innovation sense of that term, because although many generic manufacturers do engage in some innovative activity, their innovations tend to relate to the manufacturing process rather than the product itself (in other words, they are user-innovators). Nevertheless, it was suggested in Chapter 6 that generic manufacturers might choose to make a contribution to open source drug development in order to gain access to a stream of innovative drug candidates. In that case, generic manufacturers *would* be manufacturer-innovators—but their business model would still be nonproprietary. Such a nonproprietary manufacturing strategy is not seen in the open source software context because of the near-zero cost of reproducing software code. But in biotechnology, the non-negligible cost of reproducing biotechnology innovations—mentioned in Chapter 6 as a potential obstacle to open source development—might actually represent a commercial opportunity that adds to the attraction of open source from a business perspective.

Last—but emphatically not least—among possible motivations for commercial firms to contribute to open source biotechnology development is the potential for nonproprietary technologies to be used as competitive weapons. Earlier I noted that large pharmaceutical companies did not participate in the SNP consortium purely in order to gain access to a useful tool. Another important incentive for these players to contribute to the creation of a public database of human genetic markers was to avoid having to negotiate for access to the information, either among themselves or most especially with smaller biotechnology firms. Private ownership of the human genome would have been a disaster for everyone but the owner because it is a nonsubstitutable platform. As in the earlier case of ESTs, the "big companies weren't any happier than the academics that upstart genomics companies looked like cornering all the rights to valuable genome information."[42] Not wanting the relevant intellectual property to fall into the hands of small companies they could not easily control, the larger firms "placed a blocking stone on the 'Go' board [and] snuffed five or ten biotech companies that were ginning up to sell SNPs."[43]

Apart from human genome data, there are many other contexts in which this type of competitive strategy could make sense. To see the broad commercial *and* noncommercial appeal of this approach, recall from Chapter 5 that a crucial feature of open source is that it promotes free competition with respect to the technology in question. While commercial players in any industry would certainly prefer to protect *themselves* from competition, they will always want to ensure competition among their suppliers. Having a single supplier as the sole source for a critical value driver is a CEO's worst nightmare; while pharmaceutical companies would generally prefer not to share technology with one another, one thing they like even less is being beholden to the equivalent of an Intel or a Microsoft.[44] This means there is a strong motivation for commercial players to support open source development of any technology upstream of their own place in the relevant value chain.

Besides preventing competitors from gaining a choke hold on an important technology, warding off the danger of supplier lock-in, could a nonproprietary approach have other strategic applications? If open source implies free competition with respect to the relevant technology, an open source approach may be used to restructure the competitive landscape in an industry sector to the participants' commercial advantage in a variety of ways. For example, it could create the opportunity for several smaller firms to combine resources against a larger competitor, or to attract customers away from established technology providers.

The Costs of an Open Source Strategy

So far we have seen that the full range of incentives that motivate contributions to proprietary and nonproprietary technology development in software and other industries can be either empirically observed or realistically envisaged in biotechnology. We have also seen that there is scope in biotechnology and related industries for nonproprietary commercial exploitation strategies akin to those that support open source software development.

However, the existence of opportunities for commercial players to profit by contributing to open source biotechnology development is not enough on its own to establish the feasibility of the open source biobazaar. Commercial participation in nonproprietary biotechnology research and development also depends on the costs to industry participants of pursuing those opportunities in preference to a given proprietary alternative.

We saw in Chapter 3 that, for a knowledge game corporation, the point of a proprietary strategy is to liberate itself from competition, with all the attendant evils of low prices, low barriers to entry, and relentless market pressure to produce innovative, high-quality goods and services for the benefit of consumers. Of course, *any* profit-seeking firm would find the prospect of eliminating competition attractive; it is a proposition that makes the opportunity cost of free revealing seem very high. But opportunity cost depends on the opportunity lost. For those who are currently excluded from participating in biotechnology research and development by high barriers to entry and high transaction costs related to intellectual property, the opportunity cost of adopting a free-revealing strategy is effectively nil. (So is the transition cost.) Even for those who *can* afford to play, the knowledge game creates costs that must be offset against the benefits of a proprietary strategy.

What are some of these costs? First, a patent right is a right to exclude others from using a technology, not a positive right to use the technology oneself. This means that in rapidly developing fields like biotechnology where there are many unexpired patents, a patent by itself is of little value; it is also necessary to obtain permission to use other related technologies, with associated transaction costs. Second, the onus is on a patent holder to detect infringement and sue for redress; but patent suits are notoriously long and expensive, and the odds are stacked in favor of repeat players.[45] Finally, individual patents are often relatively easy to invent around. This means that the effective upper limit of the licensing fee a patent owner can charge is the estimated cost to potential licensees of developing an alternative way to achieve the same technical goal.[46] These and other costs of proprietary strategies weigh heaviest on smaller players who lack the resources necessary to acquire, maintain, and strategically deploy large patent portfolios.

Thus, apart from a very few players for whom the costs of the knowledge game are outweighed by its benefits—namely, oligopolistic profits—many biotechnology industry participants would probably be better off under the kind of competitive industry structure that open source biotechnology could bring about. If that is true, what would these innovators have to lose by adopting a nonproprietary—or more specifically, an open source—strategy? To answer this question systematically, it is helpful to consider why the various institutions that make up the biotechnology, agricultural, and pharmaceutical industries seek intellectual property protection for their respective innovations.

We have seen that pharmaceutical companies employ patents on drugs for the classical purpose of excluding competitors from a product market for as long as patent protection lasts. The commercial value of pharmaceutical patents is particularly high, especially for small-molecule drugs, because (1) chemical compounds are easy to imitate (so that it is difficult for the innovator to maintain a competitive edge in the absence of patent protection); (2) establishing infringement is a relatively simple matter (so that the patent can be effectively enforced); and (3) carrying out research and development in-house is risky and expensive (so there is a strong incentive for others to free ride). To appreciate just how lucrative this proprietary exploitation strategy can be, consider that when the patent on a blockbuster (defined as a drug with annual sales over US$1 billion) expires and multiple generic competitors enter the market, the price of the drug can drop by up to 80 percent, costing the proprietary manufacturer several million U.S. dollars per *day* in lost sales in the United States alone.[47] From the early 1980s until 2003, sales on patented blockbusters made the research-based pharmaceutical industry consistently by far the most profitable in the United States; in 2002 the combined profits for the ten drug companies in the Fortune 500 list were more than the profits for all the other 490 businesses put together.[48]

Thus, it would appear that the opportunity cost to pharmaceutical manufacturers in adopting an open source strategy for exploiting drug patents—which is clearly incompatible with their current approach—is extremely high. Yet the real opportunity cost of an open source strategy in this context is substantially lower than the apparent or perceived cost. Not all drugs are blockbusters: the probability of a new drug becoming as big a seller as Glaxo Wellcome's Zantac or Eli Lilly's Prozac, both now off-patent, is objectively very low. More importantly, despite advances in high-throughput screening, genomics, and other drug discovery techniques, the productivity of pharmaceutical research is declining steadily;[49] there are very few new drugs in the pipeline ready to take the place of these and other blockbusters as patent protection expires.[50] Open source drug development offers an alternative to the industry's present solution to this problem, which—to put it crudely—involves relying on sales of "me too" drugs while hoping that universities or the biotechnology industry will generate innovative lead compounds. The returns from sales of individual drugs under an open source regime would be lower; but so would costs and risks.

In considering the opportunity cost to pharmaceutical companies of adopting an open source strategy, it is also important to distinguish between patents on drugs and patents on research tools. We have seen that most pharmaceutical companies are not interested in generating income by selling research tools invented within the company. The reason for patenting such tools is essentially defensive: keeping them secret instead of patenting them would hinder research generally, and would leave open the possibility of independent invention and patenting by a competitor, who might then use the patent to block the first inventor's freedom to operate. In contrast to the use of patents to exclude competitors from a product market, this rationale for patenting is inherently quite compatible with an open source, possibly copyleft, approach. In any case, for pharmaceutical companies the opportunity cost of abandoning proprietary exclusivity with respect to research tools is relatively low.

Interestingly, the same cannot be said for research tools owned by the other major knowledge game players that colonize biotechnology-related value chains—namely, large agricultural or life sciences companies. The life sciences industry shares a common history with established pharmaceutical companies: both are descended in part from chemical companies founded in the late nineteenth century to exploit the chemical synthesis of newly characterized organic compounds, especially those derived from the ubiquitous industrial by-product coal tar.[51] For the reasons given earlier in relation to therapeutic compounds, agricultural chemicals are particularly well suited to the classical strategy of patenting to exclude competition. But agricultural inputs firms became involved in biotechnology research and development, not so much as a means of developing new chemical compounds, but as a way of developing an entirely new type of product—genetically modified seeds—while simultaneously extending the profitability of existing chemical inputs. By genetically engineering crops to withstand high concentrations of the firm's own pesticide or herbicide products, these firms saw the possibility of persuading growers to continue paying high prices for off-patent chemicals while at the same time generating income through repeat seed sales or licensing contracts. In this business model, patents and plant variety rights on the seeds themselves are needed to prevent competitors from selling the same seed-chemical package. But proprietary exclusivity with respect to research tools is also important, because the perceived key to generating serious profits from agricultural biotechnology is the ability to genetically engineer a range of different crops to carry the same value-adding trait; hence, Monsanto's Roundup Ready crops include not just the original soy but also maize (corn), sorghum, cotton, canola, and alfalfa.[52]

A secondary use for research tool patents on the part of large agricultural life sciences firms is as a bargaining chip to gain access to complementary research tools owned by competitors. Although this secondary motivation for owning patents may be compatible with an open source strategy in other contexts (see below), there would be a substantial opportunity cost for oligopolists in making research tools available to potential new market entrants.

What about biotechnology companies? What do their motives for patenting tell us about the likely opportunity cost of an open source exploitation strategy? In this case, the answer is not as straightforward as in the case of pharmaceutical and agricultural life sciences multinationals, because the global biotechnology industry is made up of a much larger number of companies with more heterogeneous business models.

This is especially true in agricultural biotechnology. Patenting activity on the part of for-profit local and national plant-breeding and seed firms varies widely from country to country. While patents are frequently used to exclude the competition from product markets (whether for seed or for the resulting crops), they are also sometimes obtained simply in order to comply with the rules of off-shore breeding programs that require overseas recipients to obtain patent protection in the relevant country as a condition of providing breeding material. In such a case, the opportunity cost of a nonproprietary strategy is clearly minimal.

What about medical biotechnology? In the early days of the biotechnology industry, many biotechnology companies had aspirations to become fully integrated pharmaceutical companies. In this business model, the pattern of opportunity costs is similar to that of an established pharmaceutical firm, with the following qualifications. (1) Unlike a large pharmaceutical firm, such a company is likely to have only one or two lead candidates in development at any one time. (2) Because the field of promising candidates tends to be smaller in relation to biologics than classical pharmaceutical compounds, the biotechnology company will probably be pursuing those leads in competition with others. (3) The smaller company may have no existing products, or indeed any assets at all, apart from its patent rights.[53] Thus, unlike the established pharmaceutical firm, such a company has all its eggs in one technological basket. This lack of strategic flexibility is evident in the intensity of patent conflicts in biotechnology, where patent litigation has been much fiercer than it is even among pharmaceutical firms. The reason is that for the smaller firms, patent ownership is a matter of life and death.[54]

A biotechnology company whose business is built around a single proprietary platform technology is in a similarly vulnerable position with respect to research tool patents. In this case, the typical business model is to develop a new technology platform and then form an exclusive partnership with a larger firm that has the resources to use the platform in its drug development program. An alternative and equally desirable goal is to be acquired outright by the larger firm. In either case, proprietary exclusivity is key. This is the business model that pharmaceutical companies and nonprofit research institutions sought to undermine by forming the SNP Consortium, and which would be threatened by open source research tool development.

In fact, neither of these business models has remained in favor up to the present. Only a handful of biotechnology companies managed to make a success of the fully integrated pharmaceutical company model; more recently, platform companies floundered as the first wave of platform technologies was assimilated by larger firms, leaving those companies that were yet to generate a return on investment at the mercy of highly unfavorable capital markets following the bursting of the dot-com bubble at the end of the 1990s and the general bear market of 2001 and 2002.[55] Some former platform companies have since turned to complementary markets and mixed-revenue business models in which the opportunity cost of an open source strategy in terms of loss of licensing revenue is much reduced; as we saw earlier, at least one agricultural biotechnology firm has made a modest success of exploiting a patented platform technology using an open-source-like nonproprietary strategy. In that case, the opportunity cost of not pursuing a proprietary strategy was zero because the opportunity itself did not exist.

In fact, the key opportunity cost for most biotechnology companies in pursuing such a strategy would not be a loss of revenue but a potential loss of the ability to attract investment capital. I address this issue in the next section. In the meantime, there is an even more crucial point to be made in relation to the opportunity cost of any new biotechnology business model: despite attracting thousands of start-ups, raising over US$300 billion in capital and achieving annual revenues in the tens of billions, the biotechnology industry is not and never has been profitable. Only a tiny fraction of individual biotechnology companies have ever generated positive cash flows; the sector as a whole has lost money. Further, there is no hard evidence that biotechnology has revolutionized the productivity of pharmaceutical research and development.[56] Industry participants continue to insist that profits and vastly improved productivity are just around the corner, but informed observers believe that the real problem lies in the way the industry manages and rewards risk, integrates skills and capabilities across a range of functions and disciplines, and goes about advancing critical knowledge at both organizational and industry levels.[57] These are weaknesses that would be directly addressed by an open source approach; in any case, the point here is that, from the perspective of the biotechnology industry as a whole, the opportunity cost of pursuing a different strategy is actually negative!

While the primary purpose of the present discussion is to make a case for commercial involvement in open source biotechnology research, many of the same strategic considerations arise for public-sector and private nonprofit institutions that are expected to be in some degree financially self-sufficient. It is therefore helpful to also consider the reasons why these institutions increasingly seek patent protection for their inventions.

We saw earlier that many universities, government agencies, and other nonprofit research institutions now assert ownership of employee inventions. Unable to exploit these inventions by incorporating them into products for sale, these institutions seek to license the patents to commercial companies for further development in exchange for up-front fees and ongoing royalties. While some such patents have proved exceptionally lucrative, most have failed to generate any revenue at all, at the same time contributing to problems of access to and freedom to operate with fundamental research tools. Thus, following passage of the Bayh-Dole Act in 1980, patenting by universities and public hospitals in the area of human biology increased manyfold, but only a small handful of these patents has ever generated any serious revenue.[58] Today, most university technology transfer offices around the world are not self-sufficient, let alone profitable; some cannot even meet the costs of maintaining the patents they already own. To many, the possibility that the next patent might turn out to bring in the kind of revenue generated by Cohen and Boyer's recombinant DNA technology has an almost irresistible allure.[59] Yet once again, the perceived opportunity cost of an open source approach is much higher than the actual cost.

Another reason sometimes given for university patenting is that it can contribute to regional economic development through spin-offs and science parks.[60] In Chapter 6, I noted that the most successful knowledge production networks in biotechnology exist in close geographical proximity to major universities. But empirical research suggests that the reason why universities have a positive influence on the development of regional biotechnology "clusters" has little to do with proprietary technology transfer and much to do with traditional academic free revealing. Information stickiness means that even when an innovation is freely revealed, diffusion tends to be more effective within a tighter-knit or geographically localized group of organizations, commercial and noncommercial. Thus, local linkages represent relatively transparent "channels" for information transfer, as distinct from the closed "conduits" that connect geographically distant industry participants to the main nodes.[61] This explains why universities play such an important role in the success of regions such as the Boston metropolitan area or the Bay Area around San Francisco, where large numbers of biotechnology firms have their headquarters: when the dominant nodes in an innovation network are committed to open information regimes, the entire structure is characterized by less tightly monitored ties, resulting in a local "atmosphere of innovation" that benefits all nearby organizations, even those that are not party to specific transactions.[62] Thus, there need not be any significant opportunity cost to regional economies in encouraging their education and research institutions to pursue nonproprietary exploitation strategies.

A third reason for university patenting is to promote technology transfer to industry so that publicly funded research is actually used to create useful products. This is also the main reason given for patenting and licensing by government agencies, especially in the United States. Readers will recall that technology transfer in the public interest was the rationale underpinning the Bayh-Dole Act. But the proprietary approach to technology transfer facilitated and encouraged by this legislation has arguably been deleterious to innovation overall. In a critical analysis marking twenty-five years' experience with the Bayh-Dole legislation, Sara Boettiger and Alan Bennett characterize Bayh-Dole as a "large scale experiment in how public institutions manage public assets as private goods," and argue that it has created a misalignment between the private interests of university technology transfer offices and public interests that benefit the innovation system at large and enable access to intellectual property rights for humanitarian purposes. Boettiger and Bennett call for a realignment in which public-sector institutions pay greater attention to the innovation system overall, in particular by increasing access to patented technologies for further research, granting broad access to research tools, strengthening the collaborative environment within the public sector (with a corresponding amelioration of the anticommons effect), and finding ways to manage intellectual property so as to explicitly support humanitarian applications of new technologies.[63] Thus, there are problems with the current exploitation strategy that could be addressed by an open source approach: again, the opportunity cost to universities of such an approach may be negative.

Other public-sector institutions have reasons for patenting besides licensing. In agricultural biotechnology, different kinds of national agricultural research systems (NARS) institutions take different approaches. EMBRAPA, the Brazilian equivalent of the U.S. Department of Agriculture, has the largest patent portfolio in Latin America; elsewhere in the same region, government agricultural research institutions do not patent at all, instead taking the view that government scientists are public servants and that putting a patent on their discoveries would be akin to taxing citizens twice for the same public good. In Chile and New Zealand, both economies being heavily reliant on agricultural exports, patents are used for much the same purpose as they are by commercial firms: to protect markets for agricultural products. Thus, the opportunity cost for public-sector agricultural institutions at the national level in pursuing an open source approach would vary from one country to another.

At the international level, the CGIAR centers patent partly for similar reasons to universities and some biotechnology companies—that is, to earn licensing revenue and to attract research funding from commercial firms in exchange for first options on any promising new technology. However, these centers also patent for defensive reasons, to ensure continuing access and freedom to operate for themselves and their constituencies with respect to their own technologies. (This public-sector patent rationale is not limited to agriculture: recall that one reason for the NIH's decision to file patents on ESTs was to preempt a patent "land-grab.") Here we have an incentive to patent that is perfectly consistent with open source: an open source license would achieve the desired protection without compromising these institutions' public-interest mission. Interestingly, in countries where there is currently a low number of patent applications but there is international pressure to move toward patenting, the vision of a legally protected technology commons is reportedly proving quite attractive.[64]

Finally, open source is also compatible with another major motivation for patenting by public-sector and private nonprofit institutions: to be able to use patent rights as a bargaining chip to obtain access to other people's intellectual property rights. We saw earlier that this is one strategy that these institutions share with large private-sector agricultural firms. However, unlike oligopolists, public-sector and nonprofit innovators would not necessarily have a great deal to lose in this respect by making their patent rights available on open source terms; in fact, they would have something to gain. The reason is that, as we saw in Chapter 5, open source biotechnology licenses might contain not only a standard "yank" clause terminating the license if the licensor is charged with patent infringement, but also a mutual defense clause that allows open source licensors to leverage each other's intellectual property assets for the sake of access to third-party property.

What conclusions should we draw from this overview of patent exploitation strategies in biotechnology and related industries? The key point is that different motivations to patent have different implications for the opportunity cost of adopting a nonproprietary exploitation strategy. Some existing exploitation strategies are clearly incompatible with an open source approach. But others are remarkably closely aligned with open source, implying little or no opportunity cost in adopting a fully open source strategy. Importantly, even in cases where there *appears* to be a substantial opportunity cost associated with pursuing an open source strategy, appearances can be deceptive. In the pharmaceutical industry, for example, proprietary exploitation of drug patents has sometimes been exceedingly lucrative; but blockbusters are statistical outliers, and there is no reliable prospect of replicating their extraordinary economic performance with respect to any given new product. Similarly, while some performance indicators might suggest that the proprietary exploitation strategies adopted by many small biotechnology firms have been successful, in fact these strategies have so far produced, on average, very substantial losses.

So far our discussion has concentrated on the opportunity costs of implementing a nonproprietary business strategy; let's now consider the actual or real costs. One such cost, or set of costs, relates to the dissemination of the freely revealed technology. To the extent that free revealing is adopted as a deliberate strategy and not merely a default, its success depends on ensuring uptake by users. (Of course, this is true for any form of commercial exploitation, proprietary or nonproprietary, that does not limit the technology in question to internal use.) To promote widespread adoption, innovators may choose to subsidize the acquisition, evaluation, and use of their freely revealed technologies and to engage in various forms of marketing.

A concrete example of this type of cost in the open source software context is the fact that open source licensors often invest considerable resources in preparing code for release. Commercial firms consciously factor these costs into their strategic calculations:

> There [are] . . . a lot of hidden costs associated with properly handing off code publicly, so you really have to weigh the benefit of making it available for free, versus the costs. . . . There is a lot of code

scrubbing that has to be done to be sure that it is suitable for public consumption. . . . You don't want to [just] throw your garbage in the street.[65]

In the case of software source code, typical tasks carried out in preparation for release include quality checks, checks for third-party intellectual property rights, and ensuring there are no inappropriate comments embedded in the code (such as foul language or defamatory remarks).[66] In the case of biological materials, we saw in Chapter 5 that even the transfer of a simple DNA construct requires substantial documentation; in more complex cases, considerable experimentation may be required to find the best method of transfer.

Another set of costs in any business strategy that relies on external contributions to ongoing technology development relates to the establishment of a project infrastructure. For example, the success of such a strategy depends on ensuring that new contributions are integrated with the existing technology within a reasonable time frame. This is important because, for many contributors, the point of contributing will be to have access to a leading-edge technology, as well as to see that their contributions have had some impact. Thus, the software motto "Release early and release often" applies in biotechnology too, with adjustments for different expectations as to what constitutes a reasonable turnaround.[67] To support the rapid integration of new contributions and to reinforce contributors' incentives more generally, it will probably be necessary to incur some costs in order to (1) define the technical parameters within which development is intended to take place; (2) make it easy for users to interact via the Internet or other low-cost rapid communication methods; (3) design and publish community Web pages of a quality that satisfies the immediate needs of casual visitors so that they will return to the site and, ideally, become new users or contributors; (4) offer a range of tangible or intangible rewards to sustain community participation, including explicit recognition of useful contributions; and (5) establish some mechanism for resolving disputes.[68]

Yet another class of costs associated with implementing a nonproprietary business strategy relates to commercial risks. By highlighting the risks associated with open source business models, I do not mean to suggest that they are either more or less serious than those associated with proprietary business models. All business strategies have risks; most businesses fail, whether they adopt a proprietary or a nonproprietary strategy. With forewarning, most risks can be effectively managed. However, any steps taken in mitigation ought to be regarded as costs of implementing the relevant strategy.

What kinds of risks are specifically associated with open source business strategies? One class of risks relates to the residual uncertainty of bazaar transactions. We saw in Chapter 4 that the role of any governance structure—market, firm, network, or bazaar—is to reduce transactional uncertainty through both control and incentive mechanisms. Bazaar governance is weak in both dimensions compared with other governance structures; as a result, say Demil and Lecocq, the would-be initiator or sponsor of an open source project confronts the following uncertainties: First, he or she does not initially know whether the technology will generate any interest at all on the part of potential users. Second, even if it does generate interest, there is likely to be a delay of unpredictable length between the release of the technology under an open source license and the point at which interest reaches a critical level for further development. Third, the sponsor does not know what percentage of users, if any, will choose to contribute to further development. Fourth, the nature of any contributions cannot be foretold in advance; a given contribution may or may not turn out to be relevant to the mainstream product.[69] Finally, there remains uncertainty as to the quality of contributions: in an open source project, "quality may vary from excellent to terrible."[70]

In the face of such uncertainty, why pursue an open source business strategy? First, as noted above, all business strategies have risks. The risk to a firm of dedicating all its resources to in-house research and development in an environment of great technological uncertainty is potentially much greater than the risk of a particular open source project failing. Indeed, one of the great attractions of open source development from a business perspective is that it requires minimal up-front investment and low commitment. With open source, there is no need for a firm to put all its technology eggs in one basket.

Second, as we saw in Chapter 4, all instances of the uncertainty of bazaar transactions just listed relate directly to features of bazaar governance that contribute to overall low transaction and production costs.

Third, the uncertainty of bazaar governance is mitigated in real-life open source production systems by the additional presence of firm, market, network, and collective-action–style governance. Thus, an open source sponsor would not normally simply wait for adopters to respond to the release of a new technology; instead, he or she would engage in active recruitment and marketing and use his or her position in any hierarchy to encourage adoption and contributions. In the course of a single conversation at a conference dinner table, I have watched the friend who first introduced me to open source—an academic physicist—use all of these methods. Announcing that he had just posted a new project on SourceForge, he extolled the advantages of his own program over commercially available software tools (marketing), invited a former colleague to contribute in exchange for contributions to the colleague's own project (networking), pointed out to several junior academics that getting the tool into a useful state would give them "legend status" with their supervisors and fellow students (collective-action–style recruitment), and ordered his own Ph.D. student to show the new volunteers how to get started (managerial direction). Equivalent actions might be taken in a commercial setting, though at a cost to the sponsor firm.

A second type of commercial risk associated with adopting an open source strategy relates to the fact that customer and investor perceptions of a firm's commercial offerings and future prospects may be adversely affected by the unconventional nature of the firm's business model. In part, this is merely a matter of unfamiliarity. Thus, in the case of Diversity Arrays Technology, the company's executives decided at an early stage not to describe their strategy as "open source":

> If you say "We are working with an open source model," people will say, "So what is that exactly?" [We prefer to] educate as we go.[71]

Unfamiliarity is not, however, the only concern. One reason most smaller biotechnology firms remain committed to proprietary exclusivity is that large pharmaceutical and agribusiness corporations often insist on maintaining proprietary control over inputs. Because these corporations constitute the final link in the value chain from which the revenue of smaller companies and other institutions is ultimately derived, upstream innovators who cannot guarantee an exclusive intellectual property pedigree are in danger of finding themselves without a market for their innovations. We have seen that large corporations do not always insist on exclusivity, and that in any case there is scope for nonproprietary biotechnology business strategies that do not rely on the existence of an "innovation market" (an expression that implies a proprietary manufacturing strategy). Nevertheless, the proprietary culture of larger firms does represent an obstacle to be overcome or worked around by smaller open source suppliers.

Similarly, most small biotechnology firms are keenly aware that large companies prefer to rely on other large companies for their technological infrastructure because they are suspicious that smaller firms may prove unstable and therefore unreliable. This perception may be exacerbated by the belief that an open source business is inextricably connected with an amorphous open source community—a belief I once heard expressed as a fear that the supplier would be "controlled by the Borg." Of course, the whole point of an open source license is to guarantee users and distributors of open source technologies freedom from any such external control over their business decisions. This risk can therefore be managed by educating customers about the commitment to technology freedom embodied in open source licenses. Again, however, this represents a cost to the open source business—although much of the relevant education and advocacy work has already been done in response to the same type of problem arising in the software context.[72]

The point of this example is that in commerce, appearances matter even when they are deceptive. Appearances are particularly important in the biotechnology context because the majority of biotechnology firms are dependent on capital markets, striving to meet milestones set by investors in order to qualify for further rounds of capitalization. In the absence of profits, patent positions and barriers to entry are important elements of the conventional story biotechnology firms use to attract investors. We have seen that it is possible to construct a different narrative around the benefits of open source; but such a narrative may not be well received—at least at first—by the majority of venture capitalists and others who traditionally fund private-sector biotechnology research and development.

This is not to say that venture capital funding will never be available for open source biotechnology businesses. Open source software ventures have attracted such support in the past, and as we saw in Chapter 1, the stock market has smiled upon the initial public offerings of several open source companies. But these events occurred at a time when capital markets were buoyant; as with any other business model, open source faces added challenges when the overall economic climate is unfavorable. For open source biotechnology, however, it is more important to explore other possible sources of funding. This issue is addressed in the next section.

How strongly should the risk of adverse customer and investor perceptions weigh against open source in the trade-off between proprietary and nonproprietary exploitation strategies? In answering this question, would-be open source entrepreneurs should bear in mind the *positive* associations of the open source "brand." Much of the initial hard work of defining and establishing this brand—including the work of dispelling negative misconceptions as to the stability and profitability of open source businesses—has already been done in the software context. Arguably, the value of the open source brand in biotechnology is already evident in a noncommercial setting, where nonprofit institutions have been successful in obtaining grants on the strength of their commitment to an explicitly "open source" approach (see Chapter 8). As noted in Chapter 6, in the absence of any recognized accreditation for open source biotechnology licenses, there is a danger of opportunistic exploitation and subsequent damage to the brand. Nevertheless, with careful management, the open source brand could be extended into biotechnology and its commercial value preserved and enhanced.

One way to enhance open source brand value in the biotechnology context would be to build on synergies with both the environmental and the fair-trade movements, as well as with the open source software movement and other "commons" or "public domain" efforts. "Green branding" has been shown to boost sales and enhance corporate reputations across a range of industries; Fair Trade–labeled products command premium prices in commodity markets, and businesses with an exclusively Fair Trade–labeled product line are highly profitable.[73] As I explain in the final chapter, resonances among networks dedicated to a variety of social and environmental goals could play an important role in "scaling up" the biobazaar.

Indirect Beneficiaries as Potential Sources of R&D Funding

Earlier I noted that one potential obstacle to implementing open source as a commercial strategy in biotechnology is the difficulty of attracting investment capital from traditional sources. In particular, there is a need to identify how open source biotechnology businesses, and research and development generally, could be supported in the absence of venture capital funding (which firms pursuing an unconventional business model may find difficult to attract) or funding obtained through partnerships with large firms (which might seek to impose a requirement of proprietary exclusivity).

What are some concrete examples of possible sources of indirect—that is, financial or infrastructural rather than technical—contributions to open source biotechnology? To make the argument as robust as possible, let us consider examples from the industry context in which proprietary thinking and business practices are strongest: pharmaceutical research and development. A similar argument could be made in the agricultural setting.

We have seen that the pharmaceutical industry is made up of a variety of institutional types. Besides research-based pharmaceutical companies, the industry includes biotechnology companies, contract research organizations, universities, and government and nonprofit research institutes. All of these play a role in the discovery and development of innovative drugs. But this industry also includes a range of commercial participants who, although not directly engaged in research and development, would nevertheless benefit from increased innovative activity and productivity in the sector—especially if the outcome were nonproprietary. Generic manufacturers, imitators, and parallel importers all fall into this category.[74]

Generic manufacturers are companies that specialize in manufacturing and selling compounds for which patent protection has expired. These companies benefit from selling drugs originally developed under the auspices of research-based (proprietary) pharmaceutical companies in that they are able to avoid some of the costs of obtaining regulatory approval to market a drug that has already been approved; one reason a generic industry has not flourished with respect to drugs developed using biotechnology methods is that it is rather more difficult to establish equivalence for the purpose of obtaining fast-tracked marketing approval. Yet generic manufacturers also suffer from their dependence on proprietary pharmaceutical research and development. Intense competition between generic companies to be the first on the market with the generic copy after the patent on an existing drug expires leads these companies to undertake activities that the patent holder considers infringement, resulting in expensive legal conflicts. Such conflicts also arise as a result of proprietary companies attempting to extend their monopoly through the use of "life-cycle management" or "evergreening" patents covering new forms, formulations, uses, and other subsidiary aspects of the patented drug. If generic drug manufacturers had access to a nonproprietary pharmaceutical pipeline developed using open source methods, they would be in a position to sell innovative new drugs while avoiding these costs. This is a solid commercial reason for such companies to offer financial support to open source drug developers. If such a company were also to involve itself directly in the open source production process, this would give it a first mover advantage over other generic manufacturers when the time comes to bring the drug to market. Even in a conventional research and development setting, it is recognized that a considerable portion of research management consists in choosing wisely among projects and project ideas; involvement in a range of open source projects would enable a generic manufacturer to choose wisely where to allocate its resources, while the informal nature of open source governance would permit it to maintain flexibility regarding its commitment to any one project.

As for the question of regulatory costs, while an open source drug would have the disadvantage (surely a net commercial advantage) of being new to the market and therefore more expensive to get approved, it should be noted that patents are not the only source of exclusivity in this context. In the United States, a period of effective marketing exclusivity based on the protection of clinical trial data begins at the time a drug is approved by the FDA, even if there is no patent in effect, and lasts three to five years, depending on the nature of the drug.[75] Thus, even in the absence of patent exclusivity, there would be a significant incentive for generic manufacturers to incur the expense of having a drug approved for sale.

Imitator companies are another class of drug manufacturers whose business model is nonproprietary. Often disparagingly referred to as "pirates," although their activities are not illegal, they produce patented drugs during the term of the patent, avoiding infringement by carrying out their manufacturing and sales activities in countries where there is no effective patent protection for pharmaceuticals. The number of locations where these companies can legally operate is now reduced as a result of the TRIPS agreement; clearly, it would be to their advantage to support open source drug development as an alternative source of drugs that they could legally manufacture and distribute. Similarly, parallel importers—traders who exploit the significant differences in drug prices from one country to another by buying where prices are low and selling where they are high—are under siege from proprietary companies seeking to maintain price differentials and could benefit by supporting nonproprietary development.

I have spoken of these as distinct business models, but patent expert Philip Grubb points out that in fact they are strategies that are often combined within the one business model.[76] The fact that many proprietary pharmaceutical companies are also active in the generics business, generics companies sometimes act as imitators, and both generic and imitator companies act as parallel importers means that all these companies have some leeway to shift the emphasis of their activities under the selective pressure of emerging open source competition, as described in Chapter 8.

Another group of commercial entities that are not directly engaged in pharmaceutical research and development but could benefit, at least in theory, by supporting open source production consists of pharmacy benefit management companies or health maintenance organizations (HMOs).[77] These are for-profit companies that administer pharmaceutical benefits schemes for employers, unions, health insurers, and government agencies; in the United States they purchase drugs on behalf of about 200 million Americans. Being bulk purchasers, these companies are able to negotiate with drug companies and retailers for the best prices. While there have been some accusations in recent years of collusion with drug companies, such companies, if properly constituted to prevent conflicts of interest, would have an interest in maintaining competition among manufacturers and keeping drug prices low—hence, in supporting open source development. Outside the United States, the same purchasing role is played by nonprofit government agencies that might be prepared to support open source development as a means to cheaper nonproprietary drugs.

So far we have considered for-profit corporations as potential sources of investment in open source innovation. But governments have always been a major source of funding for biomedical, agricultural, and other biotechnology-related research; extending government funding farther downstream is one way of addressing research and development productivity shortfalls in the biotechnology industry. It is important to realize that it is not necessary, in order to establish the feasibility of the biobazaar, to show that it would be entirely independent of public subsidies. Biotechnology and related industries are already heavily subsidized: in the pharmaceutical context, the public picks up a large proportion of the tab for new drugs, paying several times over through funding of basic research, tax subsidies for research and development, and monopoly prices at the point of sale (whether over the counter at the pharmacy or through private or public health insurance).[78] Thus, part of the answer to the question of where the money would come from for open source drug development is, "The same place it already comes from"—public coffers. Money spent in support of an open source alternative to the present system would be money better spent, because a greater percentage would go to technology development as distinct from profits and marketing costs, and the process itself would be cheaper and more efficient.[79]

Another important potential source of nonprofit funding for open source biotechnology is venture philanthropies. These are mostly privately funded, not-for-profit entities that focus on advancing treatments for specific diseases. They include the Bill & Melinda Gates Foundation (for research on AIDS and infectious diseases in developing countries), the Michael J. Fox Foundation for Parkinson's Research, the Multiple Myeloma Research Foundation, and the Prostate Cancer Foundation. These organizations operate in a manner similar to for-profit venture capital funds, but they have longer time horizons, and their goal is to make a therapeutic difference, not a profit.[80]

If all of the entities described in this section are plausible sources of investment in open source biotechnology, why have they not been tapped in the past? The answer is that historically, the expertise needed to develop new drugs was found exclusively in pharmaceutical companies. Just over a century ago, at the dawn of the modern era of drug research and development, no other institutions had the capacity to support the necessary applied interdisciplinary research and development.[81] But at the start of the twenty-first century, the pharmaceutical industry is no longer a monoculture: drug research and development is now the concern of a variety of institutions, not all of them committed to proprietary exploitation strategies. In particular, the emergence of a competitive contract research industry—spanning the whole length of the value chain from target identification through to new molecule optimization and development—means that there is now no doubt that new drugs could be developed on a nonproprietary basis outside the established pharmaceutical companies.[82]

The fact that traditional pharmaceutical companies are no longer the only entities capable of bringing new drugs to market is illustrated by the emergence of a new type of entity, the "virtual" pharmaceutical company. A virtual pharmaceutical company is essentially a project management firm that selects promising drug candidates, awards contracts to corporations to develop them further, and generally supervises the outsourcing of all aspects of late-stage drug development.

The major recognized advantage of a virtual or fully outsourced approach to product development is that it permits companies and institutions with different strengths to work together in a way that minimizes risk and enhances the efficient allocation of resources.[83] It has been argued that this approach is particularly suited to pharmaceutical research and development for three reasons: (1) a wide variety of scientific disciplines and skills are needed to advance a project from "bench to bedside"; (2) specialized service providers offer expertise in all these areas; and (3) pharmaceutical research and development is essentially about the production of knowledge, whether in the form of patent specifications or in the form of an information package for submission to regulatory authorities.[84]

In all these respects, the advantages of the virtual approach are the advantages of a bazaar approach. In fact, as pharmaceutical industry insider Bernard Munos points out, many of the practical problems that would be raised by open source drug development are already being successfully addressed through public-private partnerships (PPPs) such as the Medicines for Malaria Venture (MMV), established in 1999 to discover and develop new and affordable antimalarials (www.mmv.org).[85] Munos explains that such ventures operate as virtual pharmaceutical companies, receiving project proposals from anyone who cares to make a submission, as in an open source software development project. In the MMV example, proposals are screened by an expert committee, and those that are selected for funding are coordinated by members of staff acting as project managers. The actual research and development is outsourced to a range of institutions, including universities, large pharmaceutical companies, biotechnology firms, and research institutes, and paid for using funds donated by public and private sponsors. At each phase of development (target validation, identification and optimization of potential drug candidates, preclinical and clinical development), the expert committee reviews the data and makes a decision whether or not the project should proceed.[86]

To Munos, the success of PPPs confirms that open source as a development methodology for new drugs is inherently feasible.[87] Interestingly, the key difficulty in implementing the virtual pharma model—perhaps the reason it has not been widely replicated outside the nonprofit arena—is the problem of coordinating diverse contracts among multiple research partners in different disciplines. As entrepreneurial scientist David Cavalla explains, changes in direction and iterative methods of working are essential to innovation, and conventional contractual relationships are not well suited to managing such an unpredictable set of circumstances.[88] This is exactly the type of situation in which the strengths of bazaar governance, reinforced by open source licenses as an alternative to conventional partnership agreements, may outweigh the strengths of what is essentially a network approach.

This completes our examination of the feasibility of the biobazaar. The goal of this chapter has been to demonstrate that there exist incentives and opportunities for commercial involvement in open source biotechnology research and development, that the costs of adopting an open source strategy do not necessarily outweigh the benefits, and that there exist a range of possible sources of funding in addition to traditional sources.

The potential to involve commercial as well as noncommercial actors is a key feature of the open source biobazaar. This is because the principal advantage of open source over the prevailing system of knowledge production in biotechnology lies in its ability to tap into multiple sources of innovation without surrendering to proprietary restrictions on technology freedom.

It is true that in biotechnology the knowledge game, with its emphasis on private ownership of the tools of knowledge production, is heavily entrenched. Whether that could ever change is the subject of the next chapter.

8

Biotechnology's Open Source Revolution

Chapters 6 and 7 demonstrated that the raw materials necessary for the evolution of open source biotechnology are already present in biotechnology and related industries. But it might be objected that this analysis is unrealistic because it fails to take into account the proprietary predisposition of many key industry participants.

In arguing for the feasibility of commercial participation in open source biotechnology research and development, I have no wish to understate the difficulties of implementing an open source business strategy. Nor do I seek to minimize incumbents' likely resistance to the emergence of a new, nonproprietary mode of production. There is an old joke about a motorist who becomes lost in the countryside and stops by the side of the road to ask a local for directions back to the city. The local pauses, scratches his head and then drawls, "Well . . . if I were you, I wouldn't start from here."

The purpose of this chapter is to demonstrate that it *is* possible to get to the open source biobazaar from here, despite the current prevalence of proprietary business models.

The first step in this argument is to address the question of industry culture. Is it so strongly antithetical to open source as to constitute an insuperable barrier to the involvement of commercial actors in the biobazaar?

The second step is to identify some current initiatives that may turn out to be catalysts in the shift to a new industry equilibrium—

an equilibrium in which nonproprietary exploitation strategies are more widely adopted by both commercial and noncommercial players.

The final step is to offer some reflections as to how such initiatives might enroll people and resources to achieve the scale and momentum necessary to effect real social and economic change.

Industry Culture and the Proprietary Versus Nonproprietary Trade-Off

In Chapter 7, a commercial firm's choice of business strategy was treated as if it were the outcome of a conscious calculation in which the benefits of a nonproprietary approach are traded off against actual and opportunity costs. This approach could be justified solely on heuristic grounds, as a way to illuminate the likely commercial advantages and disadvantages of open source biotechnology. But the concept of a trade-off is also central to open source software business leaders' descriptions of their own business plans; it is something they suggest other business people should actually *do*.

Thus, the message that consistently emerges from formal and informal discussions of open source business strategies in the software context is that deciding whether to implement an open source strategy requires a careful assessment of one's overall business plan. Such an assessment, it is suggested, should take into account all revenue-generating opportunities, proprietary and nonproprietary, in order to determine where a technology's true value lies. In this calculation, every technology or intellectual asset that does not contribute significantly to the firm's exclusionary strategy is available for instigating external innovation from which it can derive ancillary benefits. The calculation may be quite subtle; for example, it may involve weighing up the strengths of different intellectual property assets, such as patents versus trademarks.[1]

Skeptical readers might feel that importing this idea of a strategic trade-off into our discussion of the commercial feasibility of the biobazaar introduces an air of artificiality. This is partly because, in reality, strategic thinking is constrained by industry culture, cognitive bias, and a host of other nonrational influences. For example, ticket holders in a lottery typically overestimate their chances of hitting the jackpot; similarly, biotechnology innovators are likely to overestimate the probability of winning at the knowledge game. Any strategic trade-off might therefore be expected to be weighted in favor of a proprietary approach.

A more fundamental objection, however, is that the overwhelming majority of business decisions are taken without reference to *any* conscious calculation, biased or otherwise. On this view, it is naïve to advocate a new strategy in the expectation that business leaders will go to the trouble of (1) identifying the relevant opportunities and (2) implementing a policy to take advantage of them; it expects too much of a system that is actually somewhat mindless. Thus, it has been observed that while many major corporations in biotechnology and related industries have an intellectual property *budget,* most have no intellectual property *strategy:* corporate intellectual property policy is effectively delegated to functionaries who are paid simply to minimize risk and acquire as many patents as possible.[2]

In fact, this tendency toward corporate conservatism is not unique to biotechnology and related industries. Nor is it limited to large corporations. Most business decision-makers would prefer to turn a blind eye to strategic issues—especially if they are fraught with technical, legal, and commercial uncertainty, as questions about the future of a new technology must inevitably be. Many managers try to avoid the pain of deciding which technologies are differentiators and which are enablers in the context of their own business models by thinking about the question in very black and white terms:

> You either think a technology is so completely commodity and boring . . . that you are just going to go and buy it and never even think about looking at it: it is just some cheap thing and you don't give a

> damn. At the other end of the polarity you say, "This [technology] is totally strategic to our business, and it is so strategic that we have to develop it ourselves."[3]

Formulating an open source strategy requires rather more finesse because it entails finding the middle ground between these two extremes: it is about taking something that is freely available (a "commodity") and building value into it to make it more profitable in the context of a particular business model.

In consequence, even if an open source strategy promises to improve a company's bottom line, it may not be adopted unless and until there is no other option. This has certainly been the case with respect to open source software:

> Just the fact that Linux is ten times cheaper than proprietary UNIX—you'd think that in a competitive world that would be enough. But it's not. . . . "What about risks," and "I really like how my vendor takes care of me" and a whole bunch of other reasons mean that a *ten to one* difference in cost does not guarantee success in the marketplace. . . . In Amazon's case, one day the person . . . in charge of the technology platform was looking at the revenue growth curve that the CEO was trying to shoot for . . . and he noted that there would be a point where the IT costs were going to cross the revenue curve. He went to the CEO's office and said, "Where these lines cross is where the critics are right: Amazon really will never make a profit. But if I move to Linux, these lines will never cross." Jeff Bezos said, "OK, go do it." [The point is that] many people don't operate based on what is optimal. They operate based on whether this will kill me to do it or whether it will kill me not to do it.[4]

If corporate intellectual property and technology policy are effectively knee-jerk reactions dictated by industry and organizational culture—that is, if "culture eats strategy every day for lunch"—then the default for many commercial players in biotechnology and related industries will be the proprietary approach described in Chapter 3. Pharmaceutical companies, for example, traditionally "live, breathe and die on patent positions"; not sharing intellectual property is "deeply ingrained."[5] For established firms, this proprietary culture is reinforced by organizational inertia: the design and staffing of any institution's research and development activities reflect implicit biases about the sources of innovation, so any strategy that involves externalizing some aspect of technology development that has hitherto been conducted in-house may require significant organizational changes.[6]

It might seem, then, that to suggest the possibility of an open source revolution in biotechnology is to ignore the realities of capitalism. Yet business models do change; industries do evolve over time. In his book *The Wealth of Networks,* Yochai Benkler explains that one reason proprietary exploitation strategies have come to dominate information production systems is that the availability of intellectual property protection exerts a kind of selective pressure on the ecology of strategies that information producers use to capture a return on their investment.[7] Clearly, a proprietary approach is most likely to be profitable if strong protection is available. In biotechnology, strong intellectual property rights increase the potential gains to a proprietary strategy, thereby increasing the opportunity cost of a nonproprietary approach. As we saw in Chapter 3, the structural power of large corporations—players of the knowledge game—reinforces this effect by "ratcheting up" intellectual property protection, further intensifying the pressure toward proprietary business practices.[8]

The case of open source software shows that the proprietary emphasis of business models within an industry can also evolve in the opposite direction. In software the odds were stacked against the widespread adoption of open source as a business strategy, just as they are in biotechnology. Yet, as we saw in Chapter 1, open source software has become a mainstream commercial phenomenon: a stock market success, a major earner for a number of large corporations, and a serious competitive threat to proprietary incumbents.

In fact, evolutionary economists have long recognized that the history of capitalism is one of radical and sometimes revolutionary reconfigurations of markets, in which the creation of new business models repeatedly overturns accepted patterns and principles.[9] Disruptive innovation occurs not only at the level of goods and services, but by the introduction of new methods of production and the creation of new organization systems within an industry. These new combinations have structural implications; they destabilize the commercial environment, rendering apparently entrenched ideas, technologies, skills, and infrastructure obsolete.

Further, the mechanism by which markets are transformed involves the agency of entrepreneurs: individuals and corporations whose decision-making "horizons" differ from those of their contemporaries within a particular industry. The theory of entrepreneurial innovation suggests that it does not matter whether *most* participants in biotechnology and related industries are prepared to challenge the core proprietary assumption that any uncompensated spillover of knowledge generated through private investment must necessarily reduce the innovator's profit from that investment. What matters is that *some* may shed their proprietary blinkers and attempt to breathe life into a new business model. It is perfectly realistic to speculate that, if they succeed, the structure of the market within the relevant technological niche could be quite drastically altered.

Importantly, given the strength of proprietary culture among commercial players in biotechnology and related industries, such entrepreneurs need not themselves be profit seekers. In principle, any industry participant with the capacity to innovate could trigger a shift in the ecology of exploitation strategies in a given sector toward a new, less proprietary equilibrium. As we shall see in the next section, most existing open source entrepreneurs in the biotechnology context are nonprofit entities of one kind or another.

A further subtlety is that nonproprietary and proprietary exploitation strategies are not necessarily mutually exclusive. We saw in Chapter 4 that an open source strategy may be employed within an overall business model that also employs proprietary strategies. It is also possible to apply different approaches with respect to the same technology—either simultaneously, as in the case of dual licensing, or in sequence, as in the case of a technology that is first kept as a trade secret, then patented, and then licensed on open source terms. If a sequential strategy is adopted, the critical question is when to let go of proprietary exclusivity. In fact, this is a question that must be answered in relation to any patented technology, irrespective of the type of exploitation strategy, because patent protection requires ongoing maintenance throughout the patent term. Many owners of patented research tools allow protection to lapse over the economic life of the tool; in such a case, open source licensing could be seen as an alternative to the complete abandonment of intellectual property rights.

What might the chain of events look like that would lead to an open source revolution in biotechnology? We saw in Chapter 4 that free-revealing or nonproprietary exploitation strategies are specifically designed to enhance use value—for example, by making the tool more user-friendly, technically superior, cheaper, and/or easier to obtain. If an open source technology has greater intrinsic value to users than competing proprietary technologies, it may begin to attract them away from established tools. Because open source technologies depend for their evolution on peer-review-type mechanisms, their appeal to users tends to snowball once user numbers reach a critical mass. In the absence of either strong network effects in favor of the established technology or high fixed costs that make it expensive for users to switch technologies midstream (neither of which would exist if the technology happened to be one that addresses a hitherto unserved user need), substantial market success for the nonproprietary technology is a genuine possibility.

At this stage the only real threat to the continued viability of an open source technology would be a strong competitive reaction on the part of industry incumbents. However, in this regard, open source technologies have a competitive advantage not shared by new proprietary technologies: they are resistant to many of the countermeasures industry incumbents normally employ against new entrants. As software developers Petr Hrebek and Tim Boudreau point out, an open source project has no stock, no owners, no board of directors; it cannot be bought out; it cannot be undercut in a price war because its products are already available for free.[10] More broadly, in the knowledge game, oligopolists are accustomed to dealing with competitors by lobbying for rules and regulations that rig the game in their favor. While this approach has been somewhat effective against open source in the software context,[11] ultimately its usefulness is limited.

The reason is that, through clever adaptation of conventional intellectual property licensing and the development of new business models, open source production has managed to integrate itself into mainstream legal and economic institutions. Not only does this offer a way to preserve nonproprietary, peer-based knowledge production alongside proprietary research and development without requiring massive injections of public funding, it also means that the same rules that strengthen proprietary control also tend to strengthen open source. Open source licenses—in particular, copyleft licenses—are like switches inserted into an electrical circuit to reverse the direction of the current: they convert strong intellectual property rights into strong protection for a growing technological commons. By adapting institutions and rules designed to support proprietary strategies to serve its own ends, open source achieves some degree of built-in protection from knowledge game backlash.

The upshot is that once a viable open source technology appears in the marketplace, players following a proprietary strategy have a limited range of possible responses.

The first is to attempt to compete on quality. If proprietary competitors can maintain a perception among users that their technology is intrinsically better than the nonproprietary version, they may succeed in slowing or blocking the adoption of that version.[12] One way to do this is to invest aggressively in new research and development in order to stay ahead in the quality stakes. Obviously, this outcome is beneficial for technology users, although from the perspective of the proprietary competitor it may be difficult to sustain because of the lower production costs and greater efficiency of bazaar-style production.

Given that what matters is users' *perception* of quality rather than actual quality, another possible strategy for incumbents is to engage in a marketing war, including spreading "fear, uncertainty, and doubt" (FUD) about the open source technology. This approach has been adopted with some success by proprietary technology owners in the software context.[13] The lesson for proponents of open source development is that they need to be prepared to develop effective marketing strategies, an issue to which we turn later in this chapter.

A third response to the emergence of an open source competitor is to follow the old adage, "If you can't beat 'em, join 'em." For knowledge game runners-up, given a choice between a proprietary monopoly held by a rival company and the level playing field of the bazaar, open source may appear as the lesser of two evils. If the shift toward open source production in a sector comes to seem inevitable, in principle it would be to such a company's advantage to lead the change and gain a first-mover advantage rather than to become a reluctant follower. Of course, established players would prefer not to change strategy if they can avoid it. However, given sufficient pressure from an open source competitor, they may choose to compromise by adopting a "hybrid" strategy.

Earlier in this section I highlighted the possibility of combining proprietary and nonproprietary strategies in a single business model or strategic plan. Here I am referring to something different—namely, relaxing one or more of the key principles of open source licensing to produce a true proprietary/nonproprietary cross. A discussion of hybrid exploitation strategies is beyond the scope of this book; the point here is simply that such strategies might be employed as a transitional step away from the knowledge game and toward a more competitive industry structure.[14]

Ultimately, economic modeling of the contest between proprietary and nonproprietary technologies in the software context—specifically, Microsoft Windows versus Linux—suggests that if market forces are permitted to determine the final balance of technologies in a given market niche, the most likely outcome is a mixed ecology in which neither achieves complete dominance.[15]

This result reflects assumptions about the costs of switching from an established technology that are not necessarily transferable from one technology setting to another. However, what counts in establishing the feasibility of the open source biobazaar is the commercial *viability* of an open source alternative, not its overall market share. Provided user (or more precisely, developer) numbers are sufficient to sustain bazaar production, continuing adherence to proprietary business models on the part of some industry participants need not exclude open source development of appropriate technologies for small or nonexistent markets. Thus, a recent article by Bernard Munos, an employee of the large pharmaceutical firm Eli Lilly, explicitly advocates the adoption of an open source approach to drug development as a solution to the problem of dwindling research and development productivity in that industry; Munos argues that pharmaceutical companies "stand to gain from co-opting the open-source model and allowing it to flourish in 'co-opetition' with traditional R&D, to handle the diseases or R&D steps for which it is best suited."[16]

It might be supposed from some of the examples given previously that only the most basic ("fundamental," "upstream") biotechnologies are eligible candidates for open source development. Given that much upstream research takes place in a noncommercial setting, this would suggest a rather limited application for open source biotechnology as a commercial strategy. But this supposition is inaccurate.

It *is* true that one of the situations in which a nonproprietary approach is especially likely to be appropriate is when the innovation in question is a fundamental enabling technology or a scientific resource that represents a nonsubstitutable standard. Proponents of open source software sometimes use a dendritic metaphor to explain the distinction between "enabling" and "differentiating" programs. In this metaphor, the Linux kernel might be the roots and trunk of the tree; software libraries and the Apache server software might form the branches; and applications that are not essential and not depended on by any other software are the leaf nodes.[17] Another way to express the same idea is to describe technologies as forming a "stack," in which a discrete bottom layer forms a platform for the next layer up, and so on.[18] The general principle is that the leaves on the tree (or the top layer of the stack) can be proprietary, but the trunk (or bottom layers) should be nonproprietary.

This generalization makes sense from the perspective of encouraging ongoing innovation in the field as a whole. But from the perspective of an individual innovator considering whether or not to freely reveal his or her innovation under an open source license, the analogies above are problematic: in practice, it is often impossible to draw an objective distinction between basic and applied technologies.[19] In agricultural biotechnology, for example, the same material—germplasm—represents simultaneously the most basic of research tools and the end point of the value chain. Even where the technology in question is not a living thing, one firm's differentiator may be another's enabler: for example, a dedicated biotechnology firm's "product" is a pharmaceutical company's "research tool."

Thus, in biotechnology, you don't have a tree: you have a shrub.[20] You don't have a stack: you have soup. (Actually, the same is sometimes true in software, where vendors often have trouble deciding whether theirs is a "niche" or a "platform" product.[21]) This diversity of commercial perspectives on any given technology means that even downstream technologies may be eligible for open source development *by some subset of industry participants.* As noted in Chapter 7, commercial players have an incentive to support open source development of any technology that lies upstream of their own place in the relevant value chain. What I am now emphasizing is that "upstream" is a relative, not an absolute, term.

What about technologies that lie at the very end of the value chain? Even here, end consumers who themselves have no capacity for biotechnology research and development may nevertheless have an interest in supporting open source, because (1) open source development is inherently cheaper than proprietary development, (2) it ensures competition among producers, and (3) it permits producers to tailor end products to consumer needs. Thus, while large pharmaceutical companies may not have much interest in open source drugs (as distinct from research tools), *purchasers* of pharmaceutical products certainly would. Similarly, while large agribusiness firms may have little interest in open source seeds, farmer collectives would. As we saw in the last chapter, the capacity for end consumers and others to support open source development *indirectly,* as a source of capital rather than through research and development activities, turns out to be an important part of the answer to the question of where the money will come from to sustain open source biotechnology.

Catalyzing Open Source Biotechnology

The foregoing discussion is not, of course, intended to suggest that the biobazaar will inevitably become a commercial phenomenon on the software scale. Rather, it describes a "happy path" (see Chapter 2) whereby the plausibility of an open source technology achieving market success—as many have done in the software context—is readily apparent. However, as this chain of events begins with the choice of a nonproprietary exploitation strategy on the part of either a commercial or noncommercial innovator, we must now ask whether there are any signs that this has begun to happen.

One of the earliest detailed proposals for adapting open source software licensing to biotechnology was the "General Public License for Plant Germplasm" (GPLPG). The proposal was first published on the Internet in 1999 by Tom Michaels, then a professor of plant agriculture at the Ontario Agricultural College, University of Guelph.[22] As the name suggests, the proposal was modeled after the software General Public License, or GPL. Michaels's idea was to have a means of disseminating plant germplasm that could coexist with other release protocols, like seed certification systems, and with property rights including plant variety protection and plant patents.

Michaels explains his motivation as follows.[23] Plant germplasm is the indispensable raw material for a plant-breeding program. Breeders have traditionally had relatively unrestricted access to plant germplasm, and have used selective breeding to iteratively improve plant varieties over many generations. Over the past century, attempts to protect owners of particular varieties and compensate developing nations for their contributions introduced some bureaucratic restrictions, but in general these did not constitute a serious impediment to ongoing development of new varieties. The advent of genetically modified organisms (GMOs) dramatically altered this situation. Gene constructs introduced into crop plants, coding for traits such as insect and herbicide resistance, came with patents that made these plants off-limits to other breeders, even if the breeder had no interest in the transgene itself. Breeders developing GMOs could access the corn and soybean germplasm developed through publicly funded plant breeding at universities, government institutes, and international research centers, insert a patented gene they had developed, and restrict others from accessing the resulting germplasm. Over time this process had the potential to severely restrict the basic materials for plant breeding. Michaels's intention in proposing a GPL equivalent for plant germplasm was to keep these materials broadly accessible.

Structurally, the proposal was designed to be a straightforward adaptation of the software GPL. According to Michaels, this goal ought to be achievable with respect to plant germplasm because the technology in question is more or less entirely embodied in the germplasm itself. This means that the distribution of the software GPL with the source code to which it applies can in principle be mimicked directly with a materials transfer agreement (MTA). In this proposal, the equivalent of a software "derivative work" was simply defined as any new germplasm descended from germplasm distributed under the GPLPG. Derivatives could be produced by standard cross-breeding, natural variation, or any form of genetic modification.

The MTA contained four main requirements. First, anyone could release germplasm under the terms of proposed license, unless the material was already encumbered. Second, a sample of all germplasm released under the GPLPG had to be deposited in an existing publicly accessible collection of plant gene resources. The license explicitly allowed the owner to charge for the physical act of reproducing and distributing the germplasm, and also to charge a royalty to a third party for the right to reproduce and distribute the unmodified germplasm. Third, the license allowed any germplasm descended from a GPLPG strain to be released and distributed freely, provided that the resulting material was also designated as GPLPG and full pedigree information accompanied the product. Finally, it was prohibited to release and distribute novel germplasm based wholly or in part on GPLPG plant germplasm except as expressly provided under the license. Inability to comply with that license, which could occur if material was included in a new plant germplasm that had a license that made redistribution under the GPLPG illegal, would mean that the derivative work in question could not be released at all.

This arrangement is a classic copyleft scheme designed to ensure that released material, including improvements, will always remain accessible. In principle, says Michaels, such a scheme could have been applied to patented germplasm, although the proposal itself was not accompanied by a patent license or any other license draft. Michaels felt it was important to ensure that the particulars of the MTA did not interfere with existing plant breeders' rights and that it should be clearly apparent to any prospective user how the terms interacted with plant patents. It also had to work easily with the entrenched processes for transferring plant materials so that adopting it was not too inconvenient to breeders.[24]

Michaels's proposal was presented to the Bean Improvement Cooperative in early 1999 and the Expert Committee on Grain Breeding in Ottawa a year later. Michaels brought it to the attention of his peers, and received positive feedback from them, but did not aggressively advertise the idea beyond his personal contacts. Some of the spirit of the GPLPG has been adopted by the International Treaty on Plant Genetic Resources for Food and Agriculture, within the provisions establishing a multilateral system of access and benefit-sharing.[25] The policy for the multilateral system states that it is possible to develop and sell commercial products based on materials in the system without compulsory sharing of profits with the developers of those materials, *provided* the resulting product is made available without restriction to others for further research and breeding.[26] This provision is highly reminiscent of the GPLPG.

At the time of this writing (January 2007), despite the positive feedback that Michaels's idea has generated within the immediate research community, no plant germplasm has yet been released under the terms of the GPLPG. This experience parallels that of the vast majority of open source software projects, which garner little community involvement unless and until they gain a "champion" who is willing and able to advocate for the project, as distinct from making a purely technical contribution. As we saw in previous chapters, even if a project is deemed highly worthwhile by the community, there is rarely much action until it attracts a critical mass of active user-developers. With respect to the GPLPG, and open source biotechnology initiatives more generally, there is scope for institutional actors such as the Consultative Group on International Agricultural Research or other private or public nonprofit bodies to provide this leadership.

Another open source biotechnology proposal dating from 1999 is more widely known, though it was never actually implemented. The context of this proposal was the public-private race to complete the draft sequence of the human genome, first mentioned in Chapter 2. Public-sector researchers in the genome race saw themselves as handicapped by their commitment to make their own data publicly available, because it gave their private-sector competitors an opportunity to free ride on their efforts. Specifically, there was a concern that Celera and others would mix their own data with the published sequence information and then appropriate the combined result. Tim Hubbard, head of the Sanger Institute's data analysis group, was reportedly surfing the Internet one night when it occurred to him that the software GPL could be adapted as a data license that would prevent this.[27] Over the next month or so, a draft license was produced. Ultimately, however, it was decided that any kind of constraint on data use was too restrictive to be acceptable to the wider research community. As noted in Chapter 5, in that case open source was deemed not to be open enough.[28]

Nevertheless, a similar approach *was* subsequently adopted by the International HapMap Project, a private-public collaboration established to create a haplotype map of the human genome. (A haplotype is a set of closely linked alleles—for example, genes or DNA polymorphisms, such as SNPs—that tend to be inherited together as a unit.) For a time, researchers had to agree to a set of copyleft-style click-wrap conditions in order to gain access to haplotype mapping information. These conditions were consciously modeled on the GPL, and that influence was acknowledged on the website setting out the project's data access policy.[29] The purpose was to discourage database users from filing patent applications that would block other users' access to the data.

The HapMap click-wrap policy was abandoned in December 2004, ostensibly because it had fulfilled its purpose. However, a number of problems with the arrangement have since been widely recognized and discussed. The most serious related to the ownership of the data on which the copyleft-style obligation was predicated; while software source code is undoubtedly eligible for copyright protection, provided that it satisfies the relevant legislative requirements, the HapMap claim that haplotype mapping information was protected by copyright was on shakier legal ground.[30] In the absence of intellectual property rights, only those who had agreed to the contractual terms of the click-wrap license were obliged to observe restrictions on patenting. In an attempt to close this loophole, the license incorporated provisions seeking to prevent users of the data from releasing it to anyone not bound by the same license terms. In consequence, publications based on the data could not include the data itself and therefore could not be adequately peer-reviewed.[31]

The HapMap license was further compromised by the clumsiness of its attempts to reconcile the goal of preserving public access to and use of HapMap data with the perceived need to permit users to file for patent protection on specific haplotypes that might matter for future product development.[32] In other words, the design of the copyleft "hook" in this case was not entirely successful either from a drafting perspective—the relevant provisions were somewhat ambiguous—or from the perspective of satisfying prospective users. There is an obvious tension between the incentives of public-sector data users, who ultimately want their additions to the database to result in low-cost, innovative therapies, and those of commercial players acting in pursuit of proprietary business models. Private concerns have lobbied to avoid this issue in other community resource projects by softening the language considerably, using phrasings that encourage users to share results (and by implication, profits) rather than making that process a legal requirement.[33]

Despite criticism, the existence of the HapMap policy demonstrates that both public- and private-sector actors—the HapMap project being a public-private collaboration along the lines of the SNP Consortium—are willing to borrow the idea of an open source license in order to compel users of a given technology to release improvements and other downstream developments without legal encumbrance. Along with that encouraging trend come two important lessons.

First, in order to facilitate genuine open-source-style collaboration and remove encumbrances to ongoing research, it is important to have intellectual property protection. Simply releasing the data into the public domain allows it to be rendered inaccessible by future patents, and trying to restrict dissemination using a confidentiality agreement causes difficulties and resentment. Once intellectual property rights are secure, then the choice is available to implement either a fully controllable, or a guaranteed open and accessible, data stream.

Second, ambiguity in the goals of a policy will result in a technology being licensed in a way that is attractive neither to industry players focused on generating proprietary exclusivity nor to those who would prefer to build a free market for the technology. Hybrid approaches are clearly possible for certain technologies where there are clear boundaries between the open-sourced technology and the products developed from that technology, but in general, and for any single element in the technology chain, the two options most likely to be exploited are a fully proprietary model or a fully open model. Trying to hold on to proprietary exclusivity with one hand while reaching for unencumbered research and development with the other will often prove unworkable.

A third biotechnology initiative consciously modeled on the open source approach is the Tropical Diseases Initiative (TDI), an open source drug discovery scheme proposed by lawyers Stephen Maurer and Arti Rai and computational biologist Andrej Sali.[34] A common objection to the feasibility of open source biotechnology, particularly in the biomedical context, is that the returns from open source business models would be insufficient to get new therapeutic products over the hurdles of regulatory approval and into circulation. In the conventional model of drug development, that burden is carried by large pharmaceutical corporations, who are supposed to channel the profits from patented blockbusters back into the more expensive aspects of research and development. The problem the TDI is intended to solve is that, as we saw in Chapters 2 and 3, this model does not work well for drugs that are never going to have a large market either because the diseases they treat do not affect many people or because the people who are affected are too poor to pay for treatment.

In the TDI scheme, identification of drug targets and candidates is coordinated by means of the bazaar, with research projects being designed to take maximum advantage of computational methods in order to minimize the need for more expensive wet lab experimentation. Later stages of drug development are coordinated by virtual pharmaceutical companies—that is, nonprofit venture capital organizations that select promising drug candidates, award contracts to corporations to develop them further, and generally supervise the outsourcing of all aspects of late-stage drug development. As proposed, TDI is not linked with a specific open source license; the intention is for researchers who make use of this approach to develop their own licenses.

Supporters of the TDI proposal argue that an open source approach to early-stage drug development would reduce total costs in three ways. First, it would draw on highly trained volunteer labor, employing nonmonetary rewards of the kind described earlier in this chapter and employing spare capacity in established laboratories in a manner reminiscent of the SETI@home project's employment of underused and highly distributed computing power to search for signs of extraterrestrial life.[35] Second, sponsors could avoid overpaying research and development costs. As noted above, these are difficult to estimate; the longer sponsors can delay having to put a price on research and development, the more accurate the estimate will be. Third, because the results of the discovery effort would be available on open source terms, any company could manufacture the drug. Virtual pharmaceutical companies would award contracts to the lowest bidders, and in the absence of a patent monopoly, price competition would keep the cost of finished products to a minimum.[36]

Although TDI has few publicized active projects outside its partnership with the Synaptic Leap (discussed below), community reactions to this initiative have been favorable. Importantly, as the name indicates, the TDI proposal focuses on finding cures for tropical diseases, but there is nothing in the logic of the proposal that would disqualify open source drug development for diseases that are prevalent in rich countries. The efficiency of the distributed target search process and the availability of virtual pharmaceutical companies as integrators are not affected by the "profitability" of a disease; Chapter 7 details a range of ways in which later stages of the research and development process could be financed. Naturally, there is industry resistance to the idea of extending open source drug development outside the realm of tropical disease. However, the ecology of the pharmaceutical industry is now sufficiently complex, and the process of drug development sufficiently modular, to evolve toward genuine competition.[37]

As noted above, at present the most visibly active of the TDI's operations are those coordinated by the Synaptic Leap (TSL), a nonprofit organization launched in 2005 with the goal of helping biomedical scientists to collaborate openly on developing treatments for neglected diseases.[38] The premise of this initiative is that before publication in a peer-reviewed journal, scientific collaboration tends to be confined to personal networks—that is, limited to interactions among colleagues within the same laboratory or among those who have established trust-based relationships through university, conference connections, or previous work experience. TSL's goal is to extend the range of possible collaborations via the Internet, permitting self-selection of contributors to collaborative efforts based on an open information and communications infrastructure. At the same time, TSL aims to extend the scope of collaboration to include results and observations—"data" in the broadest sense—that would not be considered reliable or significant enough for publication through traditional channels (recall Hilgartner and Brandt-Rauf's spectrum of data reliability from Chapter 2).

Almost from its inception, TSL has been closely affiliated with the Tropical Diseases Initiative as both partner and service provider. TDI members are active as TSL board members and volunteers, providing leadership on scientific and governance matters, while TSL's focus is on building and maintaining a website and tools that can be used to enhance collaborative research.[39] This symbiosis provides an opportunity for iterative learning-by-doing on both sides: TSL's involvement with the TDI provides a focus for its initial efforts and supplies a broad vision as to how scientific discoveries resulting from the collaborative process it aims to foster may be translated into new disease treatments, while TDI's efforts are facilitated by tools that TSL develops in response to specific needs.

TSL consists of a network of online research communities, each of which is dedicated to a specific disease. This disease focus contrasts with other efforts at online scientific collaboration using blogging (TSL's tool of choice for collaborations in which ideas are posted and owned by individual users). These tend to be organized along disciplinary lines; for example, the Useful Chemistry Blog at Blogspot specializes in open chemistry experiments with a focus on malaria and other tropical diseases. The goal-oriented organization of TSL's research communities is intended to complement such efforts by bringing together ideas and resources across disciplinary boundaries.

TSL's research communities are led by volunteers. To date, all are academic scientists with recognized expertise in a relevant field of research.[40] The responsibilities of a community leader include writing an introduction page for the research community that will inspire others to pitch in; identifying online news sources, tools, and resource links relating to the relevant disease; "evangelizing" to recruit new participants; monitoring site content and helping connect resources to needs; and providing constructive criticism on TSL processes and tools.[41]

What does TSL look like from the perspective of a research scientist wishing to extend his or her work on a particular disease? On login, the scientist is offered an introduction to the community and its goals, together with a list of menu items including current projects (a hierarchy of project books and pages allowing users to organize, attach files, and discuss a given project), research tools (links to relevant Web resources), RSS news feeds, community posts, and a gallery to which he or she may upload images to illustrate a post. According to the most optimistic vision of what might happen[42] next, the scientist chooses to post a blog describing a new project idea. He or she may then obtain feedback, relevant citations, or even volunteer resources from other community members via comments posted to the original blog. If the project looks promising, the scientist may apply for a research grant, either alone or together with those whose comments have helped to strengthen the proposal, highlighting the proven benefits of open collaboration in this specific context. If the grant is awarded, the core research team begins collaborating via the Web, posting laboratory notes and other materials not normally made public at this stage of the process. Other scientists are able to keep track of the process and offer their own ideas; additional resources may be forthcoming as the project evolves.

If and when the project reaches a significant milestone, its leader may decide to seek publication in a peer-reviewed journal. The paper is co-authored by the project leader and any significant participants and peer-reviewed in the traditional manner. If the milestone amounts to the discovery of a new drug candidate, the researchers can recommend it to a virtual pharmaceutical company for clinical trials, as in the TDI model articulated by Maurer, Rai, and Sali. The virtual pharma or another funding agency may then award the necessary grants to outsource clinical trials to a contract research organization—and ultimately, if all goes well, to outsource the manufacturing of a new pharmaceutical product.

Founded by Ginger Taylor, a commercial software industry veteran who saw a new application for her experience managing corporate intranet portals, TSL is still in the start-up phase—in other words, the foregoing description is of an untested "happy path." The initiative deployed its first pilot research community, for malaria, in mid-November 2005. In 2006 it added communities focused on schistosomiasis and tuberculosis; as of the end of 2006, efforts were under way to establish a Chagas community. A number of other, mostly tropical, diseases have also been identified as potential foci based on high DALY (disability adjusted life years) burdens combined with low profitability. Current plans include the development of a "Gene Wiki" pilot for the parasite *Plasmodium falciparum,* which accounts for nearly 90 percent of malaria deaths.

With one of three community leaders based outside the United States, TSL has already gone some way toward establishing a global presence. However, its active membership is still very small, and its viability as a collaborative platform, let alone as a means of filling virtual pharma's product pipeline, remains to be proven.

By concentrating on these issues, TSL is bringing scientific exchange into the Internet era, thereby addressing one of the three major differences between the traditional and the open source versions of the biobazaar canvassed in Chapters 6 and 7. Like the TDI, though, TSL does not attempt to incorporate the intellectual property licensing aspects of the open source model. These initiatives are focusing on problems with large social and low monetary value; in such a setting, it is reasonable to expect that openness may be sustained by goodwill rather than legal protection. Current participants have, in Taylor's words, decided to "make it public domain and hope for the best."[43] Large pharmaceutical companies have shown interest in these initiatives precisely because basic science in the public domain is a useful resource. The idea of intellectual property protection that guarantees public accessibility throughout the development chain may be less attractive to these players.

A less focused but far more ambitious project than TSL is the Science Commons, an offshoot of the hugely successful Creative Commons initiative.[44] Conceptually, it targets all the same problems at a multidisciplinary level. Current projects include Scholar's Copyright, which promotes the freedom to archive and reuse scholarly works on the Internet; a Biological Materials Transfer project, which aims to develop and deploy standard, modular contracts to lower the costs of transferring physical biological materials; and the Neurocommons, an open-content, open-software knowledge management platform for biological research.

A further initiative known as the BioBricks Foundation (BBF) seeks to coordinate the production of standardized synthetic biological systems.[45] Overall, Synthetic Biology aims to produce modular elements that can be assembled into functional biological systems. Such an ambitious program requires enormous collaborative efforts, and in an attempt to ensure this work continues without restriction, the BBF is searching for ways to license this technology in an open source or open access fashion, currently in coordination with the Science Commons.[46] The president of BBF, Drew Endy, has also been affiliated with the Molecular Sciences Institute (MSI) in Berkeley, another group committed to releasing their technology in a publicly accessible fashion, and who are also investigating open source biotechnology licenses. Roger Brent and Rob Carlson from the MSI wrote about the concept in late 2000, and the MSI hosted the first-ever open source biology workshop in early 2003.[47]

Another recent attempt to emulate the principles of open source software licensing in a biotechnology setting is the BIOS (Biological Innovation for Open Society) initiative of CAMBIA (the Center for Application of Molecular Biology in International Agriculture). CAMBIA is an autonomous not-for-profit research organization located in Canberra, Australia. Founded in 1994 by a small group including CEO and chairman Dr. Richard Jefferson, it is an unusual operation in that it combines wet-lab development of biotechnology research tools with intellectual property informatics and policy development. The institute is financed by grants from philanthropic organizations such as the Rockefeller Foundation, by national and international research funding bodies, by official development assistance, and by license revenue from its own patented technologies.[48]

In 2005, CAMBIA launched its BIOS initiative, which incorporates three distinct components. The first is the "Patent Lens," a searchable database containing European Patent Office, USPTO, and Patent Cooperation Treaty documents, together with ancillary information and tutorials. The second is "BioForge," a portal for protocol sharing, comments on patents, and discussion tools in both public and secure environments. Modeled on SourceForge.net, BioForge is intended to evolve into a collaborative technology development platform and virtual hangout for members of the BIOS "community." The third and final aspect of the BIOS initiative is "Biological Open Source," or "BiOS" licensing. This is the element that "intended to extend the metaphor and concepts of Open Source to biotechnology and other forms of innovation in biology."[49]

At present (January 2007) there are two BiOS licenses, one for plant enabling technologies (PET) and one for genetic resources indexing technologies (GRIT). Each must be read in conjunction with its own Technology Support Services Subscription Agreement (PET TSSS and GRIT TSSS). All of these instruments are currently in version 1.3 and are available on the BIOS website at http://www.bios.net/daisy/bios/home.html. The website also includes license information at the FAQ "About BiOS Licenses," and hosts a license discussion forum.

The idea behind BiOS is to make some of CAMBIA's own intellectual property available as the seed for collaborative, Internet-enabled development of biotechnology research tools. Although the BiOS licenses explicitly invoke the language of open source, Jefferson notes that the parallels between the BiOS licenses and open source software licenses are "very parlous and we should be careful of using these as more than metaphors."[50] Both existing licenses are copyleft-style licenses (though note the caveat below) based on technology protected by patents. The seed technology from CAMBIA is made accessible for a negotiable fee, and improvements to that technology must be granted back to CAMBIA. Subject to these and other constraints, products based on this technology may be patented and commercialized.

One difficulty in translating open source software licenses into patent-based biotechnology is that an appropriate definition of "improvement" is hard to formulate. The breadth of the BiOS interpretation—and hence, the definition of technologies that must be granted back—has led, along with more technical concerns about reporting requirements, to some reluctance to adopt the license in university environments.[51] This breadth makes BiOS-licensed technology akin to GPL software, which cannot be distributed as part of a proprietary product. While such a GPL-style license enforces a very open toolset, those restrictions can prevent the technology from being used at all. This is particularly true in an environment where the ability to coexist with other proprietary licenses without overriding them (as BiOS does explicitly) is often a practical necessity. Perhaps a better starting point for an open source biotechnology license would be the Lesser GNU Public License (LGPL), which is designed specifically for that purpose. It operates by defining the technology that requires innovators to release their improvements very carefully as a modular component, thereby allowing it to be released in coexistence with proprietary material.

The BiOS licenses also contain a grant-back structure that differs from most copyleft licenses in ways that could limit their adoption. Licenses are granted from CAMBIA to a licensee, and all improvements are granted back to CAMBIA, which may then license them to other researchers. This network, with CAMBIA at the center, is enforced by restrictions on licensees sublicensing the technology. By contrast, a standard GPL- or LGPL-style license would require that any improvements be licensed on the same terms as the original license, without giving any party a privileged position in the distribution. For reasons explained in Chapter 5, the traditional copyleft obligation is not to the licensor per se—though he or she may be responsible for enforcing it—but to all potential users of the follow-on innovation.

Another licensing proposal inspired by open source as an approach to intellectual property management, as distinct from purely a development methodology, has been put forward by postdoctoral researcher Amy Kapczynski and her colleagues at Yale University.[52] The proposal covers "Equitable Access" licensing, designed to improve access to biomedical innovations in low-income and middle-income countries, and "Neglected Disease" licensing, intended to facilitate research beneficial to people suffering from neglected diseases.

Under an Equitable Access license, a university would, for a fair royalty payment, grant to another party (such as a commercial firm) a nonexclusive license to use its patented technology to produce an item for sale in poorer countries—as well as at least some rich countries. In return, the licensee would agree to grant back to the university any improvements it might make to the technology and to cross-license any other rights the licensee holds that might be used to block production of the item in question. These rights would be extended automatically on the same terms to any third party notifying an intention to produce the item for sale in low- to middle-income countries. The license would not require licensees to sell the item in poorer countries at any particular price, but the price would be kept down by the possibility of competition from new licensees. From the licensee's perspective, the opportunity to commercialize the licensed technology in rich country markets would compensate for the other, less favorable aspects of the deal.

Under a Neglected Disease license, a university licensing out its technology would retain the right to license the technology for the purpose of research on neglected diseases anywhere and for commercial purposes in poorer countries. There are various possible formulations: the licensee might be required to grant back and cross-license its own intellectual property; the research exemption might be confined to noncommercial institutions, or to a definite list of neglected diseases, or to diseases that meet a general standard (for example, as set out in U.S. legislation defining a rare condition).

The parallels between Equitable Access and Neglected Disease licensing and open source software licensing are strong. As Kapczynski and her coauthors point out, "like the licensing practices that govern free software, [Equitable Access licensing] uses proprietary rights to secure freedom for an open class of potential users, rather than to secure exclusivity for a closed class of licensees. Like the GPL, it uses intellectual property rights not to exclude and monopolize, but rather to ensure the right of third parties to access and distribute the innovation and its derivative products."[53] The proposal also incorporates the key element of competition articulated in Chapter 5, treating "all actors symmetrically vis-à-vis the resource in question."[54] The proposed licenses could be used by participants in other initiatives, such as the TDI/TSL initiatives described earlier, that do not specify a particular licensing approach. Alternatively, they could be adopted by individual intellectual property owners seeking to bring their management policies into line with global public health goals. Although the proposal focuses on universities, there is no reason in principle why other public-spirited intellectual property owners might not choose to adopt either or both forms of technology transfer.

Many of the ideas described above are synthesized by the Network for Open Scientific Innovation (NOSI), a nonprofit corporation based in Brazil. Proposed by social entrepreneur Joseph Jackson and Brazilian lawyer Mauricio Guaragna, NOSI plans initially to introduce three programs: Open Source Biotechnology, Capacity Building in Technology Management, and Synthetic Biology. It aims to provide support for legal and scientific development and training, as well as advocate for its own programs.

The fact that NOSI would be based in Brazil is of interest in that Brazil is a developing country with sufficient publicly funded expertise in biotechnology research that it has the capacity to build a critical mass of open-sourced material. It is also, unlike the United

States, a major *importer* of intellectual property. This means that not only would the health and agriculture of Brazilians benefit from access to open-sourced biotechnology innovations, but the country could save large amounts of money in the process. In other words, what some of the actors noted above in this chapter have been motivated to do from a philanthropic viewpoint, Brazil may choose to do out of self-interest.

Brazil has a history of adopting strong positions in intellectual property, including in the biotechnology area. A high-profile example was the controversial decision to break AIDS-drug patents and manufacture a publicly funded, publicly distributed anti-HIV cocktail locally.[55] Brazil has even adopted government-wide policies to use open source software, and the current president, Luiz Inacio Lula da Silva, campaigned on that basis in the general election in 2003. In this context, there are many fewer cultural barriers to the adoption of open source biotechnology than in other research and development centers such as the United States and Europe; on the contrary, open source branding may prove very useful. With its significant capacity for low-cost production, Brazil would also be well placed to manufacture and distribute cutting-edge open source biotechnologies worldwide.[56]

The Power of Strategic Modeling

Taken together, these initiatives prove the existence of nucleation points around which an open-source-inspired biobazaar might crystallize. None constitutes a mature working example of an open source biotechnology project incorporating all three of the features highlighted at the start of Chapter 5—namely, successful collaborative technology development, open source licensing, and nonproprietary commercialization. In other words, the nucleation event has not yet occurred. Nevertheless, each of these efforts has the potential to help bring about improvements in the institutional arrangements by which citizens' needs for health care and security are met—irrespective of its own success or failure.

To see why, it is necessary to appreciate the power of strategic modeling. The term *modeling* here refers to the active diffusion of ideas in which a solution to a given problem is borrowed or translated from one context to another.[57] To model open source in biotechnology is to borrow an approach that (1) is designed to address problems strikingly similar to those outlined in the first part of this book; (2) has shown itself capable of bringing about a significant transformation in the software industry; and (3) has been actively debugged over a period of years—or decades, depending on when one chooses to pinpoint its emergence in the software context.

The obvious advantage of modeling the solution to any problem is that a solution built from scratch is likely to have defects that cannot be minimized or eliminated except through experience.[58] The obvious *dis*advantage is that new contexts breed new bugs. Furthermore, features that are bugs in the eyes of one constituency may actually be benefits in the eyes of others.[59] Thus, although modeling is usually more efficient than trying to design a completely new solution, it may entail substantial work to overcome a less-than-perfect fit to local conditions. All of the open-source-style initiatives described in this chapter are engaged in that work.

The reason this modeling effort has value beyond the realization of individual initiatives' goals is that strategic modeling is a way for groups with few material resources and little structural power to bring about social and economic transformations that would appear impossible if we were to base our predictions on a simple test of strength between competing interests. In biotechnology, the contest is between those who would benefit from greater competition and enhanced participation in research and development and those who wish to preserve or strengthen proprietary exclusivity. Structurally, the latter interest group is undoubtedly the more powerful;

but it is conceivable that clever modeling could secure a win for the weaker side. This is because its effectiveness lies in framing the terms of debate: the success of any given model depends less on the power of the promoter than on the power of the model itself. Further, because the interests of the powerful often lie in maintaining the status quo, they are vulnerable to strategic modeling by weaker groups who have less to lose in drawing out the weaknesses and contradictions inherent in prevailing models.[60]

How does modeling work in a knowledge game setting? As an example, consider the incorporation of minimum standards for protection of intellectual property rights into the international trade regime administered by the World Trade Organization (WTO). Negotiated at the Uruguay Round of the General Agreement on Tariffs and Trade (GATT) in 1994—the same round that established the WTO itself—the agreement on Trade-Related Aspects of Intellectual Property Rights (TRIPS) was signed by more than one hundred nations that were net *importers* of intellectual property rights.

As regulatory scholars Peter Drahos and John Braithwaite point out, on its face this is a puzzling outcome because stronger intellectual property protection clearly favors intellectual property *exporters*—then, as now, a tiny minority of WTO member states.[61] In this case, the explanation is partly structural: that tiny minority included the United States, a huge net exporter of intellectual property whose economic power was effectively leveraged through the threat of excluding nations that did not cooperate on introducing intellectual property rights into their domestic laws from access to U.S. markets. But the important point for the present discussion is that the structural explanation is incomplete.

The idea of expanding monopoly rights by incorporating intellectual property rules into an international trade regime designed to dismantle trade monopolies and remove barriers to competition is counterintuitive, to say the least. During the 1980s the apparent probability that such a deep institutional shift could take place was extremely low. Drahos and Braithwaite record that it was explicitly regarded as bad policy by key individuals in the GATT secretariat and the European Community, had no initial support from European and Japanese businesses (let alone their governments), and was against the interests of nearly everyone—not only including African AIDS patients or schoolchildren in developing countries who would have to pay more for overseas textbooks, but also most U.S. citizens and businesses.[62]

These initial hurdles to the adoption of TRIPS were overcome through a process of modeling. The initial idea behind the TRIPS agreement came from individuals who were not themselves powerful, except in possessing the ability to persuade powerful others to adopt their vision of the future. These individuals—"model missionaries," in Braithwaite's terminology[63]—enrolled "model mercenaries" whose economic interests would be served by the model's implementation. By gradually widening the circle of support from the CEOs of large U.S. corporations such as Pfizer and IBM to include European businesses and governments, then Japanese businesses and government, these model missionaries and mercenaries were able to "render the implausible plausible," so that maximizing intellectual property privileges became the top priority of U.S. trade policy.[64] It was not until after this point was reached that structural power was brought to bear in coercive bilateral and multilateral trade negotiations by the United States and intensive lobbying by the United States, Europe, and Japan.[65]

Thus, in the recent past and in a context that overlaps with the subject area of open source biotechnology, modeling has proved capable of bringing about sweeping changes that seemed highly unlikely in prospect. In the example just given, the modeling process was supported and ultimately driven by some of the most powerful interests on the planet. But a related case illustrates the use of modeling to promote the interests of the weak. The power of a given model often comes from its ability to sustain identities: for example, some intellectual property importing nations were sold on the TRIPS model as a way to sustain their identities as technologically advanced, knowledge-based societies in the process of becoming net intellectual property exporters.[66] In consequence, weaker interests can sometimes prevail by precipitating an identity crisis and then offering new models that help resolve it.[67] An example is the identity crisis suffered by the global pharmaceutical industry in the lead-up to the Doha Declaration on the TRIPS Agreement and Public Health, when the consequences for AIDS victims in developing countries of an "overly narrow" reading of TRIPS became devastatingly apparent. In that case, the solution was the Declaration itself, adopted by the WTO Ministerial Conference of 2001 in Doha on 14 November 2001 and expressed as reaffirming the need to interpret TRIPS in light of member states' obligations to ensure access to essential medicines for their citizens.

What lessons can be drawn from these examples to assist those who seek to promote open source as a model for research and development in biotechnology? Clearly, a key aspect of successful strategic modeling is achieving scale: widening the circle of those who are aware of the model, find it persuasive, and are prepared to support its realization in a new context. (From the perspective of an individual open source biotechnology licensor, this may also be a requirement of effective marketing.) Recent research on the successful scaling-up of the Fair Trade movement suggests that one way to do this is to link existing networks by targeting key nodes in those networks and emphasizing the resonances between network goals and those that could be achieved by implementing the model.[68]

One opportunity for scaling up open source biotechnology is to build on resonances with Fair Trade itself. More than six decades old, this movement is organized at the global level into a number of distinct networks including the International Fair Trade Association (IFAT) and Fairtrade Labelling Organizations International (FLO). These networks share two interrelated goals: first, to provide opportunities for development to small producers and poor workers in developing countries through "trade, not aid"; and second, to influence the international trading system and private companies toward practices that enable just and sustainable development.[69] A key aspect of the Fair Trade agenda is the implementation of Fair Trade principles in order to generate proof by example that it is possible to trade fairly without compromising the viability of commercial enterprises. Just as supporters of open source in biotechnology (and software) must find ways to counter the perception that voluntarily giving up proprietary exclusivity is incompatible with commercial success, the Fair Trade movement has had to demonstrate that making the commitment to pay a fair minimum price for coffee, tea, bananas, and other commodities—thereby voluntarily giving up the right to exploit small farmers and producers further up the value chain—does not automatically reduce a distributor's profits. In fact, the commercial success of unconventional Fair Trade business models has been such that the movement now faces a new challenge: as major commercial players, many with a history of engaging in unethical business practices, recognize the potential benefits of using the Fair Trade label, there is a danger that their influence may compromise the movement's credibility and commitment to its original social justice–oriented goals. This is a dilemma familiar to many in the software setting, played out in the continuing tension between advocates of free versus open source software. Assuming growth and good fortune, it may eventually become an issue in the biobazaar as well.

This is not the place to explore the many parallels between open source and Fair Trade; the point is simply that these resonances are worth exploring for the sake of mutual learning and support. On the Fair Trade side, at least, this opportunity has been recognized: as the editor of one Fair Trade journal points out, the open source approach can itself be construed as a form of Fair Trade because "it empowers workers, it empowers consumers [and] it is good for local economies."[70]

Another broad social movement with potentially very strong connections to open source biotechnology is the appropriate technology (AT) movement. Adherents believe that the failure of decades of technology transfer from industrialized countries to solve problems of poverty and hunger in the developing world suggests a need for development pathways that de-emphasize growth and technological monoculture.[71] They advocate the development and use of alternative technologies that are appropriate to local user needs. Such technologies are variously called intermediate, progressive, alternative, light-capital, labor-intensive, indigenous, appropriate, low-cost, community, soft, radical, liberatory, and convivial technologies.[72]

To appreciate the connection between open source biotechnology and the AT philosophy, consider the perspective on technological innovation articulated by Austrian philosopher and anarchist Ivan Illich in his book *Tools for Conviviality.*[73] For Illich, tools are intrinsic to social relationships: individuals relate to society through the use of tools, either by actively mastering those tools or by being passively acted upon. A tool is "convivial" to the extent that it gives each person who uses it the opportunity to enrich the environment with the fruits of his or her vision. Convivial tools facilitate autonomous and creative intercourse among people and between people and their environment; by contrast, "industrial" tools allow designers to determine the meanings and expectations of users.[74] In a technological age, rationally designed convivial tools are the basis for participatory justice—that is, for justice that consists not only in equal distribution of technological outputs (for example, material goods such as drugs or seeds) but also equal control over inputs.[75] "The principal source of injustice in our epoch," argues Illich, "is political approval for the existence of tools that by their very nature restrict to a very few the liberty to use them in an autonomous way."[76]

Consider the use of agricultural biotechnology to produce technical "locks" such as hybridization and genetic use restriction technologies ("GURTs") that render seed unsuitable for replanting or suppress the expression of introduced traits in saved seed.[77] These are only the most extreme examples: the use of genetically engineered crops that may contaminate others in the vicinity also restricts the autonomy of those who would prefer to grow traditional crops; even the development of new food crops for developing countries is often a case of tools "acting upon" the intended beneficiaries instead of empowering them to define their own productive future. An Andean potato farmer may be very poor and yet not *want* a genetically engineered potato that boosts yield so as to generate a cash crop. Such a commercial existence may threaten a way of life that the farmer values more than he or she values the ability to buy industrial goods; yet a closely related technology that improves the taste of a variety the family eats every day may be very welcome.[78]

In Chapter 6, I noted that the capital intensiveness of biotechnology research and development is sometimes perceived as an obstacle to the implementation of open source. This view is linked to assumptions about the nature of biotechnology as an essentially industrial—as distinct from convivial—technology. But molecular biotechnology and other advanced technologies need not be anticonvivial. Science can be used, not to replace human initiative with highly programmed tools, but to facilitate autonomous, decentralized production.[79] New possibilities for cognitive and material advance opened up by basic discoveries in biotechnology offer a choice: we can apply our new understanding to develop tools that would propel us into a "hyperindustrial age," or we can use it to help us develop truly "modern and yet convivial tools" that "enable the layman to shape his immediate environment."[80] Such a convivial biotechnology need not be inherently expensive, because it would consist of simple tools that work with rather than against the tendency of living things to proliferate of their own accord.

One of the major insights of constructivist science and technology studies is that no system of knowledge or belief about the natural world is built independently of the social world in which it is embedded: scientific products embody beliefs not only about how the world is, but how it *ought to be*.[81] It follows that capture of fundamental enabling tools of biotechnology by a political-industrial elite in the global North must result in a collection of technological artifacts that are designed not only to perform a particular set of tasks but also to promote particular values. By this analysis, the anticonvivial nature of most agricultural biotechnologies is no accident. Rather, it is a consequence of the political settlements embodied in biotechnology itself, both at the level of individual technologies and at the level of the innovation system. Open source biotechnology is an attempt to renegotiate those settlements, based on (1) a reframing of intellectual property as a means of facilitating, rather than hindering, the production of knowledge as a public good and (2) the gradual transformation of biotechnology research and development practices toward the production of more convivial tools.

How would open source development support the production of convivial tools in biotechnology? Illich asserts that tools foster conviviality to the extent to which they can be easily used, by anybody, as often or as seldom as desired, for the accomplishment of a purpose chosen by the user. The use of such tools by one person does not restrain another from using them equally, and they do not require previous certification of the user by any formal accreditation system.[82] These are the very properties that open source licensing seeks to confer through the guarantee of "technology freedom," described in Chapter 5. Meanwhile, bazaar governance ties the rewards for knowledge creation to the diffusion of knowledge rather than its exclusive control and restores the patterns of communication through which knowledge goods "come to life in society as public goods."[83] Open source biotechnology would give those who are excluded from the organized interests of science, state, and industry the ability not merely to question the trajectory of technology development, but to affect that trajectory directly by participating in the design of the technology itself.

In offering these two examples of social movements whose goals and strategies resonate with open source biotechnology, I do not mean to exclude other possibilities. The free and open source software movement itself is an obvious potential source of support and community for those attempting to implement open source biotechnology; so is the open hardware movement, itself inspired by the AT philosophy. Activism and scholarship directed at reforming the patent system or developing cooperative intellectual property management mechanisms such as clearinghouses and patent pools are also clearly relevant. My purpose here is not to provide a comprehensive list of potential avenues for scaling up the nascent open source biotechnology movement; it is merely to suggest that proponents of open source biotechnology should try to identify and build on connections with existing networks.

For the purpose of making such connections it is useful to articulate the open source model in terms that are sufficiently clear and definite to facilitate communication, yet open enough to accommodate a range of views. (A model that retains some degree of interpretive flexibility is more likely to be effective in bringing about institutional change, because it is able to act as a rallying point for groups whose interests are only imperfectly aligned.)[84]

This book offers such a model. The vision of open source biotechnology it presents is that of traditional bazaar-style production, modified to permit the integration of contributions from a wider range of participants. In particular, the relationship between commercial and noncommercial participants in the biobazaar would be synergistic and symbiotic, not parasitic: noncommercial contributions would provide opportunities for commercial players to enhance their profits, while commercial players would bring private resources to the production of public knowledge.

We saw in earlier chapters that contributors to an open-source-inspired biobazaar would be motivated by individual and/or organizational process benefits, user benefits, and other types of commercial opportunities driven by the enhanced use value of open source tools. Their contributions would be coordinated, not by price signals or managerial directions, but via freely accessible information about the object of production itself. The distinction between users and developers would be blurred: users would be free to become developers at any level of engagement they might choose. Similarly, participation would not need to entail a substantial long-term commitment, though some contributors might choose to make such a commitment in accordance with their own incentives. Contributions would be freely revealed with no expectation of direct reciprocity; any intellectual property would be owned by contributors, but licensed to all comers on open source terms. At the minimum, such terms would need to offer a credible commitment to provide ongoing access to the technology itself and guarantee a level playing field between licensor and licensee with respect to the privileges of intellectual property ownership. Optionally, they could also offer copyleft-style assurance of access to downstream technologies. As in open source software production, bazaar governance would predominate but could coexist with firm, market, and network structures; also as in open source software production, the mix of private incentives enforced by these structures would be supplemented by public funding and by collective-action–style incentives. Finally, the biobazaar would be strongest if there were at least a prospect of forming user communities and an effective brand community to support technology development.

In Chapter 1, I noted that there are many possible ways to map the open source landscape, depending on the purpose of the map and the disciplinary background of the cartographer. I do not claim any special authority for my own version of the "biobazaar" or "open source biotechnology": other interpretations exist, and as we have seen, they differ in important respects both from the biobazaar as it is described here and from each other. If the strategic power of modeling derives from the qualities of the model itself, not all models are equal: an effective model of open source biotechnology must be both internally consistent and practically plausible. Within those parameters, however, diversity is a good thing because it provides a broad base from which robust adaptations of open source may evolve through a process of iterative or selective learning. This process is likely to be most effective if the selection is reasoned and deliberative. By providing an analytical framework within which different implementations of open source biotechnology may be distinguished and compared, I hope to encourage such deliberation.

NOTES

REFERENCES

ACKNOWLEDGMENTS

INDEX

Notes

All URLs were current at 1 June 2007, unless otherwise stated.

1. An Irresistible Analogy

1. Crystallographer Roslyn Franklin died at 37, five years before her colleagues James Watson, Francis Crick, and Morris Wilkins received the Nobel Prize for elucidating the double helical structure of the DNA molecule. Kary B. Mullis won the Nobel Prize in Chemistry in 1993 for developing the polymerase chain reaction, a technique that allows scientists to generate large numbers of copies of a single DNA molecule.
2. Bush (1945).
3. Kuhn (1970).
4. Fox (1981); Eisenberg (1987), p. 178 n. 3.
5. Stallman (1999), n. 1.
6. XEmacs website: http://www.xemacs.org.
7. Wikipedia entry on XEmacs, http://en.wikipedia.org/wiki/XEmacs.
8. Zawinski's description of how the fork happened is at http://www.jwz.org/doc/lemacs.html.
9. The link is at http://www.jwz.org/hacks/why-cooperation-with-rms-is-impossible.mp3.
10. These lyrics are available on the Free Software Foundation's own website at http://www.gnu.org/music/free-software-song.html, together with other versions, including a death metal version.
11. A copy of this photograph is at http://www.stallman.org/saintignucius.jpg.
12. Levy (2001); Weber (2004).

13. Levy (2001), p. 424.
14. Blumenthal et al. (1986); Blumenthal (1992); Blumenthal, Campbell, et al. (1996); Blumenthal, Causino, et al. (1996); Blumenthal et al. (1997); Campbell et al. (2000); Campbell et al. (2002); Walsh and Hong (2003).
15. Levy (2001), p. 425.
16. See "The Free Software Definition," Free Software Foundation, http://www.gnu.org/philosophy/free-sw.html.
17. See Stallman (1999).
18. A copy of the GNU Manifesto is available at http://www.gnu.org/gnu/manifesto.html.
19. Free Software Foundation website, http://www.fsf.org/.
20. The reason is that software was considered to be essentially a series of mathematical algorithms or mental steps—in the language of patent law, a "discovery" rather than a patentable "invention." This position has been under siege since the 1981 case of *Diamond v. Diehr* 450 US 175, in which the United States Supreme Court ordered the Patent and Trademark Office (USPTO) to grant a patent on an invention using computer software to direct the process of curing rubber. Guidelines issued by the USPTO in 1995 and finalized in 1996 interpreted subsequent court cases as extending software patentability to programs that are essentially algorithms only distantly connected to physical processes: *Computer-Related Inventions Examination Guidelines,* 61 Fed. Reg. 7478 (28 February 1996). Software patents remain highly controversial and have become something of a thorn in the side for the free and open source software movements; but this issue is not directly relevant to open source biotechnology, as any such regime in biotechnology would need to deal with patent rights from the outset (see Chapter 5). Software can also be protected under trademark legislation. Trademarks identify the source of goods, processes, or services and may be used in conjunction with patent and/or copyright protection. For a discussion of the role of trademark protection in the open source software context, see Rosen (2005), pp. 92ff.
21. State of New York et al. v. Microsoft Corporation, No. 98-1233 (D.D.C. June 12, 2002), Direct Testimony of Bill Gates, 18 April 2002, http://www.microsoft.com/presspass.trial/mswitness/002/billgates/billgates.asp (website discontinued); cited in Fitzgerald and Bassett (2003), p. 12.

22. Von Krogh and von Hippel (2003), p. 1150.
23. Fitzgerald and Bassett (2003), p. 13.
24. Eben Moglen, personal communication, March 2003.
25. Weber (2004), p. 54.
26. Linus Torvalds, ¡torvalds@klaava.Helsinki.FI¿ Free minix-like kernel sources for 386-AT Article ¡1991Oct5.054106.4647@klaava.Helsinki.FI¿ in Usenet newsgroup comp.os.minix, 5 May 1991. Quoted in Weber (2004), p. 55.
27. Weber (2004), p. 55.
28. Raymond (2000a).
29. Ibid.
30. These statistics are based on the number of registered projects and users recorded on the SourceForge.net homepage, http://sourceforge.net as of May 2, 2007.
31. On 9 December 2006 the largest number of downloads recorded for a single project on SourceForge.net was 264,456,930.
32. Bonaccorsi and Rossi (2003); von Krogh and von Hippel (2003).
33. Von Krogh and von Hippel (2003), p. 211.
34. Weber (2004), p. 62.
35. Mulgan et al. (2005), p. 10.
36. Open Source Initiative website, http://www.opensource.org.
37. See, for example, "Open Source Case for Business," Open Source Initiative website, http://www.opensource.org/advocacy/case_for_business.php.
38. Weber (2004), p. 6.
39. V. Valloppillil, "Open Source Software: A (New?) Development Methodology," annotated version available online at http://catb.org/~esr/halloween/halloween1.html.
40. SourceForge website, http://sourceforge.net/.
41. Goldman Rohm (1999).
42. Kawamoto (1999).
43. O'Reilly (2005), at p. 463.
44. See http://news.netcraft.com/archives/2007/05/01/may_2007_web_server_survey. Apache's market share peaked in November 2005 at close to 71 percent.
45. BIND9.net website, http://www.bind9.net/bind-support.
46. McMillan (2004).
47. Weber (2004), p. 6.

48. Raymond (2000a).
49. All three essays now form part of an online collection, Raymond (2006), comprising Raymond (2000a), Raymond (2000b), Raymond (2000c), and Raymond (2000d).
50. McMillan (2002), p. 5.
51. One of the richest repositories of this fast-growing and multidisciplinary literature is the collection of online papers maintained at Free/Open Source Research Community, http://opensource .mit.edu/.
52. The terms *disruptive technology* and *disruptive innovation* were introduced by Clayton M. Christensen in Christensen (1997) and Christensen and Raynor (2003).
53. Drahos (1995); see also Drahos and Braithwaite (2002), pp. 39–60.
54. Weber (2004), p. 9.
55. Ibid., p. 5.

2. The Trouble with Intellectual Property in Biotechnology

1. O'Reilly (1999).
2. Jasanoff (2005), p. 22.
3. Popper (1947).
4. I owe this point to an anonymous reviewer.
5. Etzkowitz (1989), p. 15.
6. Jasanoff (2005), pp. 225–226.
7. "Corporate Chronology," Genentech company website, http://www .gene.com/gene/about/corporate/history/timeline/index.jsp.
8. "History of the Industry," BayBIO website, http://www.baybio.org/wt/home/Industry_Statistics.
9. P. L. 96-517. The Patent and Trademark Act of 1980 and amendments included in P. L. 98-620 (1984).
10. For a critical retrospective on the Bayh-Dole legislation, see Boettiger and Bennett (2006) and Rai and Eisenberg (2003).
11. *Diamond v. Chakrabarty* (1980) 447 US 303.
12. Ibid., at p. 309.
13. For example, in 1985 the U.S. Patent and Trademark Appeals Board awarded a patent for a type of genetically engineered corn, holding that the general availability of plant patents had not been restricted by the passage of legislation granting specific plant patent and plant variety rights protection: *Ex parte Hibberd* (1985) 227 USPQ 443. In 1987 it

confirmed that, in principle, patents could be granted on nonhuman higher animals: *Ex parte Allen* (1987) 2 USPQ (2nd ser.), 1425.

14. The patent claimed "any nonhuman mammal transgenically engineered to incorporate into its genome an oncogene tied to a specific promoter." Leder et al., "Transgenic Nonhuman Animals," United States Patent No. 4,736,866, 12 April 1988.
15. Silverman (1990), p. 162.
16. Drahos and Braithwaite (2002), pp. 155, 162.
17. Biotechnology Industry Organization (2006), p. 5.
18. The photograph can be viewed online at http://www.time.com/time/poy2000/venter.html.
19. Lemonick (2000).
20. Sulston and Ferry (2002), p. 87.
21. Watson (1968).
22. Roberts (2001), referring to *Science,* 11 Oct. 1991, p. 184.
23. Roberts (1992).
24. Jasanoff (2005), p. 216.
25. Sulston and Ferry (2002), pp. 107–109.
26. Cohen (1997).
27. Sulston and Ferry (2002), p. 111.
28. Ibid., p. 109.
29. Merck and Co. Press Release (1995).
30. Heller and Eisenberg (1998).
31. Coase (1960).
32. Sulston and Ferry (2002), pp. 199–200.
33. Walsh et al. (2003).
34. Nottenburg et al. (2002), p. 392.
35. Kryder et al. (2000), pp. vi–vii, 32–33.
36. Graff, Rausser, and Small (2003), pp. 4–5.
37. Conway and Toennissen (2003).
38. *Madey v. Duke University,* 307 F.3d 1351 (Fed. Cir. 2002) at 1362.
39. For a survey of laws on experimental use in a range of jurisdictions, see Rimmer (2005).
40. The existence of a patent in the relevant jurisdiction is not, in practice, the only constraint on freedom to innovate, or even necessarily the most important. In developing countries in which patent protection does not exist, for example, scientists may still need the cooperation of overseas patentees to obtain necessary tools and information. This is

often denied, on various grounds, including mistrust of the legal system in developing countries and risk-management problems. (I owe this point to my colleague Carolina Roa-Rodriguez.)

41. See Nicol (2005).
42. Graff, Rausser, and Small (2003), p. 351.
43. Roa-Rodriguez and Nottenburg (2003).
44. This is a common flaw in university technology transfer agreements: see generally Bennett et al. (2002).
45. Von Gavel (2001), pp. 4–6.
46. Eisenberg (2001), p. 235.
47. Long (2000), p. 834.
48. Heller and Eisenberg (1998), p. 701.
49. World Intellectual Property Organization (1992), p. 67.
50. Long (2000), pp. 828–834.
51. Ibid., p. 834.
52. Eisenberg (2001), p. 236. See also Heller and Eisenberg (1998), pp. 700–701.
53. Eisenberg (2001), pp. 239–241.
54. For an explanation, see Ellis (2000).
55. Hilgartner makes this point in the context of collaboration agreements between scientists (Hilgartner 1997, p. 5).
56. See the section "The Knowledge Game" in Chapter 3.
57. Eisenberg (2001), p. 244.
58. Nielsen (2002), pp. 12–13; Mandeville (1996), pp. 71–73; Barton (1997a); Maskus and Reichman (2004).
59. See Murray and Stern (2005); Epstein and Kuhlik (2004).
60. Walsh et al. (2003); see also Straus et al. (2002) and Nicol and Nielsen (2003).
61. See generally Nottenburg et al. (2002); regarding patent pools, see Clark et al. (2000), Gaulé (2006), Verbeure, van Zimmeren, et al. (2006), Ebersole et al. (2005), Goldstein et al. (2005), Merges (2001), Grassler and Capria (2003); regarding intellectual property clearing-houses, see Graff et al. (2001), Graff and Zilberman (2001a), Graff and Zilberman (2001b), Atkinson et al. (2003), van Zimmeren et al. (2006), Horn (2003).
62. It is worth bearing in mind that the present pattern of investment is an outcome of historical contingencies as well as perceived technological potential; under different industry conditions—such as a full-scale

open source revolution—other applications might become more prominent.

63. *Primer* (2007).
64. Spillane (1999), p. 28.
65. The description that follows draws on Grubb (2004), pp. 402–403; Angell (2005), pp. 21–36; and Drews (1999), pp. 117–156.
66. Angell (2005), p. 22.
67. Grubb (2004), p. 402.
68. Ibid.
69. It has been suggested that the majority of phase IV trials are actually a way of overcoming marketing restrictions. Doctors may prescribe an approved drug for any use or at any dose, but direct-to-consumer marketing by manufacturers is limited to uses and doses approved by the FDA. The criticism is that phase IV trials offer a pretext for companies to pay doctors to prescribe drugs to patients without necessarily meeting the scientific standards required to obtain approval in the first place. See Angell (2005), pp. 29–31.
70. See http://www.hbs.edu/mba/academics/coursecatalog/1911.html.
71. Ibid.
72. The following paragraphs summarize Fernandez-Cornejo (2004), pp. 28–29. See also Shoemaker et al. (2001), Ramaswami (2002), and Andersen and Butler (2003).
73. Fernandez-Cornejo (2004), p. 28.
74. Ibid.
75. Ibid., pp. 29, 23.
76. Ibid., pp. 28–29.
77. Eisenberg (2001), p. 225.
78. Walsh et al. (2003), Straus et al. (2002), and Nicol and Nielsen (2003), respectively.
79. Eisenberg (2001), p. 234.
80. Organisation for Economic Co-operation and Development (OECD) Workshop on Collaborative Intellectual Property Mechanisms, Washington, D.C., December 2005. On the effect of diagnostic test patents, see Verbeure, van Zimmeren, et al. (2006); and van Zimmeren et al. (2006).
81. Gorry (2005).
82. Ibid.; Grody (2005).
83. Spielman and Grebmer (2004).

84. *Madey v. Duke University,* 307 F.3d 1351 (Fed. Cir. 2002) at 1362. See also Rimmer (2005).
85. See generally Nottenburg et al. (2002).
86. Graff, Rausser, and Small (2003), p. 349.
87. Ibid.
88. B. D. Wright (1998).
89. Graff, Heiman, et al. (2003).
90. Roa-Rodriguez and Nottenburg (2003).
91. B. D. Wright (1998), p. 13.
92. Merges (2001), pp. 129–130; see also Merges (1996, 2000). On patent pools in biotechnology, see Gaulé (2006).
93. Merges (2001), p. 139.
94. O'Neill (2003), p. 22, quoting Richard Jefferson.
95. Spillane (1999), p. 28; and see Knight (2003), p. 569.

3. Intellectual Property and Innovation

1. Jasanoff (2005), p. 205.
2. Constitution of the United States, Article 1, Section 8.
3. Boehm (1967), pp. 14–26.
4. Dutton (1984), pp. 17–29.
5. Drahos and Braithwaite (2002), chaps. 6–9.
6. Locke (2007); Björkman and Hansson (2006); Cullet (2005); Drahos (1996), chap. 3. For a case on property in one's own cells, see *John Moore v. The Regents of the University of California,* Supreme Court of California, 51 Cal. 3d 120, 793 P.2d 479, 271 Cal. Rptr 146 (1990).
7. Nelson and Mazzoleni (1997).
8. On free riding in the market, see Cooter and Ulen (2000), pp. 39–44.
9. Arrow (1962).
10. Kitch (1977).
11. Jasanoff (2005), p. 203.
12. Kevles (1998), p. 69.
13. Mandeville (1996), pp. 9ff.
14. Eamon (1975), pp. 335, 338–340.
15. Mill (1879), p. 41.
16. Popper (1963).
17. Polanyi (1962), p. 55.
18. Merton (1996).

19. Raymond (2000c).
20. Merton (1996), p. 45.
21. Ibid., p. 44.
22. See, for example, Eisenberg (1987, 1989).
23. Merton (1996), p. 46.
24. Ibid., p. 40.
25. Laudan (1982), p. 266; see also Collins and Pinch (1993).
26. Laudan (1982), p. 263; Mitroff (1974).
27. Mulkay (1980).
28. See Mulkay (1976).
29. Especially Kuhn (1970).
30. Jasanoff (2005), p. 19.
31. I owe this point to an anonymous reviewer.
32. The "science wars," which began in the early 1990s, involved attacks by scientists and ex-scientists, including Paul Gross, Norman Levitt, Alan Sokal, and biologist Lewis Wolpert, on literary and sociological analyses of science, followed by retaliation from the proponents of these analyses. The most notorious event in the long-running controversy was the submission of an article by physicist Alan Sokal to the postmodernist journal *Social Text,* whose editors accepted a deliberately nonsensical and meaningless discussion of quantum physics as a genuine scholarly contribution. This hoax and the discussion it triggered became known as "the Sokal Affair."
33. Giles (2006).
34. See, for example, Collins and Pinch (1993).
35. Hilgartner and Brandt-Rauf (1994); see also Hilgartner (1997).
36. Hilgartner and Brandt-Rauf (1994), p. 360.
37. Ibid., pp. 360–361. Jordan and Lynch (1998) describe how the polymerase chain reaction (PCR) technique has been adapted to different circumstances in science, medicine, industry, and criminal forensics. Their paper explores in detail the evolution of the information status of a molecular biological technique from unreliable to standardized.
38. Hilgartner and Brandt-Rauf (1994), p. 361.
39. Ibid., p. 363.
40. Ibid., pp. 358, 363–366.
41. Ibid., pp. 366–367. Note that transferred material is considered to be bailed property.
42. Mandeville (1996), pp. 50, 52–54.

43. Ibid., pp. 52, 50–51.
44. Ibid., p. 98.
45. Hilgartner (2002).
46. Mandeville (1996), p. 96.
47. Ibid., pp. 97.
48. Polanyi (1962), p. 54.
49. Eisenberg (1989), pp. 1063–65, citing works by Robert K. Merton and Warren O. Hagstrom.
50. Benkler (2002), p. 381.
51. Ibid., pp. 373, 376; and see generally Benkler (2006).
52. Kitch (1977), pp. 288–289.
53. Boyle (2001). See also Benkler, "The Battle over the Institutional Ecology of the Digital Environment," chap. 11 in Benkler (2006).
54. Drahos and Braithwaite (2002), chaps. 3 and 10.
55. Hall and Ham (1999).
56. Macdonald (2004), citing various authors.
57. Anonymous industry informant cited in Ross Gittins, "Software's Game of Mutually Assured Damage," *Sydney Morning Herald,* 31 July 2004, available at http://www.smh.com.au/articles/2004/07/30/1091080437270.html.
58. Drahos and Braithwaite (2002), pp. 48ff. See also Arora (1997).
59. Drahos and Braithwaite (2002), p. 51 and p. 53.
60. Ibid., pp. 58–59.
61. Ibid.
62. Schumpeter (1975), at pp. 82–85.
63. Muoio (1998).
64. Hannaford (2007).
65. Drahos and Braithwaite (2002), p. 166.
66. Hannaford (2007).
67. West (2003).
68. Braithwaite and Drahos (2000).
69. Drahos and Braithwaite (2002), p. 226 n. 13.
70. Ryan (1998).
71. Matthews (2002).
72. Sell (2003).
73. Much of this discussion draws on a broad-ranging study of the global impact of intellectual property rights, particularly in the developing world, conducted by the Commission on Intellectual Property Rights (CIPR) of the United Kingdom (CIPR, 2002).

74. Lewontin (1993), p. 44.
75. CIPR (2002), p. 30.
76. UNAIDS (2006a).
77. UNAIDS (2006b), pp. 6–7.
78. CIPR (2002), p. 30.
79. Ibid., pp. 30, 31.
80. Ibid., pp. 32–33.
81. Ibid., p. 32.
82. Ibid., pp. 34ff.
83. In the long term, if lower barriers to entry were to permit the growth of a thriving local drug-manufacturing industry, that industry might be expected to contribute to the development and enhancement of distribution networks in its own commercial interests. However, this would also require a local market.
84. CIPR (2002), p. 36.
85. Ibid., p. 30.
86. Ibid., p. 38.
87. Ibid., p. 37.
88. "Last Word" (2001).
89. Pray and Naseem (2003).
90. Graff, Heiman, et al. (2003).
91. Pray and Naseem (2003).
92. CIPR (2002), p. 60; Pray and Naseem (2003), pp. 4–10; "Last Word" (2001).
93. Pray and Naseem (2003), pp. 8–9.
94. CIPR (2002), pp. 68–69.
95. Ibid.
96. See the section "The Costs of an Open Source Strategy" in Chapter 7.
97. For example, Pisano (2006).
98. GMO Communiqué, Nineteenth International Congress of Genetics, Melbourne, Australia, 6–11 July 2003, available at http://www.geneticsmedia.org/gmo_communique.htm, last accessed 18 December 2004 (page now discontinued).

4. Welcome to the Bazaar

1. Powell (2001), p. 253; see also Arora et al. (2001).
2. Benkler (2002); see Williamson (1985, 1991), quoted in Demil and Lecocq (2006).

3. Raymond (2000a), at "The Early Free Unixes."
4. Powell (2001).
5. Powell (1990), pp. 301–302.
6. Ibid., p. 303–304.
7. Ibid., p. 304.
8. Von Hippel (2002).
9. Benkler (2002).
10. See Gläser (2003, 2004); Demil and Lecocq (2006).
11. Demil and Lecocq (2006), p. 1454.
12. Gläser (2003).
13. Polanyi (1962).
14. Bonaccorsi and Rossi (2003), p. 1248.
15. Demil and Lecocq (2006), p. 1455.
16. Ghosh (1998).
17. Demil and Lecocq (2006), p. 1450.
18. Ibid., p. 1455.
19. Williamson (1999), p. 1090.
20. Demil and Lecocq (2006), p. 1452.
21. Ibid., p. 1453.
22. Williamson (1979), p. 245.
23. Von Hippel and von Krogh (2001), pp. 213, 215.
24. Benkler (2002).
25. Ibid., p. 369.
26. Brian Behlendorf, personal communication, 20 March 2003.
27. Von Hippel (2002); Lakhani and von Hippel (2003).
28. The seminal work in the user innovation literature is Eric von Hippel's book *The Sources of Innovation* (1988); see also von Hippel (2005).
29. Ghosh and Prakash (2000).
30. Von Hippel (1988), pp. 3–4.
31. Ibid., p. 3.
32. Ibid., p. 4.
33. Cohen et al. (2000); Levin et al. (1987); Mansfield et al. (1981). Interestingly, the exceptions to this rule are chemicals and pharmaceuticals; the implications of this fact for the commercial sustainability of open source biotechnology are explored in Chapters 6 and 7.
34. Von Hippel (2005), p. 77.
35. Ibid. Again, we shall see in Chapter 5 that open source licensing departs slightly from this general statement, in that the principles of open

source licensing do permit the charging of license fees *provided* that the fee structure does not compromise licensees' freedom to operate—including their freedom to fork the production process. (For an explanation of this expression, see Chapter 1.)

36. Ibid., chap. 6 (pp. 77–92).
37. Raymond (2006).
38. Meyer (2003), p. 21.
39. Ibid.; Nuvolari (2004).
40. Perens (2005), heading "Who Contributes to Open Source, and How Do They Fund That?" and accompanying table.
41. Benkler (2006), p. 46; Drahos and Braithwaite (2002), pp. 115–119.
42. Benkler (2006), p. 46.
43. Demil and Lecocq (2006), p. 1456.
44. Von Krogh et al. (2003).
45. For example, Raymond (2000c).
46. Walker (2006).
47. Benkler (2004), pp. 275–276.
48. For a discussion of formal governance structures within particular open source software projects, see Weber (2004), pp. 185–189.
49. Powell (1990), p. 304; and see generally Benkler (2006), pt. 1, "The Networked Information Economy."
50. Powell (1990), p. 304.
51. The logic is that peer production is becoming more prevalent as widespread access to digital technology lowers the costs of information production and communication (Benkler 2002, p. 444). This effect is more pronounced the more readily codified the relevant information.
52. Smith (1904), bk. 5, chap. 1, pt. 3, "Of the Expence of Public Works and Public Institutions."
53. Von Hippel and von Krogh (2001).
54. Ibid.
55. Weber (2004), pp. 175–179. Von Hippel's notion of a "private-collective" hybrid (von Hippel and von Krogh 2001) is essentially the same idea, but now we see that there can be many forms of the hybrid: different mixes of private and collective regimes and different elements of both.
56. Franke and Shah (2003).
57. Weber (2004), p. 66.
58. Muniz and O'Guinn (2001).

59. Perens (2005).
60. Rosen et al. (2003), p. 39.
61. I owe this point and its wording, plus much of the preceding paragraph, to an anonymous reviewer.

5. Open Source Licensing for Biotechnology

1. The distinction between open source licensing and straightforward free revealing is discussed later in this chapter under the heading "Open Source Biotechnology Versus Simple Free Revealing."
2. These include the workhorse BLAST library and the widely used scripting language BioPerl. Other open source bioinformatics projects are listed on the website of the Open Bioinformatics Foundation, http://www.open-bio.org/wiki/Main_Page.
3. The Open Software License, the Mozilla Public License, and the Common Public License each contains an express patent license. Other open source licenses are assumed to contain an implied license to make, use, sell, offer for sale, or import the original licensed software to the extent necessary to give effect to the explicit copyright license grant. See Rosen (2005), pp. 66–67, 189.
4. Von Gavel (2001), pp. 8ff.
5. The full list is available at http://www.opensource.org/licenses/.
6. The Open Source Initiative's official policy on license proliferation is online at http://opensource2.planetjava.org/docs/policy/licenseproliferation.php.
7. Rosen (2005), p. 3.
8. Perens (1999) offers his own account of this process.
9. Ibid.
10. Ibid.
11. See "OSD Change log," http://www.opensource.org/docs/definition.php.
12. The list is archived at http://www.crynwr.com/cgi-bin/ezmlm-cgi?3.
13. Free Software Foundation, "Various Licenses and Comments about Them," http://www.gnu.org/philosophy/license-list.html.
14. See the Debian Social Contract and Free Software Guidelines, http://www.debian.org/social_contract#guidelines.
15. The debian-legal mailing list current threads and archives can be viewed at http://lists.debian.org/debian-legal/.

16. Rosen (2005), pp. 8–11.
17. Weber (2004), p. 85.
18. One common objection to the feasibility of open source licensing with respect to patented biotechnologies springs from this misconception of open source licensing as inherently anticommercial. It has been suggested that open source patent license contracts run the risk of being declared void for public policy reasons or for frustrating the goals of the patent system (Boettiger and Burk 2004; Feldman 2004). The argument is that a critical purpose of the patent system is to provide an economic incentive or reward for innovation, and that open source patent licenses might be seen as an attempt to short-circuit that incentive. A possible (and valid) counterargument is that open source licensing promotes another stated goal of the patent system—that of moving information and technology into the public domain. But a more direct answer is that open source patent licenses *do* provide an economic incentive and reward for innovation, based on the often substantial economic gains associated with free revealing.
19. Raymond (2000d).
20. A shrink-wrap license is a product license to which users are taken to agree by the act of unwrapping the product packaging, so-called because purchased software is commonly packaged in plastic shrink wrap. By analogy, a click-wrap license is one to which the licensee assents in the course of downloading, installing, or using software. Click-wrap agreements often take the form of a screen displaying the license terms and conditions that cannot be bypassed without clicking on an "I Agree" icon.
21. World Intellectual Property Organization (1992), p. 19.
22. Organisation for Economic Co-operation and Development (2006); NIH (2005, 1999). Hence, choosing between a liberal (nonexclusive, minimal fee, minimal encumbrance) proprietary license and an open source license is essentially a matter of choosing which governance structure is most appropriate; as we saw in Chapter 4, these two structures have much in common and can beneficially coexist within the same production system.
23. Rosen et al. (2003), p. 39.
24. For an up-to-date count of registered projects, see the SourceForge website home page at http://sourceforge.net/. Note that not all registered projects are active.

25. Weber (2004), pp. 179–185.
26. North and Weingast (1989).
27. Rosen (2005), chap. 5, pp. 73–102.
28. Weber (2004), pp. 30ff.
29. Rosen (2005), pp. 8–11.
30. The full Free Software Definition is available at http://www.gnu.org/philosophy/free-sw.html.
31. Weber (2004), p. 85.
32. United States Copyright Act of 1976, 17 U.S.C., 101.
33. Rosen (2005), p. 74.
34. Regarding the cost of patent protection: it is commonly, but mistakenly, assumed that this cost will always be too high for patent ownership to be compatible with free revealing. Such an assumption is based on the faulty logic that says free revealing can never be as profitable as proprietary exclusivity. We saw in Chapter 4 that this is not the case: recall, for example, the dominance of IBM's "Linux-related services" revenues in recent years, compared with revenues derived from its enormous patent portfolio. This is not to say there are no opportunity costs associated with adopting a nonproprietary strategy, open source or otherwise; such costs are discussed in detail in Chapter 7. The point is simply that the decision to pursue a nonproprietary strategy does not automatically exclude the possibility of obtaining and maintaining patent protection.
35. Sulston and Ferry (2002), pp. 211–212. In this case there would have also arisen the question what kind of intellectual property right could have been licensed, as sequence data are not automatically protected under copyright or database protection laws. See Chapter 8 for a discussion of similar difficulties arising in connection with the International Haplotype Mapping Project.
36. This is analogous not only to open source software licensing, but also to the Creative Commons objective of facilitating the dissemination of cultural material by helping copyright owners specify which rights are reserved. Creative Commons website, http://creativecommons.org.
37. Mainstream scientific journals are one avenue for defensive disclosure. There also exist journals devoted to defensive publishing, some of which are respected sources of technical information that are included as part of the Patent Cooperation Treaty minimum documentation for International Search Authorities. In addition, some large corporations

rely on their own technical disclosure bulletins. In the United States the USPTO enables applicants to request the publication of a Statutory Invention Registration (SIR) of a filed patent, which is effectively a technical disclosure of an invention for which a patent was applied for. In registering an SIR, the applicant abandons the prosecution of the patent, in exchange for the disclosure of the invention by the patent office. Esteban Burrone, "New Product Launch: Evaluating Your Freedom to Operate," World Intellectual Property Organization website, http://www.wipo.int/sme/en/documents/freedom_to_operate.html.

38. Some open source software licenses go further, incorporating yank clauses triggered by accusations of infringement not just against the licensor but against any licensor of any open source software; although the first type of device is now generally considered acceptable from an antitrust perspective, this enhancement might be questionable in a patent context. For a discussion of the relationship between open source licensing and antitrust in a biotechnology context, see generally Feldman (2004).
39. 17 U.S.C. 106.
40. See 35 U.S.C. 154(a)(1).
41. For example, see OSD clauses 5 and 6, http://www.opensource.org/docs/definition.php.
42. Rosen (2005), p. 9, principle 1.
43. OSD clauses 1 and 5, http://www.opensource.org/docs/definition.php.
44. OSD clause 7, http://www.opensource.org/docs/definition.php.
45. Weber (2004), p. 86.
46. For example, see OSD clause 1, http://www.opensource.org/docs/definition.php.
47. Lawrence Rosen, personal communication, 26 March 2003.
48. Weber (2004), p. 158.
49. Raymond (2004).
50. OSD, http://www.opensource.org/docs/definition.php.
51. Ibid.
52. Free Software Definition, http://www.gnu.org/philosophy/free-sw.html.
53. To grasp the significance of access to source code as a technical prerequisite to software freedom, think of it as like being able to open the hood of your car. Opening the hood allows you to look at the engine to check the source of a worrying noise or to perform routine maintenance. If you are a car enthusiast or a budding mechanic, it lets you tin-

ker with the engine to learn how it works and extract spare parts for use in another vehicle. If cars were routinely manufactured with their hoods soldered shut, professional servicing would cost a lot more, there would be a lot less choice among service providers, and you would have no option but to trust a professional to diagnose your problem—even if all the professionals worked for or were otherwise beholden to the parts manufacturer (for example, through a training certification system). The car analogy also makes it easy to understand why some hackers get so emotional about closed source. Telling an old-school hacker that he or she cannot look at software source code is like telling the guy who *invented* the internal combustion engine that he's not allowed to look under his own hood. Free software isn't just ideology—it's righteous indignation!

54. Sulston and Ferry (2002), p. 206.
55. Von Gavel (2001), pp. 10–11.
56. "Courts have recognized the necessity and desirability of permitting an applicant for a patent to supplement the written disclosure in an application with a deposit of biological material which is essential to meet some requirement of the statute with respect to the claimed invention. See, e.g., *Ajinomoto Co. v. Archer-Daniels-Midland Co.*, 228 F.3d 1338, 1345–46, 56 USPQ2d 1332, 1337–38 (Fed. Cir. 2000), *cert. denied,* 121 S.Ct. 1957 (2001) (explaining how deposit may help satisfy enablement requirement); *Merck and Co., Inc. v. Chase Chemical Co.*, 273 F. Supp. 68, 155 USPQ 139 (D. N.J. 1967); *In re Argoudelis,* 434 F.2d 666, 168 USPQ 99 (CCPA 1970). To facilitate the recognition of deposited biological material in patent applications throughout the world, the Budapest Treaty on the International Recognition of the Deposit of Microorganisms for the Purposes of Patent Procedure was established in 1977, and became operational in 1981. The Treaty requires signatory countries, like the United States, to recognize a deposit with any depository which has been approved by the World Intellectual Property Organization (WIPO)." *United States Manual of Patent Examining Procedure,* 8th ed., August 2001 (latest revision October 2005), sec. 2402. Available online at http://www.uspto.gov/web/offices/pac/mpep/documents/2400_2402.htm#sect2402.
57. Macdonald (2004), p. 136.
58. OSD Version 1.9, clause 2, http://www.opensource.org/docs/definition.php.

59. "2165 the Best Mode Requirement—2100 Patentability," http://www.uspto.gov/web/offices/pac/mpep/documents/2100_2165.htm. *United States Manual of Patent Examining Procedure,* 8th ed., sec. 2402. Available online at http://www.uspto.gov/web/offices/pac/mpep/documents/2400_2402.htm#sect2402.
60. Heller and Eisenberg (1998).
61. Boettiger and Burk (2004), p. 221.
62. Rosen (2005), p. 106.
63. Boettiger and Burk (2004), pp. 230–231.
64. Rosen (2005), pp. 194–196.
65. Ibid., p. 194.
66. The Mozilla Public License deals with "files containing derivative works" rather than "derivative works" more broadly: Rosen (2005), p. 143. The Community Public License excludes "separate modules of software" from the copyleft obligation: ibid., pp. 168–169.
67. 17 U.S.C. 106.
68. Boettiger and Burk (2004), p. 221.
69. Ibid.
70. Weber (2004), pp. 179–185.
71. Bruce Perens, personal communication, 25 March 2003.
72. Pettit (1996), pp. 78ff.
73. Hume (1826a). According to Hume the designer of a constitution meant to give security against misuse of power should assume that every man is a knave, but this assumption should not be made in other contexts; it is "true in politics" but "false in fact"—in fact it is "rare to meet with one, in whom all the kind affections, taken together, do not overbalance all the selfish" (Hume 1826b).
74. Pettit (1996), p. 79.
75. Weber (2004), p. 159; Vetter (2006).
76. Pettit (1996), pp. 81–85.
77. Moglen (2001).

6. Foundations of the Biobazaar

1. Kevin Sweeney, in discussion at the Open Source Biology Workshop, Molecular Sciences Institute, Berkeley, California, 27 March 2003.
2. Angell (2005), pp. 24–27.
3. Ibid., p. 30.

4. Benkler (2002), p. 383.
5. Ibid., p. 369.
6. NASA Click-workers (http://clickworkers.arc.nasa.gov/top) is a project in which members of the public volunteer to perform routine scientific analysis that requires human perception and commonsense but little training (for example, identifying and measuring craters on Mars). See Benkler (2006), pp. 69–70. Wikipedia (http://www.wikipedia.org) is a free-content, multilingual, collaboratively written online encyclopedia. The site is a Wiki, permitting anyone to edit or add to existing entries. See Benkler (2006), pp. 70–74. Project Gutenberg (http://www.gutenberg.org) is an electronic library of approximately seventeen thousand books whose copyright has expired in the United States. The books are made available to the public online for free. See Benkler (2006), pp. 80–81. SETI@home (http://setiathome.berkeley.edu) uses the spare capacity of Internet-connected computers in the Search for Extraterrestrial Intelligence (SETI). Volunteers participate by running a program that downloads and analyzes radio telescope data. See Benkler (2006), pp. 81–83.
7. Folding@home Distributed Computing, http://folding.stanford.edu.
8. Von Krogh et al. (2003), citing Wayner (2000).
9. Brian Behlendorf, personal communication, 20 March 2003.
10. Michael Tiemann, personal communication, 2 April 2003.
11. Brian Behlendorf, personal communication, 20 March 2003.
12. Bonaccorsi and Rossi (2003), p. 1252.
13. Von Hippel (2005), pp. 133ff.
14. Benkler (2002), pp. 369, 377.
15. Carlson and Brent (2000).
16. "Big Genome—Big Science?" (2001).
17. Frazier et al. (2003), p. 292.
18. Ibid.
19. Benkler (2002), pp. 435ff.
20. Ibid., p. 379.
21. Sulston and Ferry (2002), p. 78.
22. Bruce Perens, personal communication, 25 March 2003.
23. Hilgartner and Brandt-Rauf (1994).
24. Fujimura (1987).
25. Endy (2005); Baker et al. (2006).

26. Roger Brent, Open Source Biology Workshop, Molecular Sciences Institute, Berkeley, California, 27 March 2003.
27. Greg Graff, personal communication, 28 March 2003.
28. Von Krogh et al. (2003), p. 1217.
29. Drew Endy, personal communication, 14 April 2003.
30. Robert Carlson, Open Source Biology Workshop, Molecular Sciences Institute, Berkeley, California, 27 March 2003.
31. Sulston and Ferry (2002), pp. 74–75.
32. John Sulston, personal communication, 5 August 2003.
33. World Intellectual Property Organization (1992), pp. 74–76.
34. Benkler (2006), p. 352.
35. Cavalla (2003).
36. Drews (1999), pp. 229–232.
37. Porter (2001).
38. Alfred Gilman, personal communication, 31 March 2003.
39. Sulston and Ferry (2002), pp. 55, 79.
40. Brian Behlendorf, personal communication, 20 March 2003.
41. For example, World Courier, www.worldcourier.com.
42. Alfred Gilman, personal communication, 31 March 2003.
43. Benkler (2002), p. 379.
44. This example was brought to my attention by Stephen Maurer, personal communication, 19 March 2003.
45. Smith (1990).
46. Love and Hubbard (2005).
47. They suggest that the TRIPS obligation on governments to enforce U.S. patents could be modified or replaced by an obligation to spend a certain fraction of GDP on supporting health-care R&D: ibid., p. 220.
48. James (2003). This informality is not so apparent in newer licenses, and may eventually disappear altogether as open source software moves into the commercial mainstream and existing licenses are overhauled by teams of corporate lawyers.
49. For a history of the origins of copyleft licensing, see Stallman (1999). A norm entrepreneur is an individual or entity that seeks to promote or change a norm: Sunstein (1996), p. 909.
50. Hugh Hansen, personal communication, 13 August 2003.
51. See http://www.piipa.org/.
52. The NIH already promotes best-practice guidelines for biotechnol-

ogy licensing: NIH (1999, 2005). At the international level, a similar function is performed by the Organisation for Economic Co-operation and Development (OECD 2006). Science Commons (http://sciencecommons.org/) is currently engaged in the ambitious project of developing a Web-based suite of standard, modular contracts for the transfer of biological materials among nonprofit institutions and between nonprofit and for-profit institutions. See also "Model Provisions for an 'Equitable Access and Neglected Disease License,' http://www.essentialmedicine.org/EAL.pdf, implementing the proposal of Kapczynski et al. (2005).

53. See Creative Commons, "Creative Commons Licenses," http://creativecommons.org/licenses/.
54. *Module libraries* is a term borrowed from the user innovation literature on toolkit development, used here to highlight the fact that licensor-driven development of open source biotechnology licenses is itself a form of user innovation.
55. Kathy Ku, director, Stanford University Office of Technology Transfer, personal communication, 24 March 2003.
56. For critiques of the Creative Commons licensing approach, see Elkin-Koren (2005); Elkin-Koren (2006); Weatherall (2006).
57. Eisenberg (2001), pp. 242–243.
58. That is, a legislature seeking to enact appropriate law reform measures would need to balance certainty and simplicity in the formulation of the relevant provisions against broad coverage and interpretive flexibility.
59. Open Source Initiative, Policy on License Proliferation, http://opensource2.planetjava.org/docs/policy/licenseproliferation.php.
60. The draft report, soon to become final, is at http://opensource.org/osi3.0/proliferation-report. The committee proposes to guide potential licensors toward licenses that are "popular" and have the support of a developer community. However, this proposal has met with criticism: that the committee's list of "popular" licenses is not based on empirical research, that the most popular licenses may not be legally the most satisfactory, and that licenses omitted from the list may contain provisions that some potential licensors might wish to employ.
61. Latour (1996), p. 118.
62. Opderbeck (2004).
63. Powell (1990).

64. DeBresson and Amesse (1991), p. 372.
65. Granovetter (1973).
66. Owen-Smith and Powell (2004).
67. For a discussion of the leadership role in open source software projects, see von Hippel and von Krogh (2001); Bonaccorsi and Rossi (2003), esp. pp. 1249ff; Behlendorf (1999); Weber (2004), pp. 166–171. Regarding leadership in large-scale biotechnology collaborations, see Thompson (2002); "Crosstalk"(2001) (quoting Alfred Gilman as saying there should be "money in the budget for pompoms"). See also generally Sulston and Ferry (2002), describing Sir John Sulston's experiences as director of the Sanger Centre during the race to sequence the human genome.
68. An example of such a workshop is the Amplified Fragment Length Polymorphisms (AFLP) Workshop cosponsored by ICBR Education Core and BEECS Genetic Analysis Laboratory, held at the Interdisciplinary Center for Biotechnology Research, University of Florida, Gainesville, Florida, on 11–13 June 2001. The interactive relationship between research tool manufacturers and their user communities is illustrated in the case of AFLP by the following extract from a newsletter of the pig genome mapping community: "Pig AFLP primers: . . . Unfortunately, existing commercial kits . . . are designed generally for plant genomes and may not work as well with pig DNA. . . . Extensive discussions have made it clear that PE AgGen has no large interest in marketing a kit for animal AFLP, and Keygene cannot do so without violating their previous agreement. . . . Keygene has now indicated that those animal scientists wishing to use AFLP should purchase the Perkin Elmer kit (http://www2.perkin-elmer.com/ag/775601/775601.html) and then should contact Keygene (keygene@euronet.nl), which will provide you the additional primers needed for animal AFLP mapping for a nominal fee. . . . One still cannot utilize the full range of available AFLP markers . . . without making your own primers or sharing with friends" (kindly provided by Jerry Dodgson) ("Pig Genome Update," 1999, no. 36, 1 May, http://www.animalgenome.org/pigs/newsletter/nl36.html.
69. Messages to Open Source License-Discuss forum (license-discuss@opensource.org), 25–27 September 2005.
70. A number of technical standards bodies exist in the software context; but the type of standard here under discussion is social, not technical.

71. A PDF version can be downloaded from the ISEAL home page, http://www.isealalliance.org/index.cfm?nodeid+1.

7. Financing Open Source Biotechnology

1. We saw in Chapters 4 and 5 that not all nonproprietary exploitation strategies are open source: some involve straightforward free revealing. For the sake of streamlining the discussion in this chapter, I do not seek to emphasize either this distinction or the distinction between different classes of open source license. For a fuller, diagrammatic view of the strategic options available to innovators, see Hope (2007), p. 109.
2. Benkler (2006), p. 49.
3. Nuvolari (2004).
4. Meyer (2003); Zero Prestige kite-building weblog, http://www.instructables.com/group/zeroprestige/.
5. Powell (2001), pp. 252ff.
6. Raymond (2000b), at "The Mail Must Get Through."
7. Von Hippel (1994); Franke and von Hippel (2003); see also von Hippel (2005), chap. 3, esp. pp. 33–34, and chap. 5, esp. pp. 66ff.
8. Von Hippel (1994).
9. On semiconductors and printed circuit board assembly, see von Hippel (1988), pp. 19–26. On printed circuit board design, see Urban and von Hippel (1988) and von Hippel (2005), pp. 23–34. On library information systems, see Morrison (2000). On pultrusion, see von Hippel (1988), pp. 28–30, 60–63, 182–188. On wind turbines, see Douthwaite (2002), pp. 67–104. On alternative currency systems, Douthwaite (2002), pp. 130–162.
10. User innovation (or "learning selection") and plant breeding, Douthwaite (2002), pp. 164–212 (for an illuminating history of the seed industry see Kloppenburg 2005); development of scientific instruments, von Hippel (1988), pp. 133–163; agricultural equipment, Douthwaite (2002), pp. 2–42, clinical chemistry analyzers, von Hippel (1988), pp. 93–100.
11. Polanyi (1958); Rosenberg (1982); Nelson (1990); see Mandeville (1996) for a comparison of terminology.
12. Von Hippel (1994).
13. Tom Knight, personal communication, 14 April 2003.
14. Ibid.

15. Alfred Gilman, personal communication, 31 March 2003.
16. Joachim von Braun (Director General, International Food Policy Research Institute), personal communication, 7 April 2003.
17. SNP Consortium, "Frequently Asked Questions," http://snp.cshl.org/thehapmap.html.en.
18. Alfred Gilman, personal communication, 31 March 2003.
19. Sulston and Ferry (2002).
20. Ibid., pp. 119–120.
21. John Sulston, personal communication, 5 August 2003.
22. Cambrosio and Keating (1998).
23. See Rabinow (1996).
24. Grody (2005).
25. Baker et al. (2006).
26. http://www.bsi-global.com/en/Shop/Publication-Detail/?pid=000000000030139919.
27. Meyer (2003).
28. Raymond (2000b), at "Release Early, Release Often."
29. Cambrosio and Keating (1998).
30. Ibid.
31. NIH (2007).
32. The SNP Consortium, "Frequently Asked Questions," http://snp.cshl.org/thehapmap.html.en.
33. Lee Bendeckgey, personal communication, 21 March 2003.
34. Electric Genetics, a company in Cape Town, is an example: http://www.egenetics.com.
35. Stewart (2002).
36. Lee Bendeckgey, personal communication, 21 March 2003.
37. John Sulston, personal communication, 5 August 2003.
38. Andrzej Kilian, personal communication, 15 September 2006.
39. Note, however, that these firms have been criticized for donating only technology that is not commercially valuable anyway: Nottenburg et al. (2002), p. 408.
40. John Sulston, personal communication, 5 August 2003.
41. Powell et al. (1996).
42. Sulston and Ferry (2002), p. 118.
43. Roger Brent, Open Source Biology Workshop, Molecular Sciences Institute, Berkeley, California, 27 March 2003. See also Sulston and Ferry (2002), p. 199.

44. Lee Bendeckgey, personal communication, 21 March 2003.
45. See, for example, Lanjouw and Schankerman (2001).
46. Von Hippel (1988), p. 52.
47. Grubb (2004), p. 404.
48. Angell (2005), p. 11.
49. Grubb (2004), p. 403.
50. Angell (2005), p. 47.
51. Drahos and Braithwaite (2002), pp. 50–154. See also G. Dukes, *The Law and Ethics of the Pharmaceutical Industry* (Amsterdam: Elsevier, 2006), pp. 5–7.
52. Wikipedia entry "Roundup," http://en.wikipedia.org/wiki/Roundup.
53. Grubb (2004), pp. 411–412.
54. Ibid.
55. Ibid., p. 410.
56. See Pisano (2006).
57. Ibid., p. 115.
58. Drahos and Braithwaite (2002), p. 163.
59. The Cohen-Boyer patents, jointly owned by Stanford University and the University of California, earned more than US$200 million in royalties through a nonexclusive license available to all comers for a reasonable royalty payment before expiring in December 1997: Grubb (2004), p. 413.
60. Heisey et al. (2005), p. 81.
61. Owen-Smith and Powell (2004), p. 10.
62. Ibid., pp. 8–9.
63. Boettiger and Bennett (2006).
64. I owe this report to my colleague Carolina Roa-Rodriguez (personal communication, 23 May 2007), who carried out extensive fieldwork on agricultural biotechnology research systems in the countries of the Andean region during 2004 and 2005. Interest in a protected technology commons or open-source-like regime was particularly intense among scientists working in public institutions (universities and national and international agricultural research centers), who perceived such a commons as a potential means both of accessing appropriate and needed technology produced elsewhere and of generating new technology locally.
65. Rosen et al. (2003), p. 59.
66. Ibid., p. 60.
67. Raymond (2000b).

68. Von Hippel and Katz (2002), discussing the design of an effective toolkit for user innovation.
69. Demil and Lecocq (2006), p. 1457.
70. Ibid., citing McKelvey (2001), p. 221.
71. Andrzej Kilian, personal communication, 23 June 2003.
72. See Walker (2006).
73. An example is Divine Chocolate Ltd. (http://www.divinechocolate.com), a 100 percent Fair Trade chocolate company based in the U.K. and co-owned by the Ghanaian cocoa farmers cooperative Kuapa Kokoo. Established in 1998 with the support of NGO Twin Trading, The Body Shop, Christian Aid, and Comic Relief, by 2007 the business had a turnover of £9 million and had launched a sister company in the United States, also co-owned by Kuapa Kokoo.
74. Descriptions of these business models in the following discussion are drawn from Grubb (2004), pp. 405–409.
75. Angell (2005), pp. 177–178.
76. Grubb (2004), p. 409.
77. Angell (2005), p. 231; Drews (1999), p. 195.
78. Drahos (1995), p. 168.
79. Pisano (2006).
80. Ibid.
81. See Drews (1999), p. 3.
82. Cavalla (2003), p. 273; Drews (1999), pp. 19–20. Contract research organizations might even be an additional source of capital for open source drug development. At least one example exists of such a firm committing risk financing to the development process; see Cavalla (2003), p. 272.
83. Cavalla (2003), p. 273.
84. Ibid., p. 268.
85. Munos (2006).
86. Ibid., p. 3.
87. Ibid., p. 7.
88. Cavalla (2003), p. 273.

8. Biotechnology's Open Source Revolution

1. Gabriel and Goldman (2004); James (2003), p. 74.
2. I owe this point to an anonymous reviewer.
3. Michael Tiemann, personal communication, 2 April 2003.

4. Ibid.
5. Lee Bendeckgey, personal communication, 21 March 2003.
6. Von Hippel (1988), p. 118.
7. Benkler (2006), p. 50.
8. Drahos (2003).
9. Schumpeter (1975), pp. 82–83.
10. Hrebejk and Boudreau (2001).
11. See West (2003); see also Worthen (2004): "Microsoft is one of the top lobbying shops in the country, [spending] close to $10 million per year on federal-level lobbyists. . . . Microsoft has tight links with many of the most powerful and influential shapers of policy at the federal and state level [and over the past five years to 2004] has developed one of the most sophisticated lobbying networks in the country: one that . . . makes it difficult for anyone to pass technology-related legislation Microsoft opposes. . . . Microsoft has lobbied particularly hard against open source, helping kill state bills that advocate for open source in Oregon and Texas."
12. Bonaccorsi and Rossi (2003), pp. 1255–56.
13. Open Source Initiative, "Halloween I: Open Source Software (New?) Development Methodology," http://catb.org/~esr/halloween/halloween1.html#quote4.
14. For a detailed discussion of hybrid strategies, see Hecker (2000).
15. Casadesus-Masanell and Ghemawat (2003). See also Bonaccorsi and Rossi (2003), pp. 1255–56; West (2003).
16. Munos (2006).
17. Bruce Perens, personal communication, 25 March 2003.
18. Poynder (2006).
19. Eisenberg (1997).
20. Greg Graff, personal communication, 28 March 2003.
21. Spolsky (2004).
22. Michaels (1999). This page has been discontinued; a later version of the proposal (V1.2, 6 October 1999) is on file with the author.
23. Tom Michaels, personal communication, 23 November 2006. See also Michaels (1999).
24. Tom Michaels, personal communication, 23 November 2006.
25. The International Treaty on Plant Genetic Resources for Food and Agriculture (the treaty) was adopted by the thirty-first session of the Conference of the Food and Agriculture Organization of the United Na-

tions (FAO) on 3 November 2001 and entered into force on 29 June 2004. Articles 10 to 13 (Part IV) of the treaty establish a Multilateral System to facilitate access to Plant Genetic Resources for Food and Agriculture and to share, fairly and equitably, the benefits arising from the use of these resources. Article 12.4 provides that facilitated access under the Multilateral System shall be provided pursuant to a Standard Material Transfer Agreement, which was adopted by the Governing Body of the Treaty in its Resolution 1/2006 of 16 June 2006. The terms of the agreement that resemble the reciprocal terms of a copyleft-style license are contained in Article 6 ("Rights and Obligations of the Recipient").

26. FAO Commission on Genetic Resources for Food and Agriculture website, at "The International Treaty on Plant Genetic Resources for Food and Agriculture," http://www.fao.org/ag/cgrfa/itpgr.htm.
27. Cukier (2003).
28. Cukier (2003); Sulston and Ferry (2002).
29. International HapMap Project, Data Access Policy for the International HapMap Project and International HapMap Project Public Access License, http://www.hapmap.org/cgi-perl/registration, last accessed 10 February 2005. The page at this URL has been replaced since discontinuation of the registration requirement.
30. Eisenberg (2006), p. 1027.
31. Ibid., p. 15.
32. Ibid., p. 1027; Gitter (2007).
33. See, for example, GAIN Intellectual Property Policy, http://www.fnih.org/GAIN/policies.shtrnl (page discontinued; last accessed 29 January 2007).
34. Maurer et al. (2004).
35. See the SETI@home home page, http://setiweb.ssl.berkeley.edu/.
36. Maurer et al. (2004).
37. Drews (1999), pp. 13, 234.
38. Taylor (2006).
39. Ginger Taylor, personal communication, 19 November 2006.
40. Ibid.
41. G. Taylor, "Chagas Community Leader Needed," http://thesynapticleap.org/?q=node/116.
42. Taylor (2006), pp. 34–38.
43. Ginger Taylor, personal communication, 19 November 2006.

44. Science Commons, http://sciencecommons.org/.
45. http://openwetware.org/wiki/The_BioBricks_Foundation.
46. For a comprehensive analysis of the advantages and disadvantages of open source versus public domain or "open access" approaches to managing the BBF's intellectual assets, see Rai and Boyle (2007).
47. Carlson and Brent (2000); see also Carlson (2001). The workshop was held on 27 March 2003 at the MSI.
48. See http://www.cambia.org/daisy/cambia/about_cambia/592/589.html.
49. The CAMBIA BiOS License for Plant Enabling Technology Version 1.3, Recitals, first paragraph. Online at http://www.bios.net/daisy/PELicense/751.html.
50. See http://www.bioforge.net/forge/thread.jspa?messageID=378ź.
51. *PIPRA Summer Newsletter,* no. 5 (2006), p. 3. Available online at http://www.pipra.org/docs/PIPRA-Newsletter-Issue5.pdf.
52. Kapczynski et al. (2005); see esp. pp. 1090ff. "Equitable Access" and "Neglected Disease" license provisions could be combined in a single agreement: see "Model Provisions for an Equitable Access and Neglected Disease License," http://www.essentialmedicine.org/EAL.pdf.
53. Kapczynski et al. (2005), p. 1090.
54. Ibid., p. 1069.
55. Amaral (2007).
56. Mauricio Guaragna and Joseph Jackson, personal communication, 27 April 2007.
57. In the sociological sense, modeling is a process of observational learning by which a conception of action is displayed, interpreted, and copied: Braithwaite (1994), pp. 449ff.
58. Following this logic, attempts are now under way to import other models of collaboration and intellectual property management, including patent pools and clearinghouses, into a biotechnology setting. For example, on patent pools see Clark et al. (2000); Gaulé (2006); Verbeure, van Zimmeren, et al. (2006); Ebersole et al. (2005); Goldstein et al. (2005); and Grassler and Capria (2003). On intellectual property clearinghouses, see Graff et al. (2001); Graff and Zilberman (2001a); Graff and Zilberman (2001b); Atkinson et al. (2003); van Zimmeren et al. (2006); and Horn (2003). See also G. Van Overwalle, ed., *Gene Patents and Clearing Models: From Concepts to Cases* (Cambridge: Cambridge University Press, forthcoming). The overarching question whether biotechnological innovation would benefit from some form

of collaborative intellectual property management is of considerable interest to international policymakers; hence, in December 2005 the OECD sponsored an Expert Roundtable on Collaborative Intellectual Property Rights Mechanisms held in Washington, D.C. At least one clearinghouse-type institution already exists and is proving successful in this field: the Public Intellectual Property Resource for Agriculture (PIPRA), http://www.pipra.org.

59. Braithwaite (1994), p. 464.
60. Ibid., pp. 468–469.
61. Drahos and Braithwaite (2002), p. 11.
62. Ibid., pp. 192–197.
63. Braithwaite (1994), pp. 457ff.
64. Drahos and Braithwaite (2002), p. 196.
65. Ibid.
66. Ibid., p. 192.
67. Braithwaite (1994), p. 469.
68. See generally Hutchens (2007).
69. Fair Labelling Organisations International (2006), p. 6.
70. Herrick (2006), p. 3.
71. Akubue (2000), p. 33.
72. Ibid. In the developed world, the AT movement first came to prominence in the early 1970s with the publication of E. F. Schumacher's book *Small Is Beautiful* (Schumacher 1999).
73. Illich (1973).
74. Ibid., para. 98.
75. Ibid., para. 76.
76. Ibid., para. 157.
77. Nottenburg et al. (2002), pp. 3–4.
78. I owe this example, drawn from fieldwork in the Andean region, to my colleague Carolina Roa-Rodriguez.
79. Illich (1973), para. 134.
80. Ibid., para. 135.
81. Jasanoff (2005), p. 19.
82. Illich (1973), para. 101.
83. Drahos and Braithwaite (2002), p. 218.
84. Braithwaite (1994).

References

Publications

Akubue, A. 2000. "Appropriate Technology for Socioeconomic Development in Third World Countries." *Journal of Technology Studies* 26 (1): 33–43. http://scholar.lib.vt.edu/ejournals/JOTS/Winter-Spring-2000/akabue.html.

Amaral, R. 2007. "Brazil Bypasses Patent on Merck AIDS Drug." Reuters AlertNet, 4 May. http://www.alertnet.org/thenews/newsdesk/N04351721.htm.

Andersen, M. A., and L. J. Butler. 2003. "The California Seed Industry: Organization, Economic Performance, and the Structure of R&D." Paper presented at the conference Productivity, Public Goods and Public Policy: Agricultural Biotechnology Potentials, Seventh International ICABR Conference, Ravello, Italy, 29 June–3 July.

Angell, M. 2005. *The Truth about the Drug Companies: How They Deceive Us and What to Do about It.* Melbourne: Scribe.

Arora, A. 1997. "Patents, Licensing and Market Structure in the Chemical Industry." *Research Policy* 26 (1997): 391–403.

Arora, A., A. Fosfuri, and A. Gambardella. 2001. *Markets for Technology: The Economics of Innovation and Corporate Strategy.* Cambridge, Mass.: MIT Press.

Arrow, K. J. 1962. "Economic Welfare and the Allocation of Resources to Invention." In R. R. Nelson (ed.), *The Rate and Direction of Inventive Activities.* Princeton: Princeton University Press.

Atkinson, R. C., et al. 2003. "Public Sector Collaboration for Agricultural IP Management." *Science* 301 (11 July): 174–175.

Baker, D., et al. 2006. "Engineering Life: Building a Fab for Biology." *Scientific American* 294: 44–51.

Barton, J. H. 1997a. "The Balance between Intellectual Property Rights and Competition: Paradigms in the Information Sector." *European Competition Law Review* 18: 440–445.

———. 1997b. "The Economic and Legal Context of Contemporary Technology Transfer." In E. Buscaglia, W. Ratliff, and R. Cooter (eds.), *The Law and Economics of Development*, 83–100. Greenwich, Conn.: JAI Press.

Behlendorf, B. 1999. "Open Source as a Business Strategy." In DiBona, C. S. Ockman, and M. Stone (eds.), *Open Sources: Voices from the Open Source Revolution*. Cambridge, Mass.: O'Reilly. http://www.oreilly.com/catalog/opensources/book/behlendorf.html.

Benkler, Y. 2002. "Coase's Penguin, or, Linux and the Nature of the Firm." *Yale Law Journal* 112: 369.

———. 2004. "'Sharing Nicely': On Shareable Goods and the Emergence of Sharing as a Modality of Economic Production." *Yale Law Journal* 114: 273–358.

———. 2006. *The Wealth of Networks: How Social Production Transforms Markets and Freedom*. New Haven: Yale University Press.

Bennett, A., G. Graff, S. Cullen, K. Bradford, and D. Zilberman. 2002. "Public Sector Intellectual Property Management for Agricultural Research and Development." Invited presentation, International Consortium on Agricultural Biotechnology Research, International Conference, Ravello, Italy, July 2002.

"Big Genome—Big Science?" 2001. Editorial. *Nature Cell Biology* 3: E65. http://www.nature.com/ncb/journal/v3/n3/full/ncb0301_e65.html.

Biotechnology Industry Organization. 2006. "Biotechnology Industry Organization Guide to Biotechnology, 2005–2006." Washington, D.C.: Biotechnology Industry Organization.

Björkman, B., and S. Hansson. 2006. "Bodily Rights and Property Rights." *Journal of Medical Ethics* 32: 209–214.

Blumenthal, D. 1992. "Academic-Industry Relationships in the Life Sciences." *Journal of the American Medical Association* 268 (23): 3344–49.

Blumenthal, D., E. G. Campbell, M. Anderson, N. Causino, and K. Seashore Louis. 1997. "Withholding Research Results in Academic Life Science." *Journal of the American Medical Association* 277 (15): 1224–28.

Blumenthal, D., E. G. Campbell, N. Causino, and K. Seashore Louis. 1996. "Participation of Life-Science Faculty in Research Relationships with Industry." *New England Journal of Medicine* 335 (23): 1734–39.

Blumenthal, D., N. Causino, E. G. Campbell, and K. Seashore Louis. 1996. "Relationships between Academic Institutions and Industry in the Life Sciences—An Industry Survey." *New England Journal of Medicine* 334 (6): 368–373.

Blumenthal, D., M. Gluck, K. Seashore Louis, M. A. Stoto, and D. Wise. 1986. "University-Industry Research Relationships in Biotechnology: Implications for the University." *Science* 232: 1361–66.

Boehm, K. 1967. *The British Patent System*, vol. 1: *Administration*. Cambridge: Cambridge University Press.

Boettiger, S., and A. B. Bennett. 2006. "Bayh-Dole: If We Knew Then What We Know Now." *Nature Biotechnology* 24 (3): 320–323.

Boettiger, S., and D. Burk. 2004. "Open Source Patenting." *Journal of International Biotechnology Law* 1: 221–231.

Bonaccorsi, A., and C. Rossi. 2003. "Why Open Source Software Can Succeed." *Research Policy* 32 (7): 1243–58.

Boyle, James. 2001. "The Second Enclosure Movement and the Construction of the Public Domain." Paper presented at the Conference on the Public Domain, Duke University School of Law, Durham, North Carolina, 9–11 November 2001.

Braithwaite, J. 1994. "A Sociology of Modelling and the Politics of Empowerment." *British Journal of Sociology* 45: 445–478.

Braithwaite, J., and P. Drahos. 2000. *Global Business Regulation*. Cambridge: Cambridge University Press.

Burrone, E. "New Product Launch: Evaluating Your Freedom to Operate." http://www.wipo.int/sme/en/documents/freedom_to_operate.html.

Bush, V. 1945. *Science: The Endless Frontier—A Report to the President*, by Vannevar Bush, Director of the Office of Scientific Research and Development, July. Washington, D.C.: USPO. http://www.nsf.gov/about/history/vbush1945.htm.

Cambrosio, A., and P. Keating. 1998. "Monoclonal Antibodies: From Local to Extended Networks." In A. Thackray (ed.), *Private Science: Biotechnology and the Rise of the Molecular Sciences,* 165–181. The Chemical Sciences in Society Series. Philadelphia: University of Pennsylvania Press.

Campbell, E. G., J. S. Weissman, N. Causino, and D. Blumenthal. 2000. "Data Withholding in Academic Medicine: Characteristics of Faculty Denied Access to Research Results and Biomaterials." *Research Policy* 29: 303–312.

Campbell, E. G., et al. 2002. "Data Withholding in Academic Genetics: Data from a National Survey." *Journal of the American Medical Association* 287: 473.

Carlson, R. 2001. *Open Source Biology and Its Impact on Industry.* IEEE Spectrum, http://www.kurzweilai.net/meme/frame.html?main=/articles/art0613.html.

Carlson, R., and R. Brent. 2000. "DARPA Open-Source Biology Letter." Molecular Sciences Institute, October. http://www.synthesis.cc/DARPA_OSB_Letter.html.

Casadesus-Masanell, R., and P. Ghemawat. 2003. "Dynamic Mixed Duopoly: A Model Motivated by Linux vs. Windows." Strategy Unit Working Paper No. 04-012; IESE Working Paper No. D/519. http://ssrn.com/abstract=439340 or DOI: 10.2139/ssrn.439340.

Cavalla, D. 2003. "The Extended Pharmaceutical Enterprise." *Drug Discovery Today* 8 (6): 267–274.

Christensen, C. M. 1997. *The Innovator's Dilemma.* Cambridge, Mass.: Harvard Business School Press.

Christensen, C. M., and M. E. Raynor. 2003. *The Innovator's Solution.* Cambridge, Mass.: Harvard Business School Press.

Clark, J., J. Piccolo, B. Stanton, and K. Tyson. 2000. *Patent Pools: A Solution to the Problem of Access of Biotechnology Patents?* Washington, D.C.: U.S. Patent and Trademark Office.

Coase, R. H. 1960. "The Problem of Social Cost." *Journal of Law and Economics* 3 (1960): 1–44.

Cohen, J. 1997. "The Genomics Gamble." *Science* 275 (7 February): 767–772.

Cohen, W., R. Nelson, and J. Walsh. 2000. "Protecting Their Intellectual Assets: Appropriability Conditions and Why U.S. Manufacturing Firms Patent (or Not)." NBER Working Paper no. W7552 (February). Cambridge, Mass.: National Bureau of Economic Research. http://papers.nber.org/papers/W7552.

Collins, H., and T. Pinch. 1993. *The Golem: What Everyone Should Know about Science.* Cambridge: Cambridge University Press. Second ed., with new afterword, Cambridge: Canto, 1998.

Commission on Intellectual Property Rights. 2002. *Integrating Intellectual Property Rights and Development Policy.* Final Report of the Commission on Intellectual Property Rights, London. http://www.iprcommission.org/graphic/documents/final_report.htm.

Conway, G., and G. Toennissen. 2003. "Science for African Food Security." *Science* 299: 1187–88.

Cooter, R., and T. Ulen. 2000. *Law and Economics.* 3rd ed. New York: Addison Wesley Longman.

"Crosstalk: Please Check EGO at Door: Alfred Goodman Gilman." 2001. *Molecular Interventions* 1: 14–21.

Cukier, K. 2003. "Open Source Biotech: Can a Non-Proprietary Approach to Intellectual Property Work in the Life Sciences?" http://www.cukier.com/writings/opensourcebiotech.html.

Cullet, P. 2005. *Intellectual Property Protection and Sustainable Development.* New Delhi: Butterworths.

DeBresson, C., and F. Amesse. 1991. "Networks of Innovators: A Review and Introduction to the Issue." *Research Policy* 20: 363–379.

Demil, B., and X. Lecocq. 2006. "Neither Market nor Hierarchy nor Network: The Emergence of Bazaar Governance," *Organization Studies* 27 (10): 1447–66. Abstract at http://oss.sagepub.com/cgi/content/abstract/27/10/1447.

Douthwaite, B. 2002. *Enabling Innovation.* London: Zed Books.

Drahos, P. 1995. "Global Property Rights in Information: The Story of TRIPS at the GATT." *Prometheus* 13: 6–19.

———. 1996. *A Philosophy of Intellectual Property.* Dartmouth Series in Applied Legal Philosophy. Aldershot, UK: Dartmouth.

———. 2003. "The Global Intellectual Property Ratchet: Why It Fails as Policy and What Should Be Done about It." http://cgkd/anu.edu.au/menus/publications.php#drahos.

Drahos, P., and J. Braithwaite. 2002. *Information Feudalism: Who Owns the Knowledge Economy?* London: Earthscan.

Drews, J. 1999. *In Quest of Tomorrow's Medicines: An Eminent Scientist*

Talks about the Pharmaceutical Industry, Biotechnology and the Future of Drug Research. New York: Springer-Verlag.

Dutton, H. I. 1984. *The Patent System and Inventive Activity during the Industrial Revolution, 1750–1852*. Manchester: Manchester University Press.

Eamon, W. 1975. "From the Secrets of Nature to Public Knowledge: The Origins of the Concept of Openness in Science." *Minerva* 23 (3): 321–347.

Ebersole, T. J., et al. 2005. "Patent Pools and Standard Setting in Diagnostic Genetics." *Nature Biotechnology* 23: 937–938.

Eisenberg, R. S. 1987. "Proprietary Rights and the Norms of Science in Biotechnology Research." *Yale Law Journal* 97 (2): 177–231.

———. 1989. "Patents and the Progress of Science: Exclusive Rights and Experimental Use." *University of Chicago Law Review* 56: 1017–86.

———. 1997. "Patenting Research Tools and the Law." In National Research Council (ed.), *Intellectual Property Rights and Research Tools in Molecular Biology*. Washington, D.C.: National Academies Press. http://www.nap.edu/readingroom/books/property/4.html#chap4.

———. 2001. "Bargaining over the Transfer of Proprietary Research Tools: Is This Market Failing or Emerging?" In R. C. Dreyfuss, D. Zimmerman, and H. First (eds.), *Expanding the Boundaries of Intellectual Property: Innovation Policy for the Knowledge Society*, 223–249. Oxford: Oxford University Press.

———. 2006. "Patents and Data Sharing in Public Science." *Industrial and Corporate Change* 15 (6): 1013–31.

Elkin-Koren, N. 2005. "What Contracts Can't Do: The Limits of Private Ordering in Facilitating a Creative Commons." *Fordham Law Review* 74: 375–422.

———. 2006. "Exploring Creative Commons: A Skeptical View of a Worthy Pursuit." In L. Guibault and P. B. Hugenholtz (eds.), *The Future of the Public Domain: Identifying the Commons in Information Law*. The Hague: Kluwer Law International. http://papers.ssrn.com/sol3/papers.cfm?abstract_id=885466.

Ellis, J. T. 2000. "Distortion of Patent Economics by Litigation Costs." In *Proceedings of the 1999 Summit Conference on Intellectual Property, University of Washington, Seattle*. CASRIP Symposium pub. ser. no. 5. http://www.law.washington.edu/casrip/Symposium/Number5/pub5atcl3.pdf.

Endy, D. 2005. "Foundations for Engineering Biology." *Nature* 438 (24 November): 449–453.
Epstein, R. A., and B. N. Kuhlik. 2004. "Is There a Biomedical Anticommons?" *Regulation* 27 (2): 54–58.
Etzkowitz, H. 1989. "Entrepreneurial Science in the Academy: A Case of the Transformation of Norms." *Social Problems* 36 (1): 14–29.
Fair Labelling Organisations International, International Federation for Alternative Trade, Network of European World Shops, European Fair Trade Association. 2006. *Business Unusual.* Brussels: Fair Trade Advocacy Office.
Feldman, R. C. 2004. "The Open Source Biotechnology Movement: Is It Patent Misuse?" *Minnesota Journal of Law, Science and Technology* 6: 117–167.
Fernandez-Cornejo, J. 2004. "The Seed Industry in U.S. Agriculture: An Exploration of Data and Information on Crop Seed Markets, Regulation, Industry Structure, and Research and Development." Agriculture Information Bulletin no. (AIB786). Washington, D.C.: Economic Research Service, United States Department of Agriculture. http://www.ers.usda.gov/Publications/AIB786/.
Fitzgerald, B., and G. Bassett. 2003. "Legal Issues Relating to Free and Open Source Software." In B. Fitzgerald and G. Bassett (eds.), *Legal Issues Relating to Free and Open Source Software,* vol. 1 of *Essays in Technology Policy and Law,* 11–36. Brisbane: Queensland University of Technology School of Law.
Food and Agriculture Organization of the United Nations. 2001. International Treaty on Plant Genetic Resources for Food and Agriculture. http://www.fao.org/AG/cgrfa/itpgr.htm.
Fox, J. L. 1981. "Can Academia Adapt to Biotechnology's Lure?" *Chemical and Engineering News* (12 October): 39–44.
Franke, N., and S. Shah. 2003. "How Communities Support Innovative Activities: An Exploration of Assistance and Sharing among End-Users." *Research Policy* 32: 157–178.
Franke, N., and E. von Hippel. 2003. "Satisfying Heterogeneous Needs via Innovation Toolkits: The Case of Apache Security Software." *Research Policy* 32: 1199–1215.
Frazier, M. E., et al. 2003. "Realizing the Potential of the Genome Revolution: The Genomes to Life Program." *Science* 300: 290–293.
Fujimura, J. H. 1987. "Constructing 'Do-able' Problems in Cancer Re-

search: Articulating Alignment." *Social Studies of Science* 17: 257–293.

Gabriel, R. P., and R. Goldman. 2004. *Open Source: Beyond the Fairytales.* http://opensource.mit.edu/papers/gabrielgoldman.pdf.

Gaulé, P. 2006. "Towards Patent Pools in Biotechnology?" *Innovation Strategy Today* 2: 123–124.

Ghosh, R. A. 1998. "Cooking Pot Markets: An Economic Model for Trading Free Goods and Services on the Internet." *First Monday* 3 (3). http://www.firstmonday.org/issues/issue3_3/.

Ghosh, R. A., and V. V. Prakash. 2000. "The Orbiten Free Software Survey." *First Monday* 5 (7). http://www.firstmonday.org/issues/issue5_7/ghosh/.

Giles, J. 2006. "Sociologist Fools Physics Judges," *Nature* 442, 8.

Gitter, D. 2007. "Resolving the Open Source Paradox in Biotechnology: A Proposal for a Revised Open Source Policy for Publicly Funded Genomic Databases." *Houston Law Review* 43 (4) (forthcoming).

Gläser, J. 2003. "A Highly Efficient Waste of Effort: Open Source Software Development as a Specific System of Collective Production." In *New Times, New Worlds, New Ideas,* Proceedings of the 2003 Conference of the Australian Society for Sociology (TASA), Armidale, 4–6 December. http://repp.anu.edu.au/GlaeserTASA.pdf.

———. 2004. "Coordination Matters: Suggestions for a Comparative Analysis of Collective Production Systems." Paper presented at the annual meeting of the American Sociological Association, San Francisco, 14 August. http://www.allacademic.com/meta/p108317_index.html.

Goldman Rohm, W. 1999. "Inside the Red Hat IPO." *Linux Magazine.* http://www.linux-mag.com/id/348/.

Goldstein, J., et al. 2005. "Patent Pools as a Solution to the Licensing Problem of Diagnostic Genetics." *Drug Discovery World* (Spring): 86–90.

Gorry, P. 2005. "Effect of Gene Patenting in the Daily Practice of Diagnostic Testing: Past and Future." Paper presented at the OECD Expert Roundtable on Collaborative Intellectual Property Rights Mechanisms, Washington, D.C., 8–9 December.

Graff, G., A. Bennett, B. Wright, and D. Zilberman. 2001. "Intellectual Property Clearinghouse Mechanisms for Agriculture: Summary of an Industry, Academia, and International Development Round Table." *IP Strategy Today* 2001 (3): 12–30.

Graff, G., A. Heiman, C. Yarkin, and D. Zilberman. 2003. "Privatization and Innovation in Agricultural Biotechnology." *Agriculture and Resource Economics Update* (Ciannini Foundation of Agricultural Economics, Berkeley) 6 (3): 5–7.

Graff, G. D., G. Rausser, and A. A. Small. 2003. "Agricultural Biotechnology's Complementary Intellectual Assets." *Review of Economics and Statistics* 85 (May): 349–363.

Graff, G., and D. Zilberman. 2001a. "An Intellectual Property Clearinghouse for Agricultural Biotechnology." *Nature Biotechnology* 19 (December): 1179–80.

Graff, G., and D. Zilberman. 2001b. "Towards an Intellectual Property Clearinghouse for Agricultural Biotechnology." *Intellectual Property Strategy Today,* no. 3: 1–12.

Granovetter, M. 1973. "The Strength of Weak Ties." *American Journal of Sociology* 78: 1360–80.

Grassler, F., and M. Capria. 2003. "Patent Pooling: Uncorking a Technology Transfer Bottleneck and Creating Value in the Biomedical Research Field." *Journal of Commercial Biotechnology* 9 (2): 111–118.

Grody, W. W. 2005. "Perspectives on Gene Patents from the Academic Medical Center." Paper presented at the OECD Expert Roundtable on Collaborative Intellectual Property Rights Mechanisms, Washington, D.C., 8–9 December.

Grubb, P. W. 2004. *Patents for Chemicals, Pharmaceuticals and Biotechnology.* Oxford: Oxford University Press.

Hall, B., and R. Ham. 1999. "The Patent Paradox Revisited: Determinants of Patenting in the US Semiconductor Industry." NBER Working Paper no. 7062. Cambridge, Mass.: National Bureau of Economic Research.

Hannaford, S. 2007. "Oligopoly Watch." http://www.oligopolywatch.com/.

Hecker, F. 2000. "Setting Up Shop: The Business of Open-Source Software." http://hecker.org/writings/setting-up-shop.

Heisey, P. W., J. L. King, and K. D. Rubenstein. 2005. "Patterns of Public-Sector and Private-Sector Patenting in Agricultural Biotechnology." *AgBioForum* 8: 73–82.

Heller, M. A., and R. S. Eisenberg. 1998. "Can Patents Deter Innovation? The Anticommons in Biomedical Research." *Science* 280 (1 May): 698–701.

Herrick, S. 2006. "Should You Use Free Software? You Already Do!" *Just*

Things: The Fair Trade Journal of Applied Counter-Economics 1 (3): 3–4. http://greenlagirl.com/2006/12/09/just-things-13/.

Hilgartner, S. 1997. "Access to Data and Intellectual Property: Scientific Exchange in Genome Research." In National Research Council, *Intellectual Property Rights and Research Tools in Molecular Biology.* Washington, D.C.: National Academy Press. http://www.nap.edu/readingroom/books/property/4.html#chap4.

———. 2002. "Acceptable Intellectual Property." *Journal of Molecular Biology* 319 (4): 943–946.

Hilgartner, S., and S. I. Brandt-Rauf. 1994. "Data Access, Ownership, and Control: Toward Empirical Studies of Access Practices." *Knowledge: Creation, Diffusion, Utilization* 15 (4): 355–372.

Hope, J. 2007. "Open Source Licensing." In A. Krattiger et al. (eds.), *Intellectual Property Management in Health and Agricultural Innovation: A Handbook of Best Practices,* 107–118. New York: MIHR-PIPRA. http://rsss.anu.edu.au/~janeth/.

Horn, L. 2003. "Alternative Approaches to IP Management: One-Stop Technology Platform Licensing." *Journal of Commercial Biotechnology* 9: 119–127.

Hrebejk, P., and T. Boudreau. 2001. "Perspective: The Coming 'Open Monopoly' in Software." *CPNet News.com* (24 October). http://news.com.com/The+coming+open+monopoly+in+software/2010-1071_3-281588.html.

Hume, D. 1826a. "Of the Independence of Parliament." In *Essays Moral, Political, and Literary,* vol. 3 of *The Philosophical Works of David Hume,* pt. 1, essay 6. Edinburgh: Adam Black and William Tait. Facsimile at http://oll.libertyfund.org/Home3/Book.php?recordID=0221.03.

———. 1826b. "Of the Origin of Justice and Property." In *A Treatise of Human Nature,* vol. 2 of *The Philosophical Works of David Hume,* bk. 3, pt. 2, sec. 2. Edinburgh: Adam Black and William Tait. Facsimile at http://oll.libertyfund.org/Home3Book.php?recordID=0859.

Hutchens, A. 2007. "Entrepreneurship, Power and Defiance: The Globalisation of the Fair Trade Movement." Ph.D. diss., Australian National University.

Illich, I. 1973. *Tools for Conviviality.* http://opencollector.org/history/homebrew/tools.html.

James, P. C. J. 2003. "Open Source Software: An Australian Perspective." In B. Fitzgerald and G. Bassett (eds.), *Legal Issues Relating to Free and Open Source Software,* vol. 1 of *Essays in Technology Policy and Law,* 63–89. Brisbane: Queensland University of Technology School of Law.

Jasanoff, S. 2005. *Designs on Nature: Science and Democracy in Europe and the United States.* Princeton: Princeton University Press.

Jordan, K., and M. Lynch. 1998. "The Dissemination, Standardization and Routinization of a Molecular Biological Technique." *Social Studies of Science* 28: 773–800.

Kapczynski, A., S. Chaifetz, Z. Katz, and Y. Benkler. 2005. "Addressing Global Health Inequities: An Open Licensing Approach for University Innovations." *Berkeley Technology Law Journal* 20: 1031–1114.

Kawamoto, D. 1999. "VA Linux Storms Wall Street with 698 Percent Gain." *CPNetNews.com.* http://news.com.com/VA+Linux+storms+Wall+Street+with+698+percent+gain/2100-1001_3-234182.html.

Kevles, D. J. 1998. "Diamond v. Chakrabarty and Beyond: The Political Economy of Patenting Life." In A. Thackray (ed.), *Private Science: Biotechnology and the Rise of the Molecular Sciences,* 65–79. The Chemical Sciences in Society Series. Philadelphia: University of Pennsylvania Press.

Kitch, E. W. 1977. "The Nature and Function of the Patent System." *Journal of Law and Economics* 20: 265–290.

Kloppenburg, J. R. 2005. *First the Seed: The Political Economy of Plant Biotechnology,* 2nd ed. Madison: University of Wisconsin Press.

Knight, J. 2003. "Crop Improvement: A Dying Breed." *Nature* 421: 568–570.

Kryder, R. D., S. P. Kowalski, and A. F. Krattiger. 2000. "The Intellectual and Technical Property Components of Pro-Vitamin A Rice *Golden Rice:* A Preliminary Freedom-to-Operate Review," *ISAAA Briefs* No. 20.

Kuhn, T. S. 1970. *The Structure of Scientific Revolutions.* Chicago: University of Chicago Press.

Lakhani, K., and E. von Hippel. 2003. "How Open Source Software Works: 'Free' User to User Assistance." *Research Policy* 32 (6): 923–943.

Lanjouw, J. O., and M. Schankerman. 2001. "Characteristics of Patent

Litigation: A Window on Competition." *RAND Journal of Economics* 32 (1): 129–151.

"Last Word: Grassroots Innovator." 2001. *Economist Technology Quarterly,* 8 December, 27–28.

Latour, B. 1996. *Aramis or the Love of Technology.* Cambridge, Mass.: Harvard University Press. Orig. pub. 1993.

Laudan, L. 1982. "Two Puzzles about Science: Reflections on Some Crises in the Philosophy and Sociology of Science." *History of Science* 20: 253–268.

Lemonick, M. D. 2000. "Gene Mapper." *Time,* 17 December, 2.

Levin, R. C., et al. 1987. "Appropriating the Returns from Industrial Research and Development." *Brookings Papers on Economic Activity* 3: 783–831. Washington, D.C.: Brookings Institution.

Levy, S. 2001. *Hackers: Heroes of the Computer Revolution.* New York: Penguin. Orig. pub. 1984.

Lewontin, R. C. 1993. *The Doctrine of DNA Biology as Ideology.* London: Penguin Books.

Locke, J. 2007. *The Second Treatise of Government.* Adelaide: dBooks@Adelaide. http://etext.library.adelaide.edu.au/l/locke/john/l81s/. Orig. pub. 1690.

Long, C. 2000. "Proprietary Rights and Why Initial Allocations Matter." *Emory Law Journal* 49 (Summer): 823–836.

Love, J., and T. Hubbard. 2005. "Paying for Public Goods." In R. A. Ghosh (ed.), *Code: Collaborative Ownership and the Digital Economy,* 207–229. Cambridge, Mass.: MIT Press.

Luthje, C., C. Herstatt, and E. von Hippel. 2002. "The Dominant Role of 'Local' Information in User Innovation: The Case of Mountain Biking." *Research Policy* 34 (6): 951–965.

Macdonald, S. 2003. "Bearing the Burden: Small Firms and the Patent System." *Journal of Information, Law and Technology* 1. http://elj.warwick.ac.uk/jilt/03-1/macdonald.html.

———. 2004. "When Means Become Ends: Considering the Impact of Patent Strategy on Innovation." *Information Economics and Policy* 16 (1): 135–158.

Mandeville, T. 1996. *Understanding Novelty: Information, Technological Change, and the Patent System.* Norwood, N.J.: Ablex.

Mansfield, E., et al. 1981. "Imitation Costs and Patents: An Empirical Study." *Economic Journal* 91 (December): 907–918.

Maskus, K. E., and J. H. Reichman. 2004. "The Globalization of Private

Knowledge Goods and the Privatization of Global Public Goods." *Journal of International Economic Law* 7: 279–320.
Matthews, D. 2002. *Globalizing Intellectual Property Rights: The TRIPS Agreement*. London: Routledge.
Maurer, S. M., A. Sali, and A. Rai. 2004. "Finding Cures for Tropical Disease: Is Open Source the Answer?" *Public Library of Science: Medicine* 1 (3): e56. http://www.plosmedicine.org.
McKelvey, M. 2001. "The Economic Dynamics of Software: Three Competing Business Models Exemplified through Microsoft, Netscape and Linux." *Economics of Innovation and New Technology* 10 (2–3): 199–236.
McMillan, J. 2002. *Reinventing the Bazaar: The Natural History of Markets*. New York: Norton.
McMillan, R. 2004. "IDC: Linux to Take 29 Percent of 2008 Server Shipments." *IDG News Service,* 17 June. http://www.infoworld.com/article/04/06/17/HNidclinux_1.html.
McNaughton, N. 1999. "Sharing Germplasm." http://www.ravenseyeconsulting.com/raven_articles_commentary.htm.
Merck & Co., Press Release, "First Installment of Merck Gene Index Data Released to Public Databases: Cooperative Effort Promises to Speed Scientific Understanding of Human Genome," 10 February 1995. http://www.ncbi.nlm.nih.gov/Web/Whats_New/Announce/merck_feb10_95.html.
Merges, R. P. 1996. "Contracting into Liability Rules: Intellectual Property Rights and Collective Rights Organizations." *California Law Review* 84 (5): 1293–1393.
———. 2000. "Intellectual Property Rights and the New Institutional Economics." In symposium, "Taking Stock: The Law and Economics of Intellectual Property Rights." *Vanderbilt Law Review* 53 (6): 1857–77.
———. 2001. "Institutions for Intellectual Property Transactions: The Case of Patent Pools." In R. C. Dreyfuss, D. Zimmerman, and H. First (eds.), *Expanding the Boundaries of Intellectual Property: Innovation Policy for the Knowledge Society,* 123–165. Oxford: Oxford University Press.
Merton, R. K. 1996. "The Normative Structure of Science." In H. Nowotny and K. Taschwer (eds.), *The Sociology of the Sciences,* 1:38–49. Cheltenham: Edward Elgar. Orig. pub. 1942.
Meyer, P. B. 2003. "Episodes of Collective Invention." BLS Working Pa-

per no. 368. Washington, D.C.: Bureau of Labor Statistics, U.S. Department of Labor.

Michaels, T. 1999. "General Public Release for Plant Germplasm: A proposal. Version 1.1." 26 February. http://www.oac.uoguelph.ca/www/CRSC/pltag/1998-99/gnucrop2.htm (page discontinued).

Mill, J. S. 1879. "Of the Liberty of Thought and Discussion." In *On Liberty* and *The Subjection of Women,* 33–99. New York: Henry Holt and Co. Orig. pub. 1859. Facsimile at http://oll.libertyfund.org/Home3/Book.php?recordID=0277.

Mitroff, I. I. 1974. "Norms and Counter-Norms in a Select Group of the Apollo Moon Scientists: A Case Study of the Ambivalence of Scientists." *American Sociological Review* 39 (4): 579–595.

Moglen, E. 2001. "Enforcing the GNU GPL." Free Software Foundation. http://www.gnu.org/philosophy/enforcing-gpl.html.

Morrison, P. D., et al. 2000. "Determinants of User Innovation and Innovation Sharing in a Local Market." *Management Science* 46 (12): 1513–27.

Mulgan, G., O. Salem, and T. Steinberg. 2005. *Wide Open: Open Source Methods and Their Future Potential.* London: Demos. http://www.ossite.org/collaborate/resources/entries/2380245424.

Mulkay, M. 1976. "Norms and Ideology in Science." *Sociology of Scientific Information* 15 (4–5): 637–656.

———. 1980. "Interpretation and the Use of Rules: The Case of the Norms of Science." *Transactions of the New York Academy of Sciences,* 2nd ser., 39: 111–125.

Muniz, A. M., and T. C. O'Guinn. 2001. "Brand Community." *Journal of Consumer Research* 27: 412–432.

Munos, B., 2006. "Can Open-Source R&D Reinvigorate Drug Research?" *Nature Reviews Drug Discovery* 5: 723–729.

Muoio, A. 1998. "Is Bigger Better?" *Fast Company,* no. 17, September 1998. http://www.fastcompany.com/online/17/one.html.

Murray, F., and S. Stern. 2005. "Do Formal Intellectual Property Rights Hinder the Free Flow of Scientific Knowledge? An Empirical Test of the Anti-Commons Hypothesis." NBER Working Paper no. 11465. Cambridge, Mass.: National Bureau of Economic Research. http://www.nber.org/papers/W11465.

National Institutes of Health. 1999. "Principles and Guidelines for Recipients of NIH Research Grants and Contracts on Obtaining and Disseminating Biomedical Research Resources: Final Notice." 64 *Federal Register* 72090.

———. 2005. "Best Practices for the Licensing of Genomic Inventions: Final Notice." 70 *Federal Register* 18413.

———. 2007. "NIH Roadmap for Medical Research: Molecular Libraries and Imaging Overview." http://nihroadmap.nih.gov/molecularlibraries/.

Nelson, R. 1990. *What Is Public and What Is Private about Technology?* Consortium on Competitiveness and Cooperation, Working Paper no. 90-9. Berkeley: Center for Research in Management, University of California.

Nelson, R. D., and R. Mazzoleni. 1997. "Economic Theories about the Costs and Benefits of Patents." In National Research Council, *Intellectual Property Rights and Research Tools in Molecular Biology*, 1–9. Washington, D.C.: National Academy Press.

Nicol, D. 2005. "Why Australian Biotech Would Benefit from a Clearing House." Paper presented at the OECD Expert Roundtable on Collaborative Intellectual Property Rights Mechanisms, Washington D.C., 8–9 December.

Nicol, D., and J. Nielsen. 2003. *Patents and Medical Biotechnology: An Empirical Analysis of Issues Facing the Australian Industry*. Centre for Law and Genetics Occasional Paper no. 6. Hobart: Centre for Law and Genetics, University of Tasmania.

Nielsen, J. 2002. "Biotechnology Patent Licences and Anti-Competitive Conduct." In *Regulating the New Frontiers: Legal Issues in Biotechnology*. Centre for Law and Genetics Occasional Paper no. 4. Hobart: Centre for Law and Genetics, University of Tasmania.

North, D., and B. Weingast. 1989. "Constitutions and Commitment: The Evolution of Institutions Governing Public Choice in Seventeenth Century England." *Journal of Economic History* 49 (4): 803–832.

Nottenburg, C., P. G. Pardey, and B. D. Wright. 2002. "Accessing Other People's Technology for Non-Profit Research." *Australian Journal of Agricultural and Resource Economics* 46 (3): 389–416.

Nuvolari, A. 2004. "Collective Invention during the British Industrial Revolution: The Case of the Cornish Pumping Engine." *Cambridge Journal of Economics* 28: 347–363.

O'Neill, G. 2003. "Coming Soon: The DIY Plant Biotech Kit." *Australian Biotechnology News*, 6 June.

Opderbeck, D. W. 2004. "The Penguin's Genome, or Coase and Open Source Biotechnology." *Harvard Journal of Law and Technology* 18 (1): 167–227.

Open Source Initiative. "Halloween Document I" (Version 1.17). http://catb.org/~esr/halloween/halloween1.html#quote4.

———. "License Proliferation." http://opensource2.planetjava.org/docs/policy/licenseproliferation.php.

———. 2006. "Open Source Definition (Annotated)," version 1.9. http://www.opensource.org/docs/definition.php.

O'Reilly, T. 1999. "How the Web Was Almost Won." Salon.com Technology. http://www.salon.com/tech/feature/1999/11/16/microsoft_servers/.

———. 2005. "The Open Source Paradigm Shift." In J. Feller, B. Fitzgerald, S. Hissam, and K. R. Lakhani (eds.), *Perspectives on Free and Open Source Software.* Cambridge, Mass.: MIT Press.

Organisation for Economic Co-operation and Development. 2006. *Guidelines for the Licensing of Genetic Inventions.* Paris: OECD, Directorate for Science, Technology and Industry. http://www.oecd.org/document/26/0,2340,en_2649_34537_34317658_1_1_1_1,00.html.

Owen-Smith, J., and W. Powell. 2004. "Knowledge Networks as Channels and Conduits: The Effects of Spillovers in the Boston Biotechnology Community." *Organization Science* 15 (1): 5–21.

Perens, B. 1999. "The Open Source Definition." In C. DiBona, S. Ockman, and M. Stone (eds.), *Open Sources: Voices from the Open Source Revolution.* Cambridge, Mass.: O'Reilly. http://www.oreilly.com/catalog/opensources/book/perens.html.

———. 2005. "The Emerging Economic Paradigm of Open Source." *First Monday.* Special issue no. 2, *Open Source.* http://ww.firstmonday.org/issues/special10_10/perens/index.html.

Pettit, P. 1996. "Institutional Design and Rational Choice." In R. E. Goodin (ed.), *The Theory of Institutional Design,* 54–89. Cambridge: Cambridge University Press.

Pisano, G. P. 2006. "Can Science Be a Business? Lessons from Biotech." *Harvard Business Review* 84 (10): 114–125.

Polanyi, M. 1958. *Personal Knowledge: Towards a Post-Critical Philosophy.* Chicago: University of Chicago Press.

———. 1962. "The Republic of Science: Its Political and Economic Theory." *Minerva* 1: 54–74.

Popper, K. R. 1947. *The Open Society and Its Enemies*. London: Routledge.

———. 1963. *Conjectures and Refutations: The Growth of Scientific Knowledge*. London: Routledge.

Porter, M. 2001. "Strategy and the Internet." *Harvard Business Review* 79 (3): 63–80.

Powell, W. W. 1990. "Neither Market nor Hierarchy: Network Forms of Organization." *Research in Organizational Behavior* 12: 295–336.

———. 2001. "Networks of Learning in Biotechnology: Opportunities and Constraints Associated with Relational Contracting in a Knowledge-Intensive Field." In R. C. Dreyfuss, D. Zimmerman, and H. First (eds.), *Expanding the Boundaries of Intellectual Property: Innovation Policy for the Knowledge Society*, 251–266. Oxford: Oxford University Press.

Powell, W. W., K. W. Koput, and L. Smith-Doerr. 1996. "Interorganizational Collaboration and the Locus of Innovation: Networks of Learning in Biotechnology." *Administrative Science Quarterly* 41: 116–145.

Poynder, R. 2006. "Biological Open Source: Interview with Richard Jefferson." *The Basement Interviews*, no. 9 (22 September). http://poynder.blogspot.com/2006/09/interview-with-richard-jefferson.html.

Pray, C. E., and A. Naseem. 2003. *The Economics of Agricultural Biotechnology Research*. ESA Working Paper no. 03-07. Rome: Economic and Social Department, Agricultural Economics Division, Food and Agriculture Organization. http://www.fao.org/docrep/007/ae040e/ad040e00.htm.

The Primer: A Biotechnology Guide for Non-Scientists. 2007. https://biotech.learn.com/.

Rabinow, P. 1996. *Making PCR: A Story of Biotechnology*. Chicago: University of Chicago Press.

Rai, A., and J. Boyle. 2007. "Synthetic Biology: Caught between Property Rights, the Public Domain, and the Commons." *Public Library of Science Biology* 5 (3): e58.

Rai, A. K., and R. Eisenberg. 2003. "Bayh-Dole Reform and the Progress of Biomedicine." *Law and Contemporary Problems* 66 (1): 289–314.

Ramaswami, B. 2002. "Understanding the Seed Industry: Contemporary Trends and Analytical Issues." Paper presented at the 62nd Annual

Conference of the Indian Society of Agricultural Economics, New Delhi, 19–21 December.

Raymond, E. S. 2000a. *A Brief History of Hackerdom*. Version 1.24. http://www.catb.org/~esr/writings/cathedral-bazaar/hacker-history/.

———. 2000b. *The Cathedral and the Bazaar.* Version 3.0, http://www.catb.org/~esr/writings/cathedral-bazaar/cathedral-bazaar/. Previously published 1998 as E. S. Raymond, "The Cathedral and the Bazaar," *First Monday* 3 (3). http://www.firstmonday.org/issues/issue3_3/raymond/index.html.

———. 2000c. *Homesteading the Noosphere*. Version 3.0. http://www.catb.org/~esr/writings/cathedral-bazaar/homesteading/. Previously published 1998 as E. S. Raymond, "Homesteading the Noosphere," *First Monday* 3 (10). http://www.firstmonday.org/issues/issue3_10/raymond/index.html.

———. 2000d. *The Magic Cauldron*. Version 3.0. http://www.catb.org/~esr/writings/cathedral-bazaar/magic-cauldron/.

———. 2004. "Java Is Not a Bazaar: Open Letter to Sun," *Linuxworld Magazine,* 25 November. http://linux.sys-con.com/read/47210.htm.

———. 2006. *The Cathedral and the Bazaar.* http://www.catb.org/~esr/writings/cathedral-bazaar/. Previously published in book form: E. S. Raymond. 2001. *The Cathedral and the Bazaar: Musings on Linux and Open Source by an Accidental Revolutionary.* Cambridge, Mass.: O'Reilly.

Rimmer, M. 2005. "The Freedom to Tinker: Patent Law and Experimental Use." *Expert Opinion on Therapeutic Patents* 15 (2): 167–200.

Roa-Rodriguez, C., and C. Nottenburg. 2003. *Agrobacterium-Mediated Transformation of Plants,* version 3 (July). Canberra: CAMBIA. http://www.patentlens.net/daisy/patentlens/g3/tech_landscapes.html.

Roberts, L. 1992. "NIH Gene Patents, Round Two." *Science* 255 (21 February): 912–913.

———. 2001. "Controversial from the Start." *Science* 291 (16 February): 1182–88.

Rosen, L. (ed.) 2005. *Open Source Licensing: Software Freedom and Intellectual Property Law.* Upper Saddle River, N.J.: Prentice Hall.

Rosen, L., D. Schellhase, Y. Lind, and B. Lard. 2003. "Live from Silicon Valley, Views of Open Source Practitioners." In B. Fitzgerald and G. Bassett (eds.), *Legal Issues Relating to Free and Open Source Software,* vol. 1 of *Essays in Technology Policy and Law,* 37–62. Brisbane: Queensland University of Technology School of Law.

Rosenberg, N. 1982. *Inside the Black Box: Technology and Economics.* New York: Cambridge University Press.

Ryan, M. 1998. *Knowledge Diplomacy: Global Competition and the Politics of Intellectual Property.* Washington, D.C.: Brookings Institution Press.

Schumacher, E. F. 1999. *Small Is Beautiful: Economics as if People Mattered: 25 Years Later . . . with Commentaries.* Richmond: Hartley & Marks. Orig. pub. 1973.

Schumpeter, J. A. 1975. *Capitalism, Socialism and Democracy.* New York: Harper. Orig. pub. 1942.

Sell, S. 2003. *Private Power, Public Law: The Globalization of Intellectual Property Rights.* Cambridge: Cambridge University Press.

Shoemaker, R., et al. 2001. *Economic Issues in Agricultural Biotechnology.* Agriculture Information Bulletin no. AIB762. http://www.ers.usda.gov/Publications/AIB762/. Washington, D.C.: Economic Research Service, United States Department of Agriculture.

Silverman, A. 1990. "Intellectual Property Law and the Venture Capital Process." *High Technology Law Journal* 5 (1): 157–192.

Smith, A. 1904. *An Inquiry into the Nature and Causes of the Wealth of Nations.* London: Methuen.

Smith, J. S. 1990. *Patenting the Sun: Polio and the Salk Vaccine.* New York: William Morrow.

Spielman, D. J., and K. V. Grebmer. 2004. "Public-Private Partnerships in Agricultural Research: An Analysis of Challenges Facing Industry and the Consultative Group on International Agricultural Research." EPTD Discussion Paper no. 113. Washington, D.C.: International Food Policy Research Institute.

Spillane, C. 1999. "Recent Developments in Biotechnology as They Relate to Plant Genetic Resources for Food and Agriculture." FAO Commission on Genetic Resources for Food and Agriculture Backyard Study Paper no. 9. http://www.fao.org/WAICENT/FAOINFO/AGRICULT/cgrfa/docs.htm#bsp.

Spolsky, J. 2004. "Platforms." http://www.joelonsoftware.com/articles/Platforms.html.

Stallman, R. M. 1999. "The GNU Operating System and the Free Software Movement." In C. DiBona, S. Ockman, and M. Stone (eds.), *Open Sources: Voices from the Open Source Revolution,* 1–12. Cambridge, Mass.: O'Reilly. http://www.oreilly.com/catalog/opensources/book/stallman.html.

Stewart, B. 2002. "Ewan Birney's Keynote: A Case for Open Source Bioinformatics." http://www.oreillynet.com/pub/a/network/2002/01/28/bioday1.html.

Straus, J., H. Holzapfel, and M. Lindenmeir. 2002. "Empirical Survey on 'Genetic Inventions and Patent Law.'" Presented at the OECD Expert Workshop on Genetic Inventions, Intellectual Property Rights and Licensing Practices, Berlin, 2002.

Sulston, J., and G. Ferry. 2002. *The Common Thread.* London: Random House.

Sunstein, C. R. 1996. "Social Norms and Social Roles." *Columbia Law Review* 96: 903–968.

Taylor, G. 2006. "The Synaptic Leap Open Source Biomedical Resource: An Introduction." Paper presented at Science Foo Camp, August 2006. http://www.thesynapticleap.org/?q=node/122.

Thompson, N. 2002. "May the Source Be with You." *Washington Monthly,* July/August, 1–5.

UNAIDS (Joint United Nations Programme on HIV/AIDS). 2006a. "Joining Forces to Tackle TB and HIV." http://www.unaids.org/en/MediaCentre/PressMaterials/FeatureStory/20061124_TB+and+HIV_en.asp.

———. 2006b. "Report on the Global AIDS Epidemic." http://www.unaids.org/en/HIV_data/2006GlobalReport/default.asp.

Urban, G. L., and E. von Hippel. 1988. "Lead User Analyses for the Development of New Industrial Products." *Management Science* 34 (5): 569–582.

van Zimmeren, E., B. Verbeure, G. Matthijs, and G. Van Overwalle. 2006. "A Clearing House for Diagnostic Testing: The Solution to Ensure Access to and Use of Patented Genetic Inventions?" *Bulletin of the World Health Organization* 84: 337–424.

Verbeure, B., G. Matthijs, and G. Van Overwalle. 2006. "Analysing DNA Patents in Relation with Diagnostic Genetic Testing." *European Journal of Human Genetics* 14: 26–33.

Verbeure, B., E. van Zimmeren, G. Matthijs, and G. Van Overwalle. 2006. "Patent Pools and Diagnostic Testing." *Trends in Biotechnology* 24: 115–120.

Vetter, G. R. 2006. "Exit and Voice in Free and Open Source Software Licensing: Moderating the Rein over Software Users." University of Houston Law Center No. 2005-W-02. February 27. http://ssrn.com/abstract=839167.

von Gavel, S. 2001. "Biotechnology Licensing." Course material presented at Licensing Executives Society–Benelux Licensing Course 2001, Leuven, Belgium, 13–14 December.

von Hippel, E. 1988. *The Sources of Innovation.* Oxford: Oxford University Press.

———. 1994. "'Sticky Information' and the Locus of Problem Solving: Implications for Innovation." *Management Science* 40 (4): 429–439.

———. 2002. "Open Source Projects as Horizontal Innovation Networks—by and for Users." MIT Sloan Working Paper no. 4366-02. Cambridge, Mass.: Massachusetts Institute of Technology. http://opensource.mit.edu/papers/vonhippel3.pdf.

———. 2005. *Democratizing Innovation.* Cambridge, Mass.: MIT Press.

von Hippel, E., and R. Katz. 2002. "Shifting Innovation to Users via Toolkits." *Management Science* 48 (7): 821–834.

von Hippel, E., and G. von Krogh. 2001. "Open Source Software and a 'Private-Collective' Innovation Model: Issues for Organization Science." *Organization Science* 14 (2): 209–223.

von Krogh, G., S. Spaeth, and K. Lakhani. 2003. "Community, Joining, and Specialization in Open Source Software Innovation: A Case Study." *Research Policy* 32 (7): 1217–41.

von Krogh, G., and E. von Hippel. 2003. "Special Issue on Open Source Software Development." *Research Policy* 32 (7): 1149–57.

Walker, J. M. 2006. "There Is No Open Source Community." Cambridge, Mass.: O'Reilly. O'Reilly ONLamp.com (1 December). http://www.onlamp.com/pub/a/onlamp/2006/01/12/no_oss_community.html.

Walsh, J. P., A. Arora, and W. M. Cohen. 2003. "Effects of Research Tool Patents and Licensing on Biomedical Innovation." In W. M. Cohen and S. A. Merrill (eds.), *Patents in the Knowledge-Based Economy,* 285–340. Washington, D.C.: National Academies Press.

Walsh, J., and Wei Hong. 2003. "Secrecy Is Increasing in Step with Competition." *Nature* 422: 801.

Watson, J. D. 1968. *The Double Helix: A Personal Account of the Discovery of the Structure of DNA.* New York: Atheneum.

Wayner, P. 2000. *Free for All: How Linux and the Free Software Movement Undercuts the High-Tech Titans.* New York: Harper Business. http://www.wayner.org/books/ffa/.

Weatherall, K. 2006. "Would You Ever Recommend a Creative Commons

License?" Australian Intellectual Property Resources. http://www.austlii.edu.au/au/other/AIPLRes/2006/4.html.

Weber, S. 2004. *The Success of Open Source*. Cambridge, Mass.: Harvard University Press.

West, J. 2003. "How Open Is Open Enough? Melding Proprietary and Open Source Platform Strategies." *Research Policy* 32 (7): 1259–85.

Williamson, O. 1979. "Transaction Cost Economics: The Governance of Contractual Relations." *Journal of Law and Economics* 22: 233–261.

———. 1985. *The Economic Institutions of Capitalism*. New York: Free Press.

———. 1991. "Comparative Economic Organization: The Analysis of Discrete Structural Alternatives." *Administrative Science Quarterly* 36 (2): 233–261.

———. 1999. "Strategy Research: Governance and Competence Perspectives." *Strategic Management Journal* 20 (12): 1087–1108.

World Intellectual Property Organization. 1992. *Guide to the Licensing of Biotechnology*. WIPO Pub. No. 708(E). Geneva: WIPO.

Worthen, B. 2004. "Mr. Gates Goes to Washington." *CIO Magazine,* 15 September. http://www.cio.com/archive/091504/microsoft.html (page discontinued).

Wright, B. D. 1998. "Public Germplasm Development at a Crossroads: Biotechnology and Intellectual Property." *California Agriculture* 52 (6): 8–13.

Wright, S. 1998. "Molecular Politics in a Global Economy." In A. Thackray (ed.), *Private Science: Biotechnology and the Rise of the Molecular Sciences,* The Chemical Sciences in Society Series, 80–104. University of Pennsylvania Press, Philadelphia.

Websites

BioForge: http://www.bioforge.net/forge/index.jspa

BSI: http://www.bsi-global.com/

CAMBIA: http://www.cambia.org/daisy/cambia/about_cambia/592/589.html

Creative Commons: http://creativecommons.org/

GAIN Program, National Institutes of Health: http://test.fnih.org/GAIN/policies.shtml

HapMap Project International: http://snp.cshl.org/
ISEAL (International Social and Environmental Accreditation and Labelling Alliance): http://www.isealalliance.org/index.cfm?nodeid=1
Open Source Initiative: http://www.opensource.org
OpenWetWare: http://openwetware.org/wiki/Main_Page
PIIPA (Public Interest Intellectual Property Advisors): http://www.piipa.org/
PIPRA (Public Intellectual Property Resource for Agriculture): http://www.pipra.org/main/contact.htm
Science Commons: http://sciencecommons.org/
SETI@home: http://setiweb.ssl.berkeley.edu/
SourceForge: http://sourceforge.net/

Acknowledgments

Researching and writing this book has been a major part of my life for over six years. In that time I have benefited from innumerable acts of kindness and generosity on the part of family, friends, colleagues, and others. While I remain deeply grateful for all the help and encouragement I have received at every stage of this project, previous milestones have given me the opportunity to acknowledge earlier contributions. My present task is to thank those whose assistance and support rendered (just) possible the unexpectedly difficult task of converting an existing body of research into a published book.

Many of the people who helped with this phase have been involved one way or another since the project first began. They deserve special recognition for sheer stamina. One is Peter Drahos, who taught me undergraduate intellectual property law. His lectures must have planted some seeds in the shifting, muddy estuary of my brain, because it wasn't too many years later that I approached him to act as a mentor on this project. In one of our early discussions, Peter confirmed my inkling that open source biotechnology was a topic of sufficient interest to warrant a sustained exploration. He also demonstrated the restraint and generosity that have characterized all his dealings with me: when I rudely dismissed an idea that turned out to be his own, his only reaction was to observe some weeks later, in a totally different context, that it is always important to respect and appreciate other people's intellectual efforts. Though I still sometimes struggle to put this lesson into practice, I am frequently reminded of its great value by Peter's unfailingly constructive approach.

Another person whose long-term support has made an enormous difference to this book is Dianne Nicol. Responding to a long-distance query about her own research back in 1994, Di was so friendly and informative that I made a point of renewing contact with her when I began work on this project. It was Di who introduced me to the skeptical literature on intellectual property rights that forms the theoretical background to this book. She subsequently became a respected mentor and collaborator whose patience, kindness, and enthusiasm are matched only by her scholarly rigor and rare personal humility.

Andrzej Kilian and Eric Huttner, partners in the commercialization of DArT, have also been stalwart supporters of my efforts to "grok" open source. Both have been incredibly generous in sharing their ideas and experiences as biotechnology entrepreneurs. While I still hope to repay their investment of time, energy, and trust, I am fortunate that neither Andrzej nor Eric has any difficulty grasping the notion of diffuse reciprocity: their contributions were genuine gifts, no strings attached. Had it been otherwise, I would not have had the time and intellectual freedom required to complete this work. In this connection I would also like to thank Peter Wenzl for his ongoing support.

My institutional home while writing this book has been the Centre for Governance of Knowledge and Development, part of the Regulatory Institutions Network (RegNet) at the Australian National University. This wonderful group has been a great source of friendship and practical help. Particular thanks are due to Carolina Roa-Rodriguez, Warwick Neville, and Cecily Stewart for reading chapters of the draft manuscript and to Anna Hutchens, Carmen

San Miguel, Cameron Neil, and Luigi Palombi for informative and stimulating discussions in the broader subject area. Other RegNet colleagues whose camaraderie helped ameliorate the pain of spending long hours away from my young family include Jen Wood, Jeremy Farrell, and John Braithwaite. Together with Dianne Nicol, John is a co-investigator on my current Australian Research Council grant; like Di, he has been a model of patience and goodwill as I continue to dedicate a disproportionate amount of time to this aspect of our joint efforts. Elsewhere on campus, I have benefited from interactions with Matthew Rimmer of the Law Faculty, Don Scott-Kemmis of the Faculty of Economics and Commerce, and Jochen Gläser and Grit Laudel of the Research Evaluation and Policy Project. Farther afield, Andrew Christie, Kim Weatherall, and Pia Waugh have all provided useful encouragement, orientation, and feedback.

Australians have long been significant players in free and open source software development; even Tux, the Linux penguin, is rumored to trace his ancestry to a particularly vicious inmate of Canberra zoo. But while Australian scientists have also made important contributions to biotechnology research and development, Canberra is a long way from the scientific and industrial centers that constitute the heart of the global biotechnology enterprise. The sixty or so face-to-face interviews that formed the initial empirical foundation of this work were conducted in the United States in 2003, on a field trip jointly funded by the Intellectual Property Research Institute of Australia and the Australian National University. During that trip I was privileged to meet a number of people who have provided a link to ongoing research and practical initiatives in the field, including Sara Boettiger, Drew Endy, and Eric von Hippel.

On subsequent trips to the United States and Europe I made new contacts with people who informed my thinking in a variety of important ways: John Stewart, Victoria Henson-Apollonio, and Fiona Murray come to mind, but there were many others. Although this project has been generously funded since 2005 by a Discovery Project grant from the Australian Research Council, most international travel would have been impossible without both supplementary funding and a clear-cut occasion to take time out from domestic responsibilities. I am especially grateful to Christina Sampogna of the OECD Biotechnology Division for organizing an Expert Roundtable on Collaborative Intellectual Property Rights Mechanisms in Washington in December 2005; to Hugh Hansen of the Fordham University School of Law for inviting me to participate in and help organize Fordham's Fourteenth Annual Conference on International Intellectual Property Law and Policy in April 2006; and to Geertrui Van Overwalle and her lively group at the Catholic University of Leuven's Center for Intellectual Property Rights, who showed extraordinary hospitality during and after their International Workshop on Gene Patents and Clearing Models in June 2006. Others who provided accommodation and essential travelers' aids such as desk space, telephone, and Internet connectivity include Brian Wright of the University of California at Berkeley's Department of Agricultural and Resource Economics and my generous friends Nigel Snoad and Miranda Sissons. Brian's quiet, consistent encouragement has been a greater force for the successful conclusion of this project than he could possibly be aware. Nigel and Miranda welcomed me into their small Brooklyn apartment at a time when privacy and leisure were at a premium; in different ways, both helped me to keep moving forward against increasing internal resistance. I am similarly grateful to Birgit Verbeure for helping restore my self-confidence at a time of great personal strain.

While the qualitative research for this book was conducted in person as far as possible, several key informants were kind enough to tolerate the inconvenience—in some cases substantial—of remote communication methods. They are John Wilbanks of Science Commons, Ginger Taylor of The Synaptic Leap, Matt Todd of the University of Sydney's School of Chemistry, and Tom Michaels of the University of Minnesota's Department of Horticultural Science. Thanks are also due to my kindly fellow panel organizers from the Fordham conference, Charles Fish of Time-Warner/AOL and Nick Groombridge of Weil, Gotshal & Manges LLP, and to those members of the Open Source Initiative's license-discuss email list who responded to my post seeking feedback on specific open source biotechnology licenses.

Aside from the tyranny of distance, a number of extraneous factors rendered work on this book particularly slow and taxing. I began by telling my editor, Michael Fisher, that I would complete the manuscript by December 2005. Since then I have repeatedly tried his patience without, it seems, ever quite exhausting it. Despite Michael's assurances that this "always happens," I would like him to know that I appreciate his faith and forbearance. Others who deserve credit for their work on the publishing side are Anne Zarrella, Alex Morgan, and manuscript editor Wendy Nelson. Eric von Hippel, Rob Carlson, Larry Rosen, Larry Lessig, and Steve Weber provided valuable advice regarding the negotiation of a publishing contract; it was Rob who first put me in touch with Harvard University Press. Hearty thanks also go to the book's reviewers, all of whom took the time to provide detailed and constructive comments that substantially improved the quality of the final text.

Throughout the writing process I have leaned heavily on friends for emotional and other kinds of sustenance. Kathryn Dwan, Gabrielle McKinnon, Chris and Rebecca Drew, Nicky Grigg, Tim and Natalie Maddalena, and many others helped in a whole host of ways, from cooking dinner to providing kids' entertainment to absorbing yet another self-centered rant about how much I'd rather be listening to someone else's problems than going on and on about my own. Sophie Cartwright deserves a unique citation for heroically unblocking my chi on several occasions, using whatever instruments came to hand (I'm sure she won't object to my glossing over the details). Other friends and family graciously tolerated my extended withdrawals from the ebb and flow of ordinary social life; their contribution, though indirect, was indispensable. The same is true of the highly professional and wonderfully caring staff of the Australian National University Preschool and Childcare Centre.

Many authors say they could not have reached this point without the support of their families. In my case that could not be more literally true. My parents Anne and John, my sister Margaret, and my husband Joe all read chapter drafts and helped clarify ideas. Anne, John, and Joe all spent many hours feeding, clothing, changing, bathing, teaching, entertaining, and loving our boy. John and Joe helped manage the idiosyncratic IT setup necessitated by my use of voice software. John performed a seemingly infinite number of tedious clerical tasks at every stage of the project, all without lapsing into a coma even once. Anne cooked dozens of meals, folded a mountain of laundry, and gave me the benefit of both her professional editing experience and her motherly wisdom. She knew precisely how I felt about this book. At times I suspected my parents knew the answer to Virginia Woolf's famous question: "Why do you attach so much importance to this writing of books by women, when . . . it requires so much effort, leads perhaps to the murder of one's aunts, will make one almost certainly late for luncheon, and may bring one into very grave disputes with certain very good fellows?" I'm pretty sure now that they *don't* know it, which only makes their extraordinary selflessness all the more remarkable. Joe made significant intellectual contributions throughout the project, especially at the final stage; if this were a journal article, he would be a co-author. But his greatest contribution was to be himself—which is to say, always and everywhere to substitute love for fear. Meanwhile, through the darkest of dark times, Margaret and her partner David took excellent care of another person who is infinitely precious to me. Dear family, I owe you all heartfelt thanks for the rare gift of freedom to "let the line of thought dip deep into the stream."

James, I have not forgotten you. As I lift my eyes from the screen at long last, I can just make out a glint of golden hair in the sun-filled garden beyond this dim and dusty room. I rise slowly and stand blinking in the doorway, blood gradually returning to stiffened limbs. Your whole body is turned away from me towards the light; but just as you seem about to spring away, you suddenly turn and reach for my hand. "Come on, Mum," you say. "Let's run!"

Index

17 U.S.C. 101, 158
17 U.S.C. 106, 165, 182
35 U.S.C. 112, 173, 174

Academic contributors, 125
Academic free revealing. *See* Traditional biobazaar
Academic or permissive open source licenses, 155, 161, 162, 186, 216; and copyleft licenses, 180–181
Academic science: and bazaar production, 108–109, 190, 193, 205, 231; and biotechnology, 105, 190, 191, 193, 202, 231, 235; and collective action model of innovation, 231; commercialization of, 32, 76, 92, 140, 200; and commons-based peer production, 87; differences from open source biobazaar, 235–236; freedom of in early postwar period, 31–32; and modularity, 202. *See also* Traditional biobazaar
Africa, HIV/AIDS in, 97
Agreement on Trade-Related Aspects of Intellectual Property Rights. *See* Trade-Related Aspects of Intellectual Property Rights (TRIPS)
Agribusiness, 58; and control over business decisions of smaller companies, 281–282; and modularity, 203
Agricultural biotechnology, 100–102; competition in, 100–101; definition, 57–58; and development of genetically modified seeds, 269; development process, 58–59; and hybridization and genetic use restriction technologies, 327–328; industry structure and transaction costs, 64–65; knowledge game and, 101; licensing in developed and developing countries, 63–64; and low cost integration, 218; and medical biotechnology, 63; opportunity costs of open source strategy in, 270, 272, 275; ownership of intellectual property in, 63, 65; patent litigation in, 45; patents and development and commercialization of products of, 40–41; private sector and, 100–101; public funding of, 100–101, 287; public-sector institutions and patents in, 275–276; and traditional bazaar production, 191–192; and tragedy of the anticommons, 65–66, 101. *See also* General Public License for Plant Germplasm (GPLPG)

Agricultural companies, 269, 270
Agricultural extension services, 59
Agrobacterium-mediated transformation, 45, 65
Alliance for Cellular Signaling, 247
Amazon, 127
Amesse, Fernand, 231
Amplified Fragment Length Polymorphisms (AFLP) Workshop, 357n68
Angell, Marcia, 191
Anticommons tragedy. *See* Tragedy of the anticommons
Anti-patent movement in Britain, 70
Antitrust patent-based cartels: and antitrust laws, 70, 91–92; and biotechnology, 92; and competition, 93–94; market power of, 94
Apache and Linux, 302
Application technologies. *See* Niche or application technologies
Appropriate technology (AT) movement, 327–328, 330
Arrow, Kenneth, 71–72
AZT, 190–191

Back-end software, 152
Bargaining failure/breakdown. *See* Tragedy of the anticommons
Bayh-Dole Act (USA): and commercialization of life sciences, 32–33, 71; and patenting by universities and public hospitals, 273, 274; and technology transfer in public interest, 274–275
Bazaar, 17–18
Bazaar development/production, 108–109, 137; and academic science, 108–109, 190, 193, 205, 231; advantages of, 134–141; and agricultural biotechnology, 191–192; and biotechnology, 199; and copyleft, 135; and costs of quality control, 219; defined, 17–18; and drug development, 190–191; effect of incentives on production costs, 135–136; in industrial setting, 239–240; and open source software, 108–109, 142, 189; and overcapacity, 210; and private rewards, 114; and reputation, 133; and use value over sale value, 135; and virtual pharmaceutical companies, 289. *See also* Bazaar governance; Biobazaar; Traditional biobazaar
Bazaar governance, 25, 107, 189–190, 329; and biobazaar, 331; in biotechnology, 26, 190–193; and codified knowledge, 131; coexistence with firm, market, and network governance, 129–131, 331; and community, 130, 230; and control mechanisms, 112, 135, 141, 194, 279; and costs in biotechnology and in software development, 218; costs of versus costs of other modes of production, 218–219; and decentralized ordering of transactions, 109; differences from firm, market, and network governance, 18, 109–111, 129, 131, 141, 194, 331; and difficulties of conventional contractual/partnership agreements, 290; and diffuse or generalized reciprocity, 110; and free revealing, 112; and free riding, 110, 112; and horizontally networked user innovation, 212; and horizontal user innovation network, 114; incentives (*see* Bazaar incentives); and inventing around overly broad patents, 139–140; and matching human

capital with information resources, 197; motivations of agents operating within, 110; and networks in biotechnology, 231; and nonproprietary exploitation strategies, 112; open membership and self-selection in, 111, 112; and open source software development, 131–132; and price signals, 110; and production costs, 280; relational ties in, 110, 132, 133, 230, 231; and self-selection of individuals for particular tasks, 111, 112, 194; and spectrum of involvement, 113–114; and transaction costs, 111–112; and uncertainty, 279–280; and use of intellectual property rights to prevent appropriation, 112; and use of property to promote distribution and sharing of product, 111

Bazaar incentives, 112, 114, 115, 127–129, 134–135, 194, 205, 279; and biotechnology, 240; and business strategies in biotechnology, 237–265; and innovation, 239

Bean Improvement Cooperative, 306

Behlendorf, Brian, 113

Benefits from innovation, 115–116, 128; complier-centered and deviant-centered institutional design principles for promoting, 184–185, 186; freely revealed technology, 122, 128; institutional design and, 183–184; from nonproprietary exploitation strategy, 119, 128; open source licensing and, 183. *See also* Incentives; Process benefits/incentives to innovate

Benkler, Yochai: and bazaar governance, 131; on commons-based peer production, 87–88, 112–113, 114, 193, 196–197, 198, 199; on distributed production, 194; on IBM and open source software, 127; on Internet-enabled bazaar production, 201, 202, 211, 214; on laboratory funding as silo-based, 209; on low cost integration, 214, 219; on proprietary exploitation strategies, 296; on sharing practices that are not community-dependent, 130; *The Wealth of Networks,* 127, 194, 209, 296

Bennett, Alan, 274–275

Berkeley Software Distribution (BSD) License, 155, 216

"Best mode" requirement (patents), 173, 174, 175–176

Big science, 199–200

Bill & Melinda Gates Foundation, 288

BIND, 16

Biobazaar, 18, 189; ability of to access sources of innovation without restrictions on technology freedom, 291; differing models of, 331–332; motivation of participants in, 331; open source and traditional, 26, 190, 193, 235–236, 237, 315; and participation of commercial actors motivated by private incentives, 133, 237, 263–265; relationship of commercial and noncommercial contributors to, 330; speculative model of, 21–24. *See also* Open source biotechnology; Traditional biobazaar

BioBricks Foundation (BBF), 316

BioForge, 317

Bioinformatics, 143, 255, 257

Biological Innovation for Open Society (BIOS), 316–317

Biological materials, 47, 144, 160, 211, 218; cost of transfer of, 184, 278; and disclosure requirement, 173; transfer of by post, 213. *See also* Material transfer agreement (MTA); Personal property rights
Biological Open Source (BiOS) licensing, 260, 317–318
Biologics, 54
Biomedical research and development, 99. *See also* Medical biotechnology
Biosafety regulation, 102
Biotechnological innovation, 364n58
Biotechnology: academic science and, 105, 190, 191, 193, 202, 231, 235; bazaar governance in, 26, 190–193, 231; bazaar incentives and business strategies in, 237–265; and bazaar production, 199, 212; and benefit-sharing, 177; and "big science," 199–200; capital investment in, 196–197, 198–199, 218; and central user facilities, 201; and codification of information, 172, 175, 244–245; and collective action communities, 231–232; and community, 230, 232–233 (*see also* Open source biotechnology, community); patent situation in, 44–45; and computational methods, 200–201; contracting for knowledge in, 41–51; as convivial technology, 328–329; and copyleft licenses, 180; and copyright protection, 160; corporate capital for research and development in, 140; cost of regulatory approval of products of, 213–214, 286; definition, 53; democratization of, 104, 105; and disclosure requirement, 172–173, 175, 183, 187; and elitism, 200, 201; and entitlement rights in open source licensing, 172–173; and exchange of information, 218; for-profit and nonprofit innovators and, 140–141; and free revealing, 241; implications of strengthening intellectual property rights in, 96–102, 103, 106; and granularity, 203, 204, 205, 210; and heterogeneous user need, 246–247; and industrial innovation, 141; instrumental justifications for intellectual property protection in relation to, 68, 69–74, 88; intellectual property rights/protection in, 31–35, 66, 106, 245; and Internet-enabled peer production, 198; interoperability in, 249–250; knowledge production networks in and proximity to universities, 231; and low cost integration, 211–218; and mixed business models, 258–259, 272; and modularity, 202–203, 204, 205; network effects in, 249; and networks/network governance, 231; nonprofit and public sector and, 192, 193; and nonproprietary business strategies, 240, 265; and open source, 18; open source business models and commercialization of, 141; open source community in, 233; and open source licensing, 142–143, 219–220, 304 (*see also* Open source biotechnology licenses and licensing); and open source software, 188, 189, 194–195, 196, 210–211, 218, 219; and ownership rights, 160; patent-based cartels and, 92; and patents, 144, 159, 188; and patterns of ownership, 143–144; and personal property rights, 47, 144, 160; political settlements embodied in, 329; and

pool of contributors, 194–195, 196, 218; potential versus successes of, 103–104; and process benefits, 241; and productivity of pharmaceutical research and development, 272; and proprietary manufacturing strategy, 240, 241; public funding for, 132, 287; refactoring data stream in, 204–205; and screening to maximize compliance with terms of collaboration, 184; and source code, 171–172; sticky information in, 244, 245–246; and tragedy of the anticommons, 52, 60, 66–67; training for research in, 205–211, 218; translation of software freedom into, 165, 166, 228 (*see also* Biotechnology freedom); and user innovation, 243, 244, 247–248; and use value, 136–137. *See also* Agricultural biotechnology; Medical biotechnology; Molecular biotechnology

Biotechnology companies, 3, 4, 190, 272

Biotechnology freedom, 166–167, 169, 170, 186. *See also* Open source biotechnology; Software freedom

Biotechnology industry: definition, 53–54; importance of intellectual property rights in, 68; patent statistics in, 35; problems of value allocation in, 47–48, 178; profitability of, 272; and proprietary exclusivity, 239; and public perceptions, 103; relational contracting in, 107, 108; role of patents in development of, 72–73. *See also* Molecular biotechnology industry

Biotechnology licenses and licensing: legal technicalities of, 220, 225; licensing process, 47; open source (*see* Open source biotechnology licenses and licensing); and plants, 144 (*see also* General Public License for Plant Germplasm (GPLPG))

Biotechnology platform companies, 259, 271–272

Biotechnology-related information: cost of acquiring, 175; cost of exchanging, 218; exchange of, 211–212; physical capital and fixation and communication of, 197; transaction costs of transfer of, 41; uncodified, 83, 86

Blanketing (patents), 90

Blitzkrieg (patents), 90

Blockbuster: business model, 98; drugs, 267–268, 277, 310; inventions, 48, 98

Blocking patents, 44, 89; strategies for dealing with, 46, 52, 85 (*see also* Inventing around blocking patents); techniques, 90

Boettiger, Sara, 274–275

Bonaccorsi, Andrea, 109, 196

Boudreau, Tim, 299

Boyer, Herbert, 32, 77. *See also* Cohen-Boyer recombinant DNA patent

Boyle, James, 88

Bracketing (patents), 90

Braithwaite, John: on global business regulation, 94–95, 105; on import and export of intellectual property, 323, 324; *Information Feudalism,* 88; on knowledge game, 88–89, 91, 92–93, 105; on model missionaries, 324

Brand communities, 134, 232–233, 331

Brand licensing, 123

Brandt-Rauf, Sherry, 79–80, 312

Brazil, 320–321; and agricultural biotechnology patents in public sector, 275; correlation of infant mortality and real wages in, 97
Brent, Roger, 198–199
Bush, Vannevar, 2, 76
Business model (or strategic plan): open source/nonproprietary, 128. *See also* Open source business models/exploitation strategies; Proprietary business and/or exploitation strategies
Business strategies: medical biotechnology and, 270–272. *See also* Mixed business models; Nonproprietary business strategies; Open source business models/exploitation strategies

CAMBIA. *See* Center for Application of Molecular Biology in Agriculture
Capital. *See* Human capital; Physical capital; Venture capital
Capitalism, 296, 297
Carlson, Rob, 198–199
Cartels, 91
Cathedral-style development, 17, 108
Cavalla, David, 290
C. elegans genome, 212
Celera Genomics, 38–39, 257–258, 307
Center for Application of Molecular Biology in Agriculture (CAMBIA) (Australia), 65, 259–260, 316, 317, 318
Centralized coordination of research and development by patent holder, 73–74, 87
Certification signal, 121, 250–251
Chains of products in data stream, 80
Chakrabarty, Ananda, 33, 73
Chemical and pharmaceutical companies: and biotechnology, 92; business model of, 240; and control over business decisions of smaller companies, 281–282; drug development outside, 289–290; and drug development without proprietary exclusivity, 216–217; and drug research and development, 288–289; and intellectual property protection, 93, 267–268; and modularity, 203; opportunity costs to of open source strategy, 268; and payment of fees to open source biotechnology start-ups for precompetitive research and development, 256–257; profitability of, 267–268, 277; and sale of research tools, 268; and science teaching, 3, 4; and the Synaptic Leap, 315; value of patents to, 267–268. *See also* Global pharmaceutical industry; Virtual pharmaceutical companies
Clearinghouses, 364n58
Click-wrap license, 152, 349n20; and International HapMap Project, 308
Clinical trials, 56–57, 191, 214, 254
Cliques, 231
Clustering (patents), 90
Codification of information, 82–83; and biotechnology, 172, 175, 244–245; and different governance structures, 131; and dissemination of information, 83; and imitation, 83–84; and markets in information, 84–85, 86; and peer production, 347n51
Cohen, Stanley, 32, 77. *See also* Co-

hen-Boyer recombinant DNA patent
Cohen-Boyer recombinant DNA patent, 62, 273
Collaboration, articulation of terms of. *See* Terms of collaboration
Collaborative development: in biotechnology, 175; open source exploitation strategies and, 142; open source licenses and, 142, 151, 153–154, 156; and open source technology development, 229; proprietary biotechnology licenses and, 152–153; the Synaptic Leap and Tropical Diseases Initiative and, 312–313. *See also* Network collaboration
Collaborative intellectual property management, 364n58
Collaborative proprietary licenses, 152–153
Collective action, 133, 134; and academic science, 231; incentives, 133, 134, 331; subsidies to by nonprofit entities, 214, 215
Collective action communities, 133, 231–232
Collective-action–style governance, 280
Collective invention, 121–122
Collective production, 109
Collins, Harry, 77
Commercialization: of academic science, 32, 76, 92, 140, 200; of life sciences research, 3–4, 35–38, 76, 92, 200; open source business models and, 141, 236, 260; and restrictive access practices, 7–8; and scientific commons, 141. *See also* Development and commercialization theory; Privatization
Commercial research, 209–210
Commitment, and bazaar governance, 110, 331. *See also* Credible commitment; Spectrum of involvement
Common Public License, 348n3
Commons, 139, 141. *See also* Technological commons
Commons-based peer production, 87–88, 109, 112–113, 114, 193, 196–197; and codified information, 347n51; and self-selection of individuals for particular tasks, 87, 113. *See also* Bazaar development/production; Nonproprietary peer-based knowledge production
Commons tragedy. *See* Tragedy of the commons
Communism or communalism, 75–77
Community: and bazaar governance, 130, 230; and biotechnology, 230, 232–233; of cooperating hackers, 9; and incentives to innovate, 133; and open source development/production, 130, 132, 133–134, 183, 230, 233; technical (of scientists), 232. *See also* Brand communities; Collective action communities; Developer community; User communities
Community Public License, 353n66
Competition: in agricultural biotechnology, 100–101; between open source entity and proprietary company, 300–302; and biotechnology freedom, 167, 169, 170; and copyleft licenses, 179; intellectual property barriers to, 69, 91; and knowledge game, 89, 93, 102, 266, 299 (*see also* Cartels); and open source, 265; and open source biotechnology, 267; and open source

Competition *(continued)*
software licenses and licensing, 170; patent-based cartels and, 93–94; patents and, 34, 267, 269; and proprietary exclusivity, 322–323; proprietary exclusivity and, 117; in software and biotechnology industries, 19; and software freedom, 167, 168, 170; transaction costs and, 93
Competitive advantage, 253, 254, 299
Complementary marketing, 257, 258, 263, 272
Complementary products, 123–124, 256, 258, 260–261
Complier-centered strategies, 184–185, 186
Computer program, 170–171
Computer science/information technology, 4
Constructivist accounts of science, 75, 78–81, 328–329
Consultative Group on International Agricultural Research (CGIAR), 59, 101, 192, 275–276, 307
Consumers, 116, 127, 303
Contract enforcement, 49
Contracting for knowledge, 41–51
Contract research organizations, 58, 191, 289
Contract writing, 49
Contributions to open source projects, 278–279
Contributors to open source software projects: classes of, 124–127; core, 196; critical mass of, 196, 298, 307; and nonproprietary incentives, 160; pool of, 194–196; and technology freedom with respect to follow-on innovations, 216. *See also* Bazaar incentives; Contributions to open source projects
Control (in governance structures), 111, 112, 135, 141, 194, 279
Conventional contractual/partnership agreements, 290
Convivial tools, 327, 328–329
Conway, G., 42–43
Copying of technological information, 73, 83–84, 86
Copyleft, 11
Copyleft licenses, 138, 155–156, 160, 163, 186; and academic or permissive open source licenses, 180–181; and bazaar production, 135; and biobazaar, 331; and competition, 179; and data stream, 176, 180, 181; definition of downstream technologies in, 181; and derivative works, 126, 156, 168, 176, 179, 181; and distribution of downstream technology, 181, 186; and dual licensing, 125–126; and freedom of downstream technologies, 135, 138, 176–177, 178, 181, 182–183; and grant-back, 179; "hook," 11, 161, 181, 183, 309; and intellectual property, 177; and International HapMap Project, 308–309; and limits of ownership, 181–182; obligations under, 163; and obligation to disclose source code, 167; and "passing it forward," 179; and patentleft licenses, 182; and public-sector human genome project, 160; reasons for using, 161; and terms of reach-through royalties, 178. *See also* Click-wrap license; General Public License (GPL)
Copyleft-style licenses in biotechnology, 180; and Biological Open Source (BiOS) licenses, 317, 318; and General Public License for Plant Germplasm, 306

Copyright: and biotechnology, 160; and derivative works, 182; and open source software licenses, 144; and patents, 10, 165–166, 188; and proprietary software, 10; and reciprocal license terms, 182; rights of owner of, 165, 166; and software, 158; and source code, 308. *See also* Copyright Act (USA); Intellectual property
Copyright Act (USA), 165. *See also* 17 U.S.C. 101; 17 U.S.C. 106
Copyright lawyers, 222
Costs: and knowledge game, 266–267; of open source in early stage of drug development, 311; of open source strategy in biotechnology, 266, 291; and patent rights, 266. *See also* Opportunity costs; Transaction costs; Transfer costs
Court of Appeals of the Federal Circuit (CAFC), 33–34
Creative Commons, 222, 224–225, 315, 350n36
Credible commitment, 154–155, 156, 186, 331

Data, 78, 79–80, 81–82, 312. *See also* Information
Data stream: and chains of products, 80; copyleft licenses and, 176, 180, 181; model of scientific research, 79–81; and modularity, 203; and patentleft license, 181–182; refactoring of in biotechnology, 204–205
Debian Free Software Guidelines, 147, 148
Debian Linux community, 147, 148
DeBresson, Chris, 231
Decentralized versus centralized coordination of research and development, 87
Decision-making, 160, 293, 294–295
Defensive disclosure or defensive publishing, 162, 163
Demil, Benoît, 109, 112, 129, 279
Democratization of biotechnology, 104, 105
Derivative inventions/works, 182; copyleft licensing and, 126, 156, 168, 176, 179, 181; copyright law and, 182; and germplasm, 305; patent law and, 182; prospect development and, 71
Developer community, 195
Developing countries: affordability of drugs in, 99; agricultural biotechnology licensing in, 63, 64; constraints on freedom to innovate in, 339n40; and contracting for knowledge in agricultural biotechnology, 42; and germplasm, 59, 177, 304; HIV/AIDS and mortality in, 97; intellectual property rights and research on treatment of diseases in, 98; molecular biotechnology and, 42–43; research and development capacity in, 98, 99, 101; research on treatment of diseases in, 97–98
Development and commercialization theory, 71, 117
Deviant-centered strategies, 184, 186
Diamond v. Chakrabarty, 33, 73
Differentiators and enablers, 294, 302–303
Disclosure. *See* Defensive disclosure or defensive publishing
Disclosure requirement, 173–174; and biotechnology, 172–173, 175, 183, 187; and noncompliance under open source license, 185–186; and open source license, 174–175; patent law, 173, 174–175, 187
Disclosure theory, 71, 73, 117

"Discovery" versus "invention" in patent law. *See under* Patent law
Discretionary projects of employee scientists, 210, 241
Disease, 97–98. *See also* HIV/AIDS
Disruptive innovation, 19–20, 90, 93
Disruptive technology, 338n52
Dissemination of information, 74; commercial return on, 252; and degree of codification, 83; mechanisms of, 80–81; and use value, 251–252. *See also* Knowledge exchange
Distributed knowledge production, 194–195, 202
Distributors, 116, 125, 168
Diversity Arrays Technology (DArT), 259–262, 281
Divine Chocolate Ltd, 361n73
DNA: disclosure requirement in MTA relating to, 172–173; modularity and sequencing of, 202; and source code, 172; turning published sequence of into usable input for further innovation, 245–246
Domain Name System (DNS) software, 16
Dow, 64
Downstream appropriation, 161–162
Downstream technologies: copyleft licenses and, 135, 138, 176–177, 178, 181, 182–183; definition of in copyleft-style biotechnology license, 181; open source development of, 303
Drahos, Peter: on global business regulation, 94–95, 105; on import and export of intellectual property, 323, 324; *Information Feudalism*, 88; on knowledge game, 88–89, 91, 92–93, 105
Drug candidates, 55, 310
Drug development, 54–57, 190–191; conventional model of, 310; cost reductions through open source in early-stage, 311; investment in, 98; open source precompetitive collaboration and, 254; outside pharmaceutical companies, 289–290; problem of integrating contributions to, 214–216. *See also under* Open source
Drugs, 99. *See also* Blockbuster; Generic drugs
Drug targets, 55, 190, 310
Dual licensing, 162–163, 298; and open source software, 125–126, 151
DuPont, 64

eBay, 127
Economic arguments in support of intellectual property rights, 71–74, 86–87, 94
Economic incentive for innovation, 349n18
Eisenberg, Rebecca, 39–40, 60, 226
Eli Lilly, 268
Elitism, 200, 201
Email, 16–17
Empresa Brasileira de Pesquisa Agropecuária (EMBRAPA) (Brazil Ministry of Agriculture), 275
Enablement, 174, 187
Enablers, 294, 302–303
Enabling or platform technologies, 123–124; free revealing of and proprietary products, 259; nonproprietary alternatives to, 140; and open source and free revealing, 139; and open source strategy, 302. *See also* Biotechnology platform companies; Enablers

End consumers. *See* Consumers
End users. *See* Consumers
Endy, Drew, 316
Enforcement of license agreements, 184–185
Enterprise software, 152
Entitlement rights, 156, 157, 158, 164; and open source biotechnology licensing, 172–173; and open source software licenses, 170, 186
Entrepreneurs, 297
Environment movement, 283
Equitable Access licensing, 319, 320
ESTs, 37, 38, 40, 264
Evergreening patents. *See* Life-cycle management or evergreening patents
Exchange of information. *See* Knowledge exchange
Exclusive marketing rights of drug manufacturers, 215
Expert Committee on Grain Breeding (Ottawa), 306
Expert Roundtable on Collaborative Intellectual Property Rights Mechanisms, 364n58

Fairtrade Labelling Organizations (FLO), 234
Fairtrade Labelling Organizations International (FLI), 325
Fair Trade movement, 283, 325–326
Fencing (patents), 90
Field-of-use and territorial provisions in licenses, 49, 167
Fightingaids@home, 195
Firm: -based selection of individuals for particular tasks, 87, 113; and bazaar governance, 18, 109–111, 129, 131, 194, 219, 280, 331; controls and incentives, 112; definition, 108; and open source development, 131; transaction cost economics and, 107; and use of intellectual property rights to control assets, 111
Fixation of biotechnology information: costs of, 197, 199
Fleming, Alexander, 115
Flooding (patents), 90
Folding@home, 195
Follow-on innovations and innovators: and benefit-sharing with initial innovators, 177, 178; and benefits of open source technology development, 180; and copyleft licenses, 161, 176; and free revealing, 121; and technology freedom, 216
Food and Drug Administration (FDA) (USA), 56, 57, 286
Food insecurity, 96, 100, 198
Forking, 5, 169–170, 184, 346n35
For-profit entities, 55, 210, 241–243
Franchising, 123
Franklin, Roslyn, 335n1
FreeBSD, 17
Freedom: in "free speech" and "free beer," 8, 175; meaning in open source software, 31; of scientists and philosophers, 31. *See also* Software freedom; Technology freedom
Freedom to operate (FTO) analysis, 43–46
Freely revealed technology: benefits from, 122, 136; dissemination of, 277, 278; as investment in brand and reputation, 262; potential further development of, 251; and provision of services, 122–123
Free revealing, 142; and bazaar governance, 112, 231; and biobazaar, 331; and biotechnology, 241; and

Free revealing *(continued)*
broad patents on enabling or platform technologies, 139; and cumulative transactions or collective invention, 121–122; and defensive disclosure or defensive publishing, 162; definition, 119; of enabling or platform technologies, 259; and enhancement of company's reputation, 262–263; and innovation, 128; as intentional strategy, 119–120; and legal freedom (without legal encumbrance), 120; and network effects, 251; and network governance, 231; and nonproprietary business strategies, 115, 118, 150, 358n1; and open source, 136; and open source licensing, 138, 186; versus open source licensing, 158–164; and open source projects, 208; and open source software, 119, 142; opportunity costs of, 266; and other proprietary barriers, 138; and private rewards (*see under* Private rewards/incentives); profitability of, 240; and public-sector human genome project, 139; and restructuring of competitive landscape of an industry, 124; and return on private investment in innovation, 251; risks of, 161–162; and technical transparency, 120, 122; and technology transfer to commercial entities, 237; and use value, 120, 249, 251, 252, 263, 298; as viable business strategy, 128, 240. *See also* Freely revealed technology

Free riders: and bazaar governance, 110, 112; and copyleft licenses, 180; ease of copying information and, 83; problem of, 71

Free software, 8, 9; and legal freedom (without legal encumbrance), 8; and Linux, 12; and open source software, 12–13; and technical transparency, 8

Free Software Definition (FSD), 148, 157, 171

Free Software Foundation (FSF), 9, 11, 148

Free software movement, 4–6, 8–9

FUD (fear, uncertainty, and doubt), 300

Functional classes of innovators, 115–116

Gates, Bill, 10

Genentech, 32, 73

Gene patents, 61

General Public License (GPL), 11, 155, 304, 305; and Biological Open Source (BiOS) licenses, 318; compatibility of other software licenses with, 148; and Equitable Access licenses, 320. *See also* Lesser GNU Public License (LGPL)

General Public License for Plant Germplasm (GPLPG), 304, 305–306, 307

Generic drug manufacturers, 92, 215, 285, 287; and financial support for open source biotechnology, 285–286; and integration of contributions to open source drug development, 215; and open source drug development, 263–264

Generic drugs: competition from, 92, 99; and price of drugs, 99; and regulatory approval, 57, 215, 286

Genetically engineered crops, 102, 328

Genetically modified organisms (GMOs), 304–305

Genetically modified seeds, 269
Genetic resources indexing technologies (GRIT), 317
Genetic technologies, 153
Genetic testing patents, 61–62
Genetic use restriction technologies (GURTs), 327–328
Genome sequencing, 1. *See also* Human genome sequencing
Germplasm: contribution of developing countries to, 59, 177, 304; and genetically modified organisms, 304–305; MTA, 305–306; in value chain of agricultural biotechnology, 302. *See also* General Public License for Plant Germplasm (GPLPG); International Treaty on Plant Genetic Resources for Food and Agriculture
Ghosh, Rishab, 110
Gilman, Alfred, 213
Glaxo-Wellcome, 40, 268
Global business regulation, 94–95, 105
Global intellectual property law and policy, 25, 94, 95
Global pharmaceutical industry, 325
GNU Emacs, 5
GNU/Linux, 12, 114, 138
GNU Manifesto, 9
GoldenRice, 41
Governance structures, 129, 132. *See also* Bazaar governance; Firm; market; Network governance
Government laboratories, 55
Graff, Greg, 41, 45
Grant-back, 179
Granularity, 201–202, 203, 204, 205, 210
Green biotechnology. *See* Agricultural biotechnology
Green Revolution, 102
Grubb, Philip W., 55, 287
Guaragna, Mauricio, 320

Hackers, 5, 7, 9, 75
Haplotype map of human genome, 308. *See also* International HapMap Project
"Happy path" sequence of events, 43, 304, 314
Hardware vendors, 126–127
Harvard University and "oncomouse," 33
Health and life expectancy, 96–97
Health care biotechnology. *See* Medical biotechnology
Health maintenance organizations (HMOs), 287
Heller, Michael, 39–40
Heterogeneous user need, 243, 246–247
Hewlett Packard, 126–127
Hierarchical structures, 108. *See also* Firm
Hilgartner, Stephen, 79–80, 85, 312
HIV/AIDS, 97, 190–191, 321, 325
Horizontally networked user innovation, 109, 212, 243
Horizontal user innovation network, 114
Hrebek, Petr, 299
Hubbard, Tim, 217, 307–308
Human capital, 197, 199
Human genome, 264, 307–308
Human Genome Project, 200, 206–207. *See also* Public-sector human genome project
Human genome sequencing, 1, 36–37. *See also* Human Genome Project; Public-sector human genome project
Hume, David, 184
Hybridization, 327–328

Hybridoma technology, 248, 251–252
Hybrid strategy, 300–301, 309–310. *See also* Mixed business models

IBM: and intellectual property rights, 150; and open source software, 15, 126–127; profitability of its Linux-related/open source services, 127, 350n34
Illich, Ivan, 327, 329
Imitator companies, 286, 287
Incentives: and bazaar production costs, 135–136; collective-action-style, 133, 134, 331; community and in relation to innovation, 133; defined, 111–112; for innovating, 110, 115, 117, 118, 170, 178; and proprietary exclusivity, 117, 118, 128, 239, 240; for software development, 206; use value of innovation as for commercial participation in open source biotechnology, 248. *See also* Bazaar incentives; Benefits from innovation; Process benefits/incentives to innovate
Industrial innovation, 141
Industry culture. *See under* Open source biotechnology
Information: closed conduits and open channels, 231, 274; codification of (*see* Codification of information); ease of copying (*see* Copying of technological information); markets in (*see* Markets in information); nature of, 82; nonmarket mechanisms of transfer, 84, 85, 86, 104–105; privatization of, 88; transfer of by Internet and other means, 212–213. *See also* Data; Sticky information
Information exchange. *See* Knowledge exchange; Markets in information
Information markets. *See* Markets in information
Information technology. *See* Computer science/information technology
Infrastructure technologies. *See* Enabling or platform technologies
Innovation: bazaar governance and, 25; bazaar incentives and, 239; benefits from, 115–116, 128; community and incentives for, 133; constraints on in developing countries, 339n40; conventional economic view of, 84; cumulative and collective/cooperative nature of, 84, 107, 139; decentralized versus centralized/coordinated by patent holder, 74, 87; free revealing and, 128; heterogeneous user need and cost of, 243; incentives for, 110, 115, 117, 118, 170, 178; and intellectual property rights, 19–20, 68–69, 74, 84, 86–88, 90–91, 99, 102, 239; and intermediate technological inputs, 106–107; knowledge game and, 92, 93, 102; and learning networks, 242, 262; open source and promotion of, 20; open source licensing and, 349n18; patent litigation and, 90; and private rewards, 132; proprietary exclusivity and investment in, 107, 118, 128, 239, 240; serendipitous, 115, 117; socially and economically valuable, 35, 40, 103, 104, 210, 216, 239; stifling of downstream by upstream intellectual property holders, 40, 122, 170; and tragedy

of the anticommons, 40, 41; transaction cost economics and, 107. *See also* Biotechnical innovation; Horizontally networked user innovation; Industrial innovation; Sale value; User innovation
Institute for Genomic Research (TIGR), 37–38
Institutional design, 183–187
Institutional heterogeneity, 49–50
Integration, 202, 214–216, 217. *See also* Low cost integration
Intellectual property, 166, 239; and access to development resources, 163–164; barriers to competition associated with, 19, 69, 91; and bazaar governance, 112; in biotechnology, 31–35, 66; Brazil and, 321; cartels and, 91; constructivist view of science and, 78–79; and copyleft licenses, 177; economic/instrumental justifications for legal protection of, 68, 69–74, 86–88, 94, 106; fragmented ownership of, 63; implications of strengthening in biotechnology, 96–102, 103, 106; and globalization, 25, 70, 94, 95; import and export of, 321, 323, 324, 325; and information markets, 71–72, 73, 82–86, 105; and innovation, 19–20, 68–69, 74, 84, 86–88, 90–91, 93, 102, 239; and international trade agreements, 20, 64, 95, 323–324; and knowledge exchange, 28–29, 69–70; and markets for intermediate technological inputs, 106–107; and market structure, 68; Mertonian view of, 74–78; and monopoly power, 167; moral justifications for protection of, 70; and open source licenses, 149–150, 186; overarching concept of, 166; ownership of in agricultural biotechnology, 63, 65; and proprietary exclusivity, 25, 71; and proprietary licenses, 177–178; "ratcheting up" protection, 296; rationale for biotechnology industry participants obtaining, 267–277; and research on treatment of disease in developing countries, 98; and scientific data, 81–82; as selective pressure on ecology of business models, 296; and "story of science," 74–82; strategy, 294; and tragedy of the anticommons, 40; and transaction costs, 50–51, 67, 81, 92–93, 103, 136, 139; and transformation of social practices, 85; United States Constitution and, 70; use of to control assets or to prevent appropriation, 111–112. *See also* Copyright; Freedom to operate (FTO) analysis; International intellectual property regime; Patents
Intermediate technological inputs, 106–107
International Agricultural Research Centers, 59
International Fair Trade Association (IFAT), 325
International Federation of Organic Agriculture Movements (IFOAM), 234
International Haplotype Mapping Project. *See* International HapMap Project
International HapMap Project, 308–310, 350n35
International intellectual property regime, 20, 95

International Social and Environmental Accreditation and Labelling (ISEAL) Alliance, 233–234
International trade agreements, 20, 64, 95, 323–324
International Treaty on Plant Genetic Resources for Food and Agriculture, 306–307
Internet: and knowledge exchange, 213; and low cost integration, 211; and open source software, 16–17, 28; the Synaptic Leap and collaboration via, 312, 314, 315; transfer of information and, 212–213
Internet-enabled bazaar/peer production, 26, 193, 194, 198, 201, 211, 214, 235–236. *See also* Granularity; Low cost integration; Modularity
Interoperability, 120, 121, 249–250
Inventing around blocking patents, 44, 46, 52, 138, 139, 259, 266. *See also* Blocking patents
Invention-inducement theory, 71, 117. *See also* Markets in information
"Invention" versus "discovery" in patent law. *See under* Patent law
Iterative learning, 243; in drug development process, 54; and Tropical Diseases Initiative and TSL, 312–313; and use value, 260

Jackson, Joseph, 320
Jasanoff, Sheila, 72–73
Jefferson, Richard, 316, 317
Jordan, K., 343n37
Joy, Bill, 155

Kapczynski, Amy, 319, 320
Kilian, Andrzej, 259, 261–262
Kitch, Edmund, 72, 87–88
Knowledge exchange: free, 30; intellectual property rights and, 28–29, 69–70; transaction costs and, 41; via Internet, 213. *See also* Biotechnology-related information; Contracting for knowledge; Dissemination of information; Markets in information; Scientific knowledge
Knowledge game, 25, 67, 69, 88–92, 105; and agricultural biotechnology, 101; archetypal strategy of, 118; and biotechnology, 291; and competition, 89, 93, 102, 266, 299 (*see also* Cartels); and costs, 266–267; dealing with competitors in, 299; and innovation, 92, 93, 102; and open source, 96; and patents, 89, 90–91; proprietary exclusivity and, 239; and strategic modeling, 323; and transaction costs, 92–93; winners and losers, 92–96
Knowledge goods, 135
Knowledge production. *See* Nonproprietary peer-based knowledge production
Knowledge production networks, 231, 273–274
Kohler, Georg, 251
Kuhn, Thomas, 3, 77

Law reform, 226
Lead users, 196
Learning, 242; and codification of information, 82, 83; and software, 206, 242. *See also* Iterative learning; Process benefits/incentives to innovate
Learning networks, 242, 262
Lecocq, Xavier, 109, 112, 129, 279
Legal freedom (without legal encumbrance): and biotechnology freedom, 170; and free revealing, 120;

and free software, 8, 9; and open source production/licensing, 155, 158
Lesser GNU Public License (LGPL), 318
Levy, Steven, 6, 7, 8
Liberty rights, 156, 157, 158, 164, 170
Licensing agreements: and competition, 91; difficulties of drafting and enforcing, 49; institutional heterogeneity and, 49–50; patent strategies and, 50–51; restrictive provisions in, 51. *See also* Open source licenses and licensing
Licensing revenue, 89, 90; open source, 150–151, 168–169, 175–176, 346n35; open source biotechnology, 176, 184
Life-cycle management or evergreening patents, 285
Life sciences industry, 269. *See also* Agricultural biotechnology
Life sciences research, 3–4, 35–38, 76, 92, 200
Linux, 11–12; and Apache, 302; and bazaar governance, 194; and Microsoft Windows, 301; and open source revolution, 13, 17
Litigation deterrent or yank clause, 163, 276
Love, James, 217
Low cost integration, 211–218
Lula da Silva, Luiz Inacio, 321
Lynch, M., 343n37

Malnutrition, 100
Mandeville, Thomas, 82, 83, 86
Manufacturers (in user innovation theory), 115–116, 117, 240, 263
Manufacturing strategy, 117–118
Market: and bazaar governance, 18, 109–111, 129–130, 131, 194, 280, 331; controls and incentives, 112, 114; definition, 108; and direction of biotechnology research and development, 98, 99, 100, 102; for intermediate technological inputs, 106–107; and open source development, 131; and selection for production tasks, 113; transaction cost economics and, 107; and use of intellectual property rights to control assets, 111. *See also* Unserved need (small or niche markets)
Market-based selection of individuals for particular production tasks, 87, 113
Marketing exclusivity, 186
Marketing war, 300
Market positioning, 123
Markets in information, 82, 86; and codification of information, 84–85, 86; and intellectual property, 82–86; and intellectual property rights, 71–72, 73, 105; use of patent rights to dominate, 91
Market structure, 68
Massachusetts Institute of Technology (MIT), 206
"Matching funds organizations," 217
Material transfer agreement (MTA), 47, 60, 144, 305–306
Matthews, Duncan, 95
Maurer, Stephen, 310, 314
Mazzoleni, Roberto, 71
Medical biotechnology, 96–99; and agricultural biotechnology, 63; business strategies of companies in, 270–272; definition, 54; and low cost integration, 214–217; opportunity costs of open source strategy in, 270–271; patents and develop-

Medical biotechnology *(continued)* ment and commercialization of products of, 40–41; public funding of, 287
Medical treatment, 96–97
Medicines for Malaria Venture (MMV), 290
Merck, 38, 40, 262
Merges, Robert, 65–66
Merrill, Thomas, 226
Merton, Robert K., 74–78, 105
"Me too" drugs, 54, 57, 268
Michael J. Fox Foundation for Parkinson's Research, 288
Michaels, Tom, 304, 305, 306, 307
Microsoft, 10, 15, 362n11
Microsoft Windows, 301
Mill, John Stuart, 74
Milstein, Cesar, 251
Mitroff, Ian, 77
Mixed business models, 298; and biotechnology, 258–259, 272; and hybrid strategy, 300–301; provision of goods or services complementary to a company's intellectual property rights, 258, 260 (*see also* Diversity Arrays Technology (DArT))
Modeling, 322, 324–325
Model missionaries and mercenaries, 324
Model open source licenses, 145–146, 223–224
Modularity, 201, 202–203, 204, 205
Module libraries, 356n54
Moglen, Eben, 11
Molecular biotechnology, 42–43. *See also* Biotechnology
Molecular biotechnology industry, 18–19. *See also* Biotechnology industry
Molecular diagnostics, 60–63, 250
Molecular Sciences Institute (MSI), 316
Monoclonal antibodies, 248, 251–252
Monsanto, 270
Mozilla Public License, 348n3, 353n66
Mullis, Kary B., 44–45, 335n1
Multiple Myeloma Research Foundation, 288
Munos, Bernard, 289–290, 301–302
MySQL, 125

National agricultural research systems (NARS), 192, 275
National Institutes of Health (NIH), 31; and integration of contributions to clinical trials, 214; and patenting of ESTs, 37, 276. *See also* Uniform Biological Material Transfer Agreement (UBMTA)
National Science Foundation (USA), 31
Neglected Disease licensing, 319, 320
Nelson, Richard, 71
Network collaboration, 229, 261
Networked intellectual property portfolios, 91
Network effects, 110, 135; in biotechnology, 249; and certification signals, peer review, and reliability of technology, 121, 250–251; and free revealing, 251; and interoperability, 121, 250
Network for Open Scientific Innovation (NOSI) (Brazil), 320
Network governance, 108, 231; and bazaar governance, 109–111, 129, 131, 141, 194, 280, 331; and biotechnology, 231; controls and incentives, 112; costs of versus costs

of bazaar governance, 219; and open source software development, 131–132; relational ties in, 110, 132, 133, 231; and use of intellectual property rights to control assets, 111
Network relationships, 108
Networks, definition, 108. *See also* Horizontally networked user innovation; Knowledge production networks; Learning networks
"New drug application," 57
Niche or application technologies, 123
Nonexclusive grant-backs, 257
Nonexclusive proprietary license, 136, 153
Nonmarket mechanisms of transfer of information, 84, 85, 86, 104–105
Nonprofit sector: and biotechnology, 192–193; and collective action subsidies, 214, 215; and drug development, 190, 191; interdependence of with for-profit sector in biotechnology, 140–141; open source entrepreneurs in, 297; and ownership of employee inventions, 273. *See also* Private nonprofit institutions
Nonproprietary business strategies: and bazaar governance, 112; benefits from, 119, 136; and biotechnology, 240, 256, 265, 266, 276–277; complementary to proprietary business strategies (*see* Mixed business models); and free revealing, 115, 118, 128, 358n1; and open source, 150, 358 n1; and open source licenses, 186; and proprietary business strategies, 21–22, 253, 298, 300–301, 302; and return on private investment in innovation, 25, 118, 249; and software distribution, 256; and transaction costs, 242; and use value (*see under* Free revealing). *See also* Hybrid strategy; Mixed business models; Nonproprietary service-based business models; Open source business models/exploitation strategies; Proprietary business and/or exploitation strategies
Nonproprietary incentives. *See* Bazaar incentives
Nonproprietary manufacturing strategy, 263–264
Nonproprietary mode of production, 237–238
Nonproprietary peer-based knowledge production, 299
Nonproprietary service-based business models, 259–262
Norm of non-excludability, 112
Norms of science, 75–77, 105
North, Douglass, 154

Oligopolists. *See* Patent-based cartels
Open Software License, 181, 348n3
Open source, 12, 142; and bazaar governance, 25, 107 (*see also* Bazaar governance); and biotechnology, 18 (*see also* Open source biotechnology); brand (*see* Open source brand); and broad patents on enabling or platform technologies, 139; commercial viability of, 301; and community, 130; and competition, 265; and end users/consumers, 303; and Fair Trade, 326; and free revealing, 136; and knowledge game, 96; and nonproprietary exploitation strategies, 358n1; perception of/acceptance

Open source *(continued)*
by potential customers and investors, 281–282
Open source biotechnology, 201, 329–330; applicability to upstream versus downstream technologies, 302; and appropriate technology, 327; as biobazaar, 18; Brazil and, 321; capitalism and, 296–297; and commercial viability, 237–238; community, 230, 233; and convivial tools, 329; costs of, 238, 266 (*see also under* Opportunity costs); and cost of regulatory approval, 310; desirability of, 228; feasibility of, 21–24, 193, 211–212, 214, 218, 219, 221–222, 230, 235, 238, 266, 290, 301; features of, 321, 330–331; indirect contributions and capital investment, 26, 238, 284–288, 303; integration of commercial and noncommercial contributions, 26, 193, 330; intellectual property policies of universities and, 209; as modeling, 22–23, 322; and pool of contributors, 208; and potential misuse of open source label/brand, 229–230, 283; and productivity of pharmaceutical research and development, 301; and proprietary culture of industry, 27; scaling up, 325, 330; and synergies with other social movements, 283–284; trade-off between proprietary and nonproprietary exploitation strategies, 26, 293–294; and tragedy of the anticommons, 137–138; and venture capital, 256, 282–283, 284. *See also* Biobazaar; Biotechnology freedom
Open source biotechnology licenses and licensing, 26, 144–145, 282; and best-practice guidelines, 228–230; and charging of fees, 176; and diversity of approaches to developing, 228; drafting of/design options for, 143–149, 183, 187, 221, 223, 234–235; and dual licensing, 151; existing open source software licenses as models for, 145–146; and field-of-use and territorial restrictions of intellectual property rights, 167; free revealing versus, 159, 160–164; and generic open source licensing principles, 235; lawyers and other licensing experts and, 221–222; and Lesser GNU Public License (LGPL), 318; licensor-driven development of as user innovation, 356n54; and mapping copyright owners' to patent owners' rights, 165–166; Molecular Sciences Institute and, 316; potential problem of proliferation of, 227–228; versus publication/public domain approach, 159–164; and setting and enforcing of standards by open source biotechnology community, 233–234; and suite of model licenses, 223–224; the Synaptic Leap and, 315; Tropical Diseases Initiative and, 311. *See also* Biological Open Source (BiOS) licensing; Equitable Access licensing; General Public License for Plant Germplasm (GPLPG); International HapMap Project; Open source licenses and licensing
Open source brand, 229–230, 283–284
Open source business models/exploitation strategies: in biotechnology, 240–265, 281–282; and commercialization, 141–142, 236, 260; de-

cision-making about, 295; profitability of, 295; and risk (*see under* Risk); software, 260, 293; and stock market, 15–16, 282. *See also* Mixed business models; Nonproprietary business strategies
Open Source Definition (OSD), 146–148, 156, 171, 174, 175
Open source development/production: as bazaar governance supported by firm, market, and network governance, 130–131; commercial and noncommercial contributions to, 140; and community, 130, 132, 133–134, 183, 230, 233; critical mass of contributors/user-developers, 196, 298, 307; of downstream technologies, 303; of drugs, 268, 311, 312; and facilitation of precompetitive collaboration, 253–254; and leadership, 232, 307; and licensing, 142 (*see also* Open source licenses and licensing); and mainstream legal and economic institutions, 299; market and, 131; and marketing strategies, 300; and pool of contributors, 194–196, 218; and overcapacity, 210; and reliability of technology, 250; and shared principles of collaboration, 229; and unserved need (small or niche markets), 244, 310; use of intellectual property licensing strategies and new business models in, 140; and user innovation, 243. *See also* Open source projects; Open source software development
Open Source Initiative (OSI), 15; and approved licenses, 145, 230; authority of, 233; certification by, 146, 148; and Debian Free Software Guidelines, 147; and license proliferation, 227–228; and standards for licenses, 233; and use of term *open source,* 230
Open source licenses and licensing, 25–26, 136, 142, 153, 221; and abandonment of patent, 160–161, 298; and articulation of terms of collaboration, 155, 156; and biobazaar, 331; BioBricks Foundation and, 316; and biotechnology, 142–143, 219–220, 304 (*see also* Open source biotechnology licenses and licensing); certain restrictions typical of proprietary licenses not permitted under, 168; characteristics of, 149; and charging of fees, 175–176, 346n35; and collaborative development, 142, 151, 153–154, 156; and collaborative proprietary licenses, 152–153; commercial applications of, 150, 186, 236 (*see also* Free revealing; Nonproprietary business strategies); and commercial investment, 150; and credible commitment, 154–155, 156, 186; and dedicating original work to public domain, 158–159; and defensive patenting, 276; and difficulties of conventional contractual/partnership agreements, 290; and disclosure requirement in patent law, 174–175; enforcement in accordance with complier-centered institutional design, 185; and field-of-use and territorial restrictions, 167; and free revealing, 138, 186; versus free revealing, 158–164; and innovation, 20–21, 349n18; and institutional design, 183, 187, 220; and intellectual property rights, 149–150, 186;

Open source licenses and licensing *(continued)*
and involvement of lawyers in license development, 220–226; and legal freedom (without legal encumbrance), 155, 158; proliferation of, 145, 225–228; purpose of, 151, 156, 299; and reporting requirement, 167; revenue from, 150–151, 168–169, 175–176, 346n35; rights of users under, 156–158; simplicity and absence of legal technicalities in, 220; standard-setting for, 233, 234; and suite of model licenses, 223–224; technical transparency, 14, 124, 155, 159; and tragedy of the anticommons, 138; and use value over sale value, 136. *See also* Biological Open Source (BiOS) licensing; Copyleft licenses; Model open source licenses; Open source biotechnology licenses and licensing

Open source projects, 13–15; competitive advantage of, 299; contributions to (*see* Contributions to open source projects); development of open source software in, 153–154; entrepreneurial, 27; and free revealing, 208; motivation for participation in, 18, 278 (*see also* Bazaar incentives); restrictive institutional policies and participation in, 208–209; spectrum of involvement, 113; and voluntary participation and voluntary selection of tasks, 14. *See also* Open source development/production

Open source revolution, 12, 13, 298

Open source software, 330; adoption of as business strategy, 296–297; and bazaar production, 108–109, 142, 189; Brazil and, 321; classes of contributors to, 124–127; commonly used applications, 15, 127; cost of preparing code for release, 277–278; and cumulative improvements/collective invention, 121–122; in daily life, 16; development of in context of open source projects, 153–154; freedom in, 31; and free revealing, 119, 142; and free software, 12–13; incentives and, 115; and Internet and email, 16–17, 28; and modularity, 203; and network effects and interoperability, 121, 249; use of by government and business, 15; and use value, 137. *See also* Software freedom

Open source software development: and bazaar production in industry setting, 239–240; and biotechnology research and development, 188, 189, 194–195, 196, 210–211, 218, 219; and low cost integration, 211–214; and network and bazaar governance, 131–132

Open source software licenses and licensing, 144; and charging of fees, 168–169; and competition, 167, 170; and complier-centered institutional design, 184; and enterprise software, 152; and entitlement to source code (*see under* Source code); and Equitable Access and Neglected Disease licensing, 320; and freedom to fork, 169–170; mix-and-match menu approach to creating new licenses not embraced, 225–226; and monopoly power of intellectual property rights, 167; and patent-based biotechnology, 317–318; rights of licensor, 167, 168; and royalties, 168, 169; and trust in open source

label/brand, 229. *See also* Open Source Definition (OSD); Open Source Initiative (OSI); Open source licenses and licensing
Open source technologies, 298–300
Opportunity costs: to biotechnology companies of not pursuing open source strategy, 272; and free revealing, 266; and nonproprietary biotechnology business strategies, 238, 276–277; of open source strategy to agricultural biotechnology company, 270, 272; of open source strategy to biotechnology companies, 270–271, 272; of open source strategy to pharmaceutical companies, 268; of open source strategy to universities, 275; of public-sector and nonprofit institutions pursuing open source strategy, 273, 275
O'Reilly, Tim, 16
Organisation for Economic Co-operation and Development (OECD), 355n52, 364n58
Organizational learning, 242
Owners. *See* Users and owners

Parallel importers, 286, 287
Patent Act (USA). *See* 35 U.S.C. 112
Patent bargain, 173–174
Patent-based cartels, 91
Patent law: Court of Appeals of the Federal Circuit and, 33–34; and equivalent of copyright derivative works, 182; and disclosure requirement, 173, 174–175, 187; "invention" versus "discovery," 33, 72–73, 88; and living organisms, 33; and reciprocal license terms, 182. *See also* Patent Act (USA)
Patent lawyers, 222
Patentleft licenses, 181–182
Patent Lens, 316–317
Patent licensing agreements. *See* Licensing agreements
Patent pools, 364n58
Patents, 159, 226; abandonment of versus open source licensing, 160–161, 298; "best mode" requirement, 173, 174, 175–176; and biotechnology, 40–41, 144, 159, 188; blocking effect of, 44, 46, 84, 85, 86, 89; blocking techniques, 90; centralized coordination of research and development by holder of, 73–74, 87; and competition, 34, 91, 267, 269; complexity and dynamic nature of landscape in biotechnology, 44–45; and copyright, 10, 165–166, 188; costs, 92, 266; defensive, 37, 90, 268, 276; disclosure requirement, 172–175; on enabling or platform technologies, 139; on ESTs, 37, 38; gene, 61; for genetically modified organisms, 305; genetic testing, 61–62; for human gene fragments, 38, 39; and knowledge game, 89, 90–91; infringement litigation, 34, 45, 50, 90, 257, 266, 271; licensing revenue from (*see* Licensing revenue); maintenance of, 298; and nonuse of patented technology, 90–91; offensive and defensive uses of, 34–35; origin of, 70; and price of drugs, 99; private nonprofit institutions and, 273; public-sector institutions and, 273; purposes of amassing portfolio of, 89; role of in development of biotechnology industry, 72–73; and software, 336n20; and software source codes, 10; and supplier lock-in, 62, 102, 151, 265; timing of application in drug development process,

Patents *(continued)*
56; and uncodified information, 85–86; value of to pharmaceutical companies, 267–268. *See also* Blocking patents; Life-cycle management or evergreening patents; Submarine patents; Freedom to operate (FTO) analysis; Intellectual property
Patent stacking, 90
Patent strategy, 50–51
Patent trolls, 90–91
Peer production. *See* Commons-based peer production; Internet-enabled bazaar/peer production
Peer review: and evolution of technology, 298; and network effects and reliability of technology, 121, 250–251
Perens, Bruce: on classes of contributors to open source software, 124, 125, 126; and OSD, 146–147; on source software licensing as social engineering, 183
Permissive open source licenses. *See* Academic or permissive open source licenses
Personal property rights, 47, 117, 144, 160, 165, 182
Pharmaceutical companies. *See* Chemical and pharmaceutical companies
Pharmaceutical industry, 284–285
Pharmaceutical Manufacturers Association, 73
Pharmaceutical research and development, 98, 103, 268
Pharmacogenetics (or pharmacogenomics), 61, 247
Pharmacy benefit management companies, 287
Physical capital, 196–197, 198–199, 218
Pinch, Trevor, 77
Pirates. *See* Imitator companies
Plant enabling technologies (PET), 317
Plant variety rights, 144, 165, 269
Platform technologies. *See* Enabling or platform technologies
Polyani, Michael, 74, 105, 109
Polymerase chain reaction (PCR) technique, 248, 343n37
Popper, Karl, 30–31, 74
Porter, Michael, 211
Positive network externalities. *See* Network effects
Poverty, 96–97, 100, 198
Powell, Woody, 107, 108, 131, 242
Pratt, Edmund T., Jr., 95
Preclinical testing, 55–56, 190, 191
Precompetitive research and development, 253–254, 256–257
Predictive toxicology, 254
Price of drugs, 99, 102
Price signals, 108, 110
Private-collective hybrid, 347n55
Private nonprofit institutions, 273, 276
Private ownership of research results, 36–37, 76
Private rewards/incentives: and bazaar, 114, 331; and free revealing, 122–124, 125, 126, 127, 263; and governance structures, 132; and innovation, 132; and commercial actors in biobazaar, 133, 237, 263–265
Privatization, 4, 88. *See also* Commercialization
Process benefits/incentives to innovate, 115, 117, 124, 241; and bazaar production, 129; and biobazaar, 331; and biotechnology, 241, 243; for-profit entities and, 241–243; and free revealing, 129;

and software, 206, 241, 242; and trainees, 208. *See also* Learning
Profitability: of biotechnology industry, 272; of blockbuster drugs, 98, 267–268, 277; of free revealing, 240; of knowledge game, 93; of open source strategy, 295; of pharmaceutical companies, 267–268, 277; of proprietary manufacturing strategies, 118; threshold of, 210
Project infrastructure, 278–279
Property, 69, 70, 164–165; unconventional use in bazaar settings, 111. *See also* Intellectual property; Personal property rights
Proprietary add-ons, 126
Proprietary business and/or exploitation strategies, 117; 151; and nonproprietary strategies, 21–22, 240, 253, 265, 296, 298, 300–301, 302. See also Hybrid strategy; Mixed business models; Nonproprietary business strategies; Open source business models/exploitation strategies
Proprietary culture: in biotechnology, 297; and open source, 27, 281, 292; and strategic thinking, 295–296
Proprietary exclusivity: biotechnology platform technology companies and, 271; and competition, 322–323; and interference in knowledge exchange, 28–29, 118; and innovation, 25, 71, 107, 117, 118, 128, 216–217, 239, 240
Proprietary manufacturing strategy, 240–241
Prospect theory, 71–72, 84, 87–88, 107. *See also* Markets in information
Prostate Cancer Foundation, 288
Public funding: of agricultural biotechnology, 100–101, 287; and bazaar production, 140, 331; of biotechnology, 132, 287; of development of suite of model open source licenses, 223; of medical biotechnology, 287; for open source biotechnology, 287–288
Public Intellectual Property Resource for Agriculture (PIPRA), 364n58
Public Interest Intellectual Property Advisors (PIIPA), 223
Public-private collaboration. *See* International HapMap Project; SNP Consortium
Public-private partnerships (PPPs), 289–290
Public-sector human genome project: and copyleft-style licenses, 160; and free revealing, 139; race with private sector, 307–308; and user innovation, 247
Public-sector institutions: and agricultural biotechnology, 65, 192, 275–276; and drug development in developing countries, 98; opportunity costs of pursuing open source strategy, 275; and ownership of employee inventions, 208–209, 273; patenting by, 273, 275–276

Quality control, 219

Rai, Arti, 310, 314
Rausser, G., 41, 45
Raymond, Eric S.: *The Cathedral and the Bazaar,* 17, 243; on software cathedrals, 108; on enterprise software, 152; on freedom to fork, 169–170; *Homesteading the Noosphere,* 17, 75; *The Magic Cauldron,* 17; on testing reliability, 250; on “scratching an itch,” 243
Reach-through royalties, 178–179

Reciprocal licenses, 140, 179, 182. *See also* Copyleft licenses
Reciprocity, 110, 179
Red biotechnology. *See* Medical biotechnology
Red Hat Linux, 125, 261
Refactoring, 204
Regulatory approval, 57, 213–214, 215, 286, 310
Relational contracting, 107, 108
Relational ties: significance in network versus bazaar governance, 110, 132, 133, 230, 231; in open source production, 132, 134
Reputation: and compliance, 185; enhancement of through free revealing, 262–263; as incentive to innovate, 133
Research exemption, 43, 63–64
Research tools: and biotechnology platform companies, 271; communities around, 232; patenting and proprietary exclusivity, 35, 40, 50, 98, 101, 103, 105, 268–270; licensing, 50, 63, 117; open source development of, 271; valuation, 48
Return on investment in innovation: in absence of proprietary exclusivity, 25, 118, 249; in agricultural biotechnology, 58; free revealing and, 251; in open source drug development, 268; prospect development theory and, 72; user innovators and, 252–253
Revenue. *See* Licensing revenue
Risk: of adverse perception of open source, 281–283; commercial, 279; in implementing open source strategies, 279, 280, 281, 282. *See also* Uncertainty
RMS. *See* Stallman, Richard
Rosen, Lawrence, 148–149, 156–157, 169, 181
Rossi, Cristina, 109, 196
Royal Society of London for the Improvement of Natural Knowledge, 29–30
Royalties, and open source software licensing, 168, 169. *See also* Reach-through royalties
Royalty stacking, 178
Ryan, Michael, 95

Sale value, 118, 241; and bazaar production, 135; and open source licensing, 136; and use value, 118–119, 248–249
Sali, Andrej, 310, 314
Salk, Jonas, 216
Salk polio vaccine, 216–217, 254
Sanctions, 185
Sanger Centre, 207
Schumpeter, Joseph, 93
Science: decentralized organization, 87; and the Enlightenment, 29–30; and free exchange of information, 30; intellectual property and, 74–82; and pedagogical narrative, 1–3. *See also* Academic science; Big science; Commons-based peer production; Communism or communalism
Science Commons, 315, 316, 355n52
"Science wars," 343n32
Scientific knowledge, 74
Scientific research, 18, 25, 75, 84
Screening of drug candidates, 55
Seed, 58–59, 100
Self-enforcing licenses. *See* Reciprocal licenses
Sell, Susan, 95
Sendmail, 17, 126
Sequence data, 350n35

Sequential exploitation strategy, 298
Service provision, 126, 258. *See also* Mixed business models; Nonproprietary service-based business models
SETI@home, 311
Shrink-wrap license, 152, 349n20
Single nucleotide polymorphisms (SNPs), 39, 40
Slashdot.org, 134
Small, A. A., 41, 45
Smith, Adam, 132
Smith, Henry, 226
SNP Consortium, 309; and collectively constructed SNP map, 253; and open source research tool development, 271; and publicly accessible SNP database, 40. *See also* Single nucleotide polymorphisms (SNPs)
Social and economic change, 322–323
Social and economic value of innovation, 35, 40, 103, 104, 210, 216, 239
Sociologists of science, 76, 77, 78
Sociology of scientific knowledge (SSK), 78. *See also* Constructivist accounts of science
Software: comparison with biotechnology, 188; competition between proprietary and nonproprietary technologies, 301; and copyright, 158; and granularity, 210; intrinsic or process-oriented benefits of developing, 206, 241, 242; and modularity, 203; and open source business models, 260, 293; patentability of, 336n20; skill and commitment for, 206; trademark protection, 336n20; and user innovation, 244
Software freedom, 156–158, 164, 166, 175, 186; and competition, 167, 168, 170; translation into biotechnology, 165, 166, 228. *See also* Technology freedom
Software industry, 18–19
Software licensing, 9–10, 151–152
Sokal, Alan, 343n32
Source code, 10, 170, 171–172; biotechnology equivalent, 171–172; and copyright protection, 308; cost of access to, 175; entitlement to under open source license, 157, 158, 171, 174, 175, 186; patenting of, 10; proprietary restrictions on, 7
SourceForge.net, 14, 134, 146
Spectrum of involvement, 113–114. *See also* Commitment
Sponsors or initiators of open source projects: and uncertainty, 279, 280; private nonprofit organizations, 214; and research and development costs, 290, 311
Stallman, Richard, 6; and copyleft/GPL, 9–10, 11; and free software, 4–6, 8–9, 12, 28, 138–139
Standardization: in biotechnology, 250; and use value, 121, 249–250, 252
Statutory Invention Registration (SIR), 350n37
Sticky information, 243, 245; in biotechnology, 244, 245–246; and knowledge production networks, 274
Strategic modeling: and knowledge game, 323; power of, 322–325, 332; and scale, 325; and social and economic change, 322–323
Submarine patents, 61, 90
Subsistence farmers, 100

Sulston, John, 38, 247–248
Supplier lock-in, 62, 102, 151, 265
Suppliers (in user innovation theory), 116
Surrounding (patents), 90
Swanson, Robert, 32
Synaptic Leap (TSL), 311, 312–315
Synthetic biology, 204–205, 206, 316

Target validation. *See* Validation of drug targets
Task partitioning. *See* Modularity
Taylor, Ginger, 314, 315
Technical transparency: and biotechnology freedom, 170; and free revealing, 120, 122; and free software, 8; and open source production/licensing, 14, 124, 155, 159; standard of, 173
Technological commons, 299
Technology freedom, 178–179, 180, 186, 291, 329; competitive impact of, 8; and follow-on innovations, 216; pursuit of through law reform instead of open source licensing, 226
Technology Support Services Subscription Agreements (for PET and GRIT), 317
Technology transfer in public interest, 32, 274–275
Terms of collaboration, 155, 156
TIGR. *See* Institute for Genomic Research
Toennissen, Gary, 42–43
Torvalds, Linus, 11–12
Trade, 70. *See also* International trade agreements
Trademarks, 233–234, 336n20
Trade-Related Aspects of Intellectual Property Rights (TRIPS), 20, 95, 99, 286, 323, 324, 325
Traditional biobazaar, 205; in agricultural biotechnology, 191–192; and knowledge production networks, 274; and open source biobazaar, 26, 190, 193, 235–236, 237, 315, 330; and technology transfer to commercial entities, 237. *See also* Academic science. *See also under* Biobazaar
Tragedy of the anticommons, 24, 39–40, 41, 137; in agricultural biotechnology, 65–66, 101; in biotechnology generally, 52, 60, 66–67; difficulty of empirical verification, 52; in molecular diagnostics, 60–63, 250; open source as a solution, 137–139; and research on diseases in developing country, 98; transaction costs and, 39, 41, 137, 139
Tragedy of the commons, 39
Transactional uncertainty: bazaar governance and, 112, 279–280; governance structure and, 107, 131
Transaction cost economics, 107
Transaction costs: and agricultural biotechnology industry structure, 64–65; and bazaar governance, 111–112, 280; and contracting for knowledge in biotechnology, 41; and intellectual property rights, 50–51, 67, 81, 92–93, 103, 136, 139; and license proliferation, 225–226; lowered through nonproprietary strategies, 242; and tragedy of the anticommons, 39, 41, 137, 139; and uncodified information, 86
Transfer costs: and codification of information, 83–84; and sticky information, 243

Transfer pricing, 99
Transparency. *See* Technical transparency
Trolltech, 125
Tropical Diseases Initiative (TDI), 310–312, 314; and the Synaptic Leap, 311

Uncertainty: bazaar governance and, 111, 279–280; and FTO analysis, 44; mitigation of by coexistence of bazaar governance with firm, market, and network governance, 280–281; and strategic bargaining over innovation inputs, 50; and value allocation, 48, 49. *See also* Risk; Transactional uncertainty
Uncodified information, 82–83; and biotechnology, 83, 86; and imitation, 83–84; and network governance, 131; patents and, 85–86; transaction costs and, 86
Uniform Biological Material Transfer Agreement (UBMTA), 225
United States Patent and Trademark Office (USPTO), 33
Universities: and drug development process, 55; intellectual property policies of and participation in open source projects, 208–209; networks in biotechnology and, 231, 273–274; and opportunity costs of open source strategy, 275; patenting by and technology transfer to industry, 32, 273, 274–275. *See also* Academic science
UNIX, 9
Unserved need (small or niche markets), 98–99, 100; satisfaction of through open source production, 244, 310. *See also* Developing countries; Food insecurity
Useful Chemistry Blog, 313
User communities: and biobazaar, 331; and biological research tools, 232–233; Diversity Arrays Technology and, 261; and support for open source production, 134
User innovation, 110, 115–116, 240, 252–253, 263; in bioinformatics, 255; in biotechnology, 243, 244, 247–248; licensor-driven development of open source biotechnology licenses, 356n54; and software code, 243, 244. *See also* Heterogeneous user need; Horizontally networked user innovation; Sticky information
Users, 115, 167
Users and developers/producers, 110, 114, 331
Users and owners, 11, 20, 155, 156
Use value, 118–119; aspects of enhanced by free revealing, 120–121; as a driver of bazaar production, 135; and free revealing/nonproprietary strategies, 120, 249, 251, 252, 263, 298; as incentive for commercial participation in open source biotechnology, 244, 248; and open source licensing, 136–137; and standardization, 249–250, 252; versus sale value, 118–119, 248–249
Use-value-oriented business strategies, 263

Validation of drug targets, 55, 190
Value allocation, 47–48, 178
Varmus, Harold, 38
Venter, Craig, 36–37, 38–39
Venture capital: and biotechnology, 258; and open source biotechnology, 256, 282–283, 284; and open

Venture capital *(continued)*
source software, 282; patents and, 73
Venture philanthropies, 288
Virtual pharmaceutical companies, 289–290, 310–311, 314. *See also* Public-private partnerships (PPPs)
Voluntary standard-setting and conformity assessment, 233, 234
Von Hippel, Eric, 120, 130, 347n55
Von Krogh, Georg, 130, 206

Watson, Jim, 37
Weber, Steven, 6, 12, 21; on coordination of contributions to open source projects, 133; on open source licenses, 149, 153, 157–158, 168, 183, 220
Weingast, Barry, 154
Wellcome Trust, 40
Whitehead Institute, 206–207
World Trade Organization. *See* Trade-Related Aspects of Intellectual Property Rights (TRIPS)

XEmacs, 5

Yank clause. *See* Litigation deterrent or yank clause